Mathematical Methods for Engineers and Scientists 2

K.T. Tang

Mathematical Methods for Engineers and Scientists 2

Vector Analysis, Ordinary Differential Equations and Laplace Transforms

With 73 Figures and 4 Tables

 Springer

Professor Dr. Kwong-Tin Tang

Pacific Lutheran University
Department of Physics
Tacoma, WA 98447, USA
E-mail: tangka@plu.edu

ISBN-13 978-3-642-06770-9   e-ISBN-13 978-3-540-30270-4

Springer is a part of Springer Science+Business Media.

springer.com

© Springer-Verlag Berlin Heidelberg 2010

Cover design: eStudio Calamar Steinen

Printed on acid-free paper

Preface

For some thirty years, I have taught two "Mathematical Physics" courses. One of them was previously named "Engineering Analysis". There are several textbooks of unquestionable merit for such courses, but I could not find one that fitted our needs. It seemed to me that students might have an easier time if some changes were made in these books. I ended up using class notes. Actually I felt the same about my own notes, so they got changed again and again. Throughout the years, many students and colleagues have urged me to publish them. I resisted until now, because the topics were not new and I was not sure that my way of presenting them was really that much better than others. In recent years, some former students came back to tell me that they still found my notes useful and looked at them from time to time. The fact that they always singled out these courses, among many others I have taught, made me think that besides being kind, they might even mean it. Perhaps it is worthwhile to share these notes with a wider audience.

It took far more work than expected to transcribe the lecture notes into printed pages. The notes were written in an abbreviated way without much explanation between any two equations, because I was supposed to supply the missing links in person. How much detail I would go into depended on the reaction of the students. Now without them in front of me, I had to decide the appropriate amount of derivation to be included. I chose to err on the side of too much detail rather than too little. As a result, the derivation does not look very elegant, but I also hope it does not leave any gap in students' comprehension.

Precisely stated and elegantly proved theorems looked great to me when I was a young faculty member. But in later years, I found that elegance in the eyes of the teacher might be stumbling blocks for students. Now I am convinced that before the student can use a mathematical theorem with confidence, he must first develop an intuitive feeling. The most effective way to do that is to follow a sufficient number of examples.

This book is written for students who want to learn but need a firm hand-holding. I hope they will find the book readable and easy to learn from.

Learning, as always, has to be done by the student herself or himself. No one can acquire mathematical skill without doing problems, the more the better. However, realistically students have a finite amount of time. They will be overwhelmed if problems are too numerous, and frustrated if problems are too difficult. A common practice in textbooks is to list a large number of problems and let the instructor to choose a few for assignments. It seems to me that is not a confidence building strategy. A self-learning person would not know what to choose. Therefore a moderate number of not overly difficult problems, with answers, are selected at the end of each chapter. Hopefully after the student has successfully solved all of them, he will be encouraged to seek more challenging ones. There are plenty of problems in other books. Of course, an instructor can always assign more problems at levels suitable to the class.

Professor I.I. Rabi used to say "All textbooks are written with the principle of least astonishment". Well, there is a good reason for that. After all, textbooks are supposed to explain the mysteries and make the profound obvious. This book is no exception. Nevertheless, I still hope the reader will find something in this book exciting.

On certain topics, I went farther than most other similar books. For example, most textbooks of mathematical physics discuss viscous damping of an oscillator, in which the friction force is proportional to velocity. Yet every student in freshman physics learnt that the friction force is proportional to the normal force between the planes of contact. This is known as Coulomb damping. Usually Coulomb damping is not even mentioned. In this book, Coulomb damping and viscous damping are discussed side by side.

Volume I consists of complex analysis and matrix theory. In this volume, we discuss vector and tensor analysis, ordinary differential equations and Laplace transforms. Fourier analysis and partial differential equations will be discussed in volume III. Students are supposed to have already completed two or three semesters of calculus and a year of college physics.

This book is dedicated to my students. I want to thank my A and B students, their diligence and enthusiasm have made teaching enjoyable and worthwhile. I want to thank my C and D students, their difficulties and mistakes made me search for better explanations.

I want to thank Brad Oraw for drawing many figures in this book, and Mathew Hacker for helping me to typeset the manuscript.

I want to express my deepest gratitude to Professor S.H. Patil, Indian Institute of Technology, Bombay. He has read the entire manuscript and provided many excellent suggestions. He has also checked the equations and the problems and corrected numerous errors.

The responsibility for remaining errors is, of course, entirely mine. I will greatly appreciate if they are brought to my attention.

Tacoma, Washington   *K.T. Tang*
December 2005

Contents

Part II Differential Equations and Laplace Transforms

Part I

Vector Analysis

1

Vectors

Vectors are used when both the magnitude and the direction of some physical quantity are required. Examples of such quantities are velocity, acceleration, force, electric and magnetic fields. A quantity that is completely characterized by its magnitude is known as a scalar. Mass and temperature are scalar quantities.

A vector is characterized by both magnitude and direction, but not all quantities that have magnitude and direction are vectors. For example, in the study of strength of materials, stress has both magnitude and direction. But stress is a second rank tensor, which we will study in a later chapter.

Vectors can be analyzed either with geometry or with algebra. The algebraic approach centers on the transformation properties of vectors. It is capable of generalization and leads to tensor analysis. Therefore it is fundamentally important in many problems of mathematical physics.

However, for pedagogical reasons we will begin with geometrical vectors, since they are easier to visualize. Besides, most readers probably already have some knowledge of the graphical approach of vector analysis.

A vector is usually indicated by a boldfaced letter, such as $\mathbf{V}$, or an arrow over a letter $\vec{V}$. While there are other ways to express a vector, whatever convention you choose, it is very important that vector and scalar quantities are represented by different types of symbols. A vector is graphically represented by a directed line segment. The length of the segment is proportional to the magnitude of the vector quantity with a suitable scale. The direction of the vector is indicated by an arrowhead at one end of the segment, which is known as the tip of the vector. The other end is called the tail. The magnitude of the vector is called the norm of the vector. In what follows, the letter V is used to mean the norm of $\mathbf{V}$. Sometimes, the norm of $\mathbf{V}$ is also represented by $|\mathbf{V}|$ or $\|\mathbf{V}\|$.

1.1 Bound and Free Vectors

There are two kinds of vectors; *bound vector* and *free vector*. Bound vectors are fixed in position. For example, in dealing with forces whose points of application or lines of action cannot be shifted, it is necessary to think of them as bound vectors. Consider the cases shown in Fig. 1.1. Two forces of the same magnitude and direction act at two different points along a beam. Clearly the torques produced at the supporting ends and the displacements at the free ends are totally different in these two cases. Therefore these forces are bound vectors. Usually in statics, structures, and strength of materials, forces are bound vectors; attention must be paid to their magnitude, direction, and the point of application.

A *free vector* is completely characterized by its magnitude and direction. These vectors are the ones discussed in mathematical analysis. In what follows, *vectors are understood to be free vectors* unless otherwise specified.

Two free vectors whose magnitudes, or lengths, are equal and whose directions are the same are said to be equal, regardless of the points in space from which they may be drawn. In other words, a vector quantity can be represented equally well by any of the infinite many line segments, all having the same length and the same direction. It is, therefore, customary to say that a vector can be moved parallel to itself without change.

1.2 Vector Operations

Mathematical operations defined for scalars, such as addition and multiplication, are not applicable to vectors, since vectors not only have magnitude but also direction. Therefore a set of vector operations must be introduced. These operations are the rules of combining a vector with another vector or a vector with a scalar. There are various ways of combining them. Some useful combinations are defined in this section.

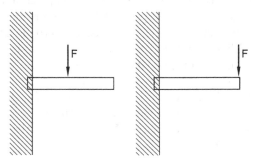

Fig. 1.1. Bound vectors representing the forces acting on the beam cannot be moved parallel to themselves

1.2.1 Multiplication by a Scalar

If c is a positive number, the equation

$$\mathbf{A} = c\mathbf{B} \qquad (1.1)$$

means that the direction of the vector $\mathbf{A}$ is the same as that of $\mathbf{B}$, and the magnitude of $\mathbf{A}$ is c times that of $\mathbf{B}$. If c is negative, the equation means that the direction of $\mathbf{A}$ is opposite to that of $\mathbf{B}$ and the magnitude of $\mathbf{A}$ is c times that of $\mathbf{B}$.

1.2.2 Unit Vector

A *unit vector* is a vector having a magnitude of one unit. If we divide a vector $\mathbf{V}$ by its magnitude V, we obtain a unit vector in the direction of $\mathbf{V}$. Thus, the unit vector $\mathbf{n}$ in the direction of $\mathbf{V}$ is given by

$$\mathbf{n} = \frac{1}{V}\mathbf{V}. \qquad (1.2)$$

Very often a hat is put on the vector symbol $(\widehat{\mathbf{n}})$ to indicate that it is a unit vector. Thus $\mathbf{A} = A\widehat{\mathbf{A}}$ and the statement "n is an unit vector in the direction of $\mathbf{A}$" can be expressed as $\mathbf{n} = \widehat{\mathbf{A}}$.

1.2.3 Addition and Subtraction

Two vectors $\mathbf{A}$ and $\mathbf{B}$ are added by placing the tip of one at the tail of the other, as shown in Fig. 1.2. The sum $\mathbf{A} + \mathbf{B}$ is the vector obtained by connecting the tail of the first vector to the tip of the second vector. In Fig. 1.2a, $\mathbf{B}$ is moved parallel to itself, in Fig. 1.2b $\mathbf{A}$ is moved parallel to itself. Clearly

$$\mathbf{A} + \mathbf{B} = \mathbf{B} + \mathbf{A}. \qquad (1.3)$$

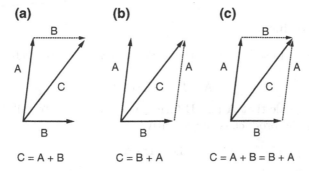

Fig. 1.2. Addition of two vectors: (a) connecting the tail of $\mathbf{B}$ to the tip of $\mathbf{A}$; (b) connecting the tail of $\mathbf{A}$ to the tip of $\mathbf{B}$; (c) parallelogram law which is valid for both free and bound vectors

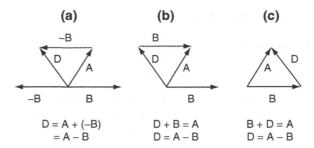

Fig. 1.3. Subtraction of two vectors: (a) as addition of a negative vector; (b) as an inverse of addition; (c) as the tip-to-tip vector which is the most useful interpretation of vector subtraction

If the two vectors to be added are considered to be the sides of a parallelogram, the sum is seen to be the diagonal as shown in Fig. 1.2c. This *parallelogram rule* is valid for both free vectors and bound vectors, and is often used to define the sum of two vectors. It is also the basis for decomposing a vector into its components.

Subtraction of vectors is illustrated in Fig. 1.3. In Fig. 1.3a subtraction is taken as a special case of addition

$$\mathbf{A} - \mathbf{B} = \mathbf{A} + (-\mathbf{B}).$$
(1.4)

In Fig. 1.3b, subtraction is taken as an inverse operation of addition. Clearly, they are equivalent. The most often and the most useful definition of vector subtraction is illustrated in Fig. 1.3c, namely $\mathbf{A} - \mathbf{B}$ is the *tip-to-tip* vector $\mathbf{D}$, starting from the tip of $\mathbf{B}$ directed towards the tip of $\mathbf{A}$.

Graphically it can also be easily shown that vector addition is associative

$$\mathbf{A} + (\mathbf{B} + \mathbf{C}) = (\mathbf{B} + \mathbf{A}) + \mathbf{C}.$$
(1.5)

If $\mathbf{A}$, $\mathbf{B}$, and $\mathbf{C}$ are the three sides of a parallelepied, then $\mathbf{A} + \mathbf{B} + \mathbf{C}$ is the vector along the longest diagonal.

1.2.4 Dot Product

The dot product (also known as the *scalar product*) of two vectors is defined to be

$$\mathbf{A} \cdot \mathbf{B} = AB \cos \theta,$$
(1.6)

where θ is the angle that $\mathbf{A}$ and $\mathbf{B}$ form when placed tail-to-tail. Since it is a scalar, clearly the product is commutative

$$\mathbf{A} \cdot \mathbf{B} = \mathbf{B} \cdot \mathbf{A}.$$
(1.7)

Geometrically, $\mathbf{A} \cdot \mathbf{B} = AB_A$ where B_A is the projection of $\mathbf{B}$ on $\mathbf{A}$, as shown in Fig. 1.4. It is also equal to BA_B, where A_B is the projection of $\mathbf{A}$ on $\mathbf{B}$.

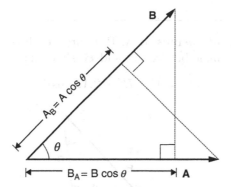

Fig. 1.4. Dot product of two vectors. $\mathbf{A} \cdot \mathbf{B} = AB_A = BA_B = AB\cos\theta$

If the two vectors are parallel, then $\theta = 0$ and $\mathbf{A} \cdot \mathbf{B} = AB$. In particular,

$$\mathbf{A} \cdot \mathbf{A} = A^2, \tag{1.8}$$

which says that the square of the magnitude of any vector is equal to its dot product with itself.

If $\mathbf{A}$ and $\mathbf{B}$ are perpendicular, then $\theta = 90°$ and $\mathbf{A} \cdot \mathbf{B} = 0$. Conversely, if we can show $\mathbf{A} \cdot \mathbf{B} = 0$, then we have proved that $\mathbf{A}$ is perpendicular to $\mathbf{B}$.

It is clear from Fig. 1.5 that

$$A(B + C)_A = AB_A + AC_A.$$

This shows that the distributive law holds for the dot product

$$\mathbf{A} \cdot (\mathbf{B} + \mathbf{C}) = \mathbf{A} \cdot \mathbf{B} + \mathbf{A} \cdot \mathbf{C}. \tag{1.9}$$

With vector notations, many geometrical facts can be readily demonstrated.

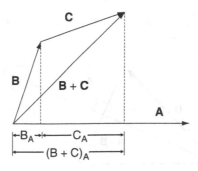

Fig. 1.5. Distributive law of dot product of two vectors. $\mathbf{A} \cdot (\mathbf{B} + \mathbf{C}) = \mathbf{A} \cdot \mathbf{B} + \mathbf{A} \cdot \mathbf{C}$

Example 1.2.1. **Law of cosines.** If A, B, C are the three sides of a triangle, and θ is the interior angle between **A** and **B**, show that

$$C^2 = A^2 + B^2 - 2AB\cos\theta.$$

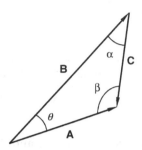

Fig. 1.6. The law of cosine can be readily shown with dot product of vectors, and the law of sine, with cross product

Solution 1.2.1. Let the triangle be formed by the three vectors **A**, **B**, and **C** as shown in Fig. 1.6. Since $\mathbf{C} = \mathbf{A} - \mathbf{B}$,

$$\mathbf{C}\cdot\mathbf{C} = (\mathbf{A} - \mathbf{B})\cdot(\mathbf{A} - \mathbf{B}) = \mathbf{A}\cdot\mathbf{A} - \mathbf{A}\cdot\mathbf{B} - \mathbf{B}\cdot\mathbf{A} + \mathbf{B}\cdot\mathbf{B}.$$

It follows

$$C^2 = A^2 + B^2 - 2AB\cos\theta.$$

Example 1.2.2. Prove that the diagonals of a parallelogram bisect each other.

Solution 1.2.2. Let the two adjacent sides of the parallelogram be represented by vectors **A** and **B** as shown in Fig. 1.7. The two diagonals are $\mathbf{A} - \mathbf{B}$ and $\mathbf{A} + \mathbf{B}$. The vector from the bottom left corner to the mid-point of the diagonal $\mathbf{A} - \mathbf{B}$ is

$$\mathbf{B} + \frac{1}{2}(\mathbf{A} - \mathbf{B}) = \frac{1}{2}(\mathbf{A} + \mathbf{B}),$$

which is also the half of the other diagonal $\mathbf{A} + \mathbf{B}$. Therefore they bisect each other.

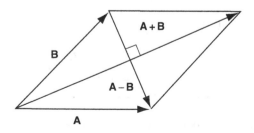

Fig. 1.7. Diagonals of a parallelogram bisect each other; diagonals of a rhombus ($A = B$) are perpendicular to each other

Example 1.2.3. Prove that the diagonals of a rhombus (a parallelogram with equal sides) are orthogonal (perpendicular to each other).

Solution 1.2.3. Again let the two adjacent sides be $\mathbf{A}$ and $\mathbf{B}$. The dot product of the two diagonals (Fig. 1.7) is

$$(\mathbf{A} + \mathbf{B}) \cdot (\mathbf{A} - \mathbf{B}) = \mathbf{A} \cdot \mathbf{A} + \mathbf{B} \cdot \mathbf{A} - \mathbf{A} \cdot \mathbf{B} - \mathbf{B} \cdot \mathbf{B} = A^2 - B^2.$$

For a rhombus, $A = B$. Therefore the dot product of the diagonals is equal to zero. Hence they are perpendicular to each other.

Example 1.2.4. Show that in a parallelogram, the two lines from one corner to the midpoints of the two opposite sides trisect the diagonal they cross.

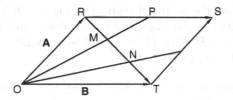

Fig. 1.8. Two lines from one corner of a parallelogram to the midpoints of the two opposite sides trisect the diagonal they cross

Solution 1.2.4. With the parallelogram shown in Fig. 1.8, it is clear that the line from O to the midpoint of RS is represented by the vector $\mathbf{A} + \frac{1}{2}\mathbf{B}$. A vector drawn from O to any point on this line can be written as

$$\mathbf{r}(\lambda) = \lambda \left(\mathbf{A} + \frac{1}{2}\mathbf{B} \right),$$

where λ is a real number which adjusts the length of OP. The diagonal RT is represented by the vector $\mathbf{B} - \mathbf{A}$. A vector drawn from O to any point on this diagonal is

$$\mathbf{r}(\mu) = \mathbf{A} + \mu \left(\mathbf{B} - \mathbf{A} \right),$$

where the parameter μ adjusts the length of the diagonal. The two lines meet when

$$\lambda \left(\mathbf{A} + \frac{1}{2}\mathbf{B} \right) = \mathbf{A} + \mu \left(\mathbf{B} - \mathbf{A} \right),$$

which can be written as

$$(\lambda - 1 + \mu) \mathbf{A} + \left(\frac{1}{2}\lambda - u \right) \mathbf{B} = 0.$$

This gives $\mu = \dfrac{1}{3}$ and $\lambda = \dfrac{2}{3}$, so the length of RM is one-third of RT. Similarly, we can show the length NT is one-third of RT.

Example 1.2.5. Show that an angle inscribed in a semicircle is a right angle.

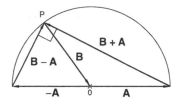

Fig. 1.9. The circum-angle of a semicircle is a right angle

Solution 1.2.5. With the semicircle shown in Fig. 1.9, it is clear the magnitude of $\mathbf{A}$ is the same as the magnitude of $\mathbf{B}$, since they both equal to the radius of the circle $A = B$. Thus $(\mathbf{B} - \mathbf{A}) \cdot (\mathbf{B} + \mathbf{A}) = B^2 - A^2 = 0$. Therefore $(\mathbf{B} - \mathbf{A})$ is perpendicular to $(\mathbf{B} + \mathbf{A})$.

1.2.5 Vector Components

For algebraic description of vectors, we introduce a coordinate system for the reference frame, although it is important to keep in mind that the magnitude and direction of a vector is independent of the reference frame. We will first use the rectangular Cartesian coordinates to express vectors in terms of their components. Let $\mathbf{i}$ be a unit vector in the positive x direction, and $\mathbf{j}$ and $\mathbf{k}$ be unit vectors in the positive y and z directions. An arbitrary vector $\mathbf{A}$ can be expanded in terms of these *basis vectors* as shown in Fig. 1.10:

$$\mathbf{A} = A_x \mathbf{i} + A_y \mathbf{j} + A_z \mathbf{k}, \tag{1.10}$$

where A_x, A_y, and A_z are the projections of $\mathbf{A}$ along the three coordinate axes, they are called components of $\mathbf{A}$.

Since $\mathbf{i}, \mathbf{j}$, and $\mathbf{k}$ are mutually perpendicular unit vectors, by the definition of dot product

$$\mathbf{i} \cdot \mathbf{i} = \mathbf{j} \cdot \mathbf{j} = \mathbf{k} \cdot \mathbf{k} = 1, \tag{1.11}$$

$$\mathbf{i} \cdot \mathbf{j} = \mathbf{j} \cdot \mathbf{k} = \mathbf{k} \cdot \mathbf{i} = 0. \tag{1.12}$$

Because the dot product is distributive, it follows that

$$\mathbf{A} \cdot \mathbf{i} = (A_x \mathbf{i} + A_y \mathbf{j} + A_z \mathbf{k}) \cdot \mathbf{i}$$
$$= A_x \mathbf{i} \cdot \mathbf{i} + A_y \mathbf{j} \cdot \mathbf{i} + A_z \mathbf{k} \cdot \mathbf{i} = A_x,$$

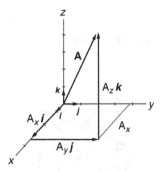

Fig. 1.10. Vector components. **i**, **j**, **k** are three unit vectors pointing in the direction of positive x-, y- and z-axis, respectively. A_x, A_y, A_z are the projections of **A** on these axes. They are components of **A** and $\mathbf{A} = A_x\mathbf{i} + A_y\mathbf{j} + A_z\mathbf{k}$

$$\mathbf{A} \cdot \mathbf{j} = A_y, \qquad \mathbf{A} \cdot \mathbf{k} = A_z,$$

the dot product of **A** with any unit vector is the projection of A along the direction of that unit vector (or the component of **A** along that direction). Thus, (1.10) can be written as

$$\mathbf{A} = (\mathbf{A} \cdot \mathbf{i})\mathbf{i} + (\mathbf{A} \cdot \mathbf{j})\mathbf{j} + (\mathbf{A} \cdot \mathbf{k})\mathbf{k}. \tag{1.13}$$

Furthermore, using the distributive law of dot product and (1.11) and (1.12), we have

$$\mathbf{A} \cdot \mathbf{B} = (A_x\mathbf{i} + A_y\mathbf{j} + A_z\mathbf{k}) \cdot (B_x\mathbf{i} + B_y\mathbf{j} + B_z\mathbf{k})$$
$$= A_xB_x + A_yB_y + A_zB_z, \tag{1.14}$$

and

$$\mathbf{A} \cdot \mathbf{A} = A_x^2 + A_y^2 + A_z^2 = A^2. \tag{1.15}$$

Since $\mathbf{A} \cdot \mathbf{B} = AB\cos\theta$, the angle between **A** and **B** is given by

$$\theta = \cos^{-1}\frac{\mathbf{A} \cdot \mathbf{B}}{AB} = \cos^{-1}\left(\frac{A_xB_x + A_yB_y + A_zB_z}{AB}\right). \tag{1.16}$$

Example 1.2.6. Find the angle between $\mathbf{A} = 3\mathbf{i} + 6\mathbf{j} + 9\mathbf{k}$ and $\mathbf{B} = -2\mathbf{i} + 3\mathbf{j} + \mathbf{k}$.

Solution 1.2.6.

$$A = (3^2 + 6^2 + 9^2)^{1/2} = 3\sqrt{14}; \quad B = \left((-2)^2 + 3^2 + 1^2\right)^{1/2} = \sqrt{14},$$

$$\mathbf{A} \cdot \mathbf{B} = 3 \times (-2) + 6 \times 3 + 9 \times 1 = 21$$

$$\cos\theta = \frac{\mathbf{A} \cdot \mathbf{B}}{AB} = \frac{21}{3\sqrt{14}\sqrt{14}} = \frac{7}{14} = \frac{1}{2},$$

$$\theta = \cos^{-1}\left(\frac{1}{2}\right) = 60°.$$

Example 1.2.7. Find the angle θ between the face diagonals $\mathbf{A}$ and $\mathbf{B}$ of a cube shown in Fig. 1.11.

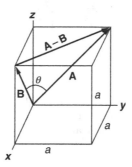

Fig. 1.11. The angle between the two face diagonals of a cube is $60°$

Solution 1.2.7. The answer can be easily found from geometry. The triangle formed by $\mathbf{A}$, $\mathbf{B}$ and $\mathbf{A}-\mathbf{B}$ is clearly an equilateral triangle, therefore $\theta = 60°$. Now with dot product approach, we have

$$\mathbf{A} = a\mathbf{j} + a\mathbf{k}; \quad \mathbf{B} = a\mathbf{i} + a\mathbf{k}; \quad A = \sqrt{2}a = B,$$

$$\mathbf{A} \cdot \mathbf{B} = a \cdot 0 + 0 \cdot a + a \cdot a = a^2 = AB \cos\theta = 2a^2 \cos\theta.$$

Therefore

$$\cos\theta = \frac{a^2}{2a^2} = \frac{1}{2}, \quad \theta = 60°.$$

Example 1.2.8. If $\mathbf{A} = 3\mathbf{i} + 6\mathbf{j} + 9\mathbf{k}$ and $\mathbf{B} = -2\mathbf{i} + 3\mathbf{j} + \mathbf{k}$, find the projection of $\mathbf{A}$ on $\mathbf{B}$.

Solution 1.2.8. The unit vector along $\mathbf{B}$ is

$$\mathbf{n} = \frac{\mathbf{B}}{B} = \frac{-2\mathbf{i}+3\mathbf{j} + \mathbf{k}}{\sqrt{14}}.$$

The projection of $\mathbf{A}$ on $\mathbf{B}$ is then

$$\mathbf{A} \cdot \mathbf{n} = \frac{1}{B}\mathbf{A} \cdot \mathbf{B} = \frac{1}{\sqrt{14}}(3\mathbf{i}+6\mathbf{j}+9\mathbf{k}) \cdot (-2\mathbf{i}+3\mathbf{j} + \mathbf{k}) = \frac{21}{\sqrt{14}}.$$

Example 1.2.9. The angles between the vector $\mathbf{A}$ and the three basis vectors $\mathbf{i}, \mathbf{j}$, and $\mathbf{k}$ are, respectively, α, β, and γ. Show that $\cos^2\alpha + \cos^2\beta + \cos^2\gamma = 1$.

Solution 1.2.9. The projections of **A** on $\mathbf{i}, \mathbf{j}, \mathbf{k}$ are, respectively,

$$A_x = \mathbf{A} \cdot \mathbf{i} = A \cos \alpha; \quad A_y = \mathbf{A} \cdot \mathbf{j} = A \cos \beta; \quad A_z = \mathbf{A} \cdot \mathbf{k} = A \cos \gamma.$$

Thus

$$A_x^2 + A_y^2 + A_z^2 = A^2 \cos^2 \alpha + A^2 \cos^2 \beta + A^2 \cos^2 \gamma = A^2 \left(\cos^2 \alpha + \cos^2 \beta + \cos^2 \gamma \right).$$

Since $A_x^2 + A_y^2 + A_z^2 = A^2$, therefore

$$\cos^2 \alpha + \cos^2 \beta + \cos^2 \gamma = 1.$$

The quantities $\cos \alpha$, $\cos \beta$, and $\cos \gamma$ are often denoted l, m, and n, respectively, and they are called the *direction cosine* of **A**.

1.2.6 Cross Product

The vector cross product written as

$$\mathbf{C} = \mathbf{A} \times \mathbf{B} \tag{1.17}$$

is another particular combination of the two vectors **A** and **B**, which is also very useful. It is defined as a vector (therefore the alternative name: *vector product*) with a magnitude

$$C = AB \sin \theta, \tag{1.18}$$

where θ is the angle between **A** and **B**, and a direction perpendicular to the plane of **A** and **B** in the sense of the advance of a right-hand screw as it is turned from **A** to **B**. In other words, if the fingers of your right hand point in the direction of the first vector **A** and curl around toward the second vector **B**, then your thumb will indicate the positive direction of **C** as shown in Fig. 1.12.

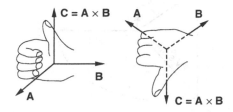

Fig. 1.12. Right-hand rule of cross product $\mathbf{A} \times \mathbf{B} = \mathbf{C}$. If the fingers of your right hand point in the direction of the first vector **A** and curl around toward the second vector **B**, then your thumb will indicate the positive direction of **C**

With this choice of direction, we see that cross product is anticommutative

$$\mathbf{A} \times \mathbf{B} = -\mathbf{B} \times \mathbf{A}. \tag{1.19}$$

It is also clear that if $\mathbf{A}$ and $\mathbf{B}$ are parallel, then $\mathbf{A} \times \mathbf{B} = 0$, since θ is equal to zero.

From this definition, the cross products of the basis vectors ($\mathbf{i}$, $\mathbf{j}$, $\mathbf{k}$) can be easily obtained

$$\mathbf{i} \times \mathbf{i} = \mathbf{j} \times \mathbf{j} = \mathbf{k} \times \mathbf{k} = 0, \tag{1.20}$$

$$\mathbf{i} \times \mathbf{j} = -\mathbf{j} \times \mathbf{i} = \mathbf{k},$$
$$\mathbf{j} \times \mathbf{k} = -\mathbf{k} \times \mathbf{j} = \mathbf{i},$$
$$\mathbf{k} \times \mathbf{i} = -\mathbf{i} \times \mathbf{k} = \mathbf{j}. \tag{1.21}$$

The following example illustrates the cross product of two nonorthogonal vectors. If $\mathbf{V}$ is a vector in the xz-plane and the angle between $\mathbf{V}$ and $\mathbf{k}$, the unit vector along the z-axis, is θ as shown in Fig. 1.13, then

$$\mathbf{k} \times \mathbf{V} = V \sin \theta \mathbf{j}.$$

Since $|\mathbf{k} \times \mathbf{V}| = |\mathbf{k}|\,|\mathbf{V}| \sin \theta = V \sin \theta$ is equal to the projection of $\mathbf{V}$ on the $\mathbf{xy}$-plane, the vector $\mathbf{k} \times \mathbf{V}$ is the result of rotating this projection 90° around the $\mathbf{z}$ axis.

With this understanding, we can readily demonstrate the distributive law of the cross product

$$\mathbf{A} \times (\mathbf{B} + \mathbf{C}) = \mathbf{A} \times \mathbf{B} + \mathbf{A} \times \mathbf{C}. \tag{1.22}$$

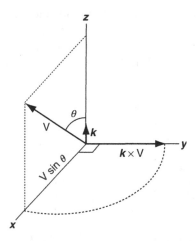

Fig. 1.13. The cross product of $\mathbf{k}$, the unit vector along the z-axis, and $\mathbf{V}$, a vector in the xz-plane

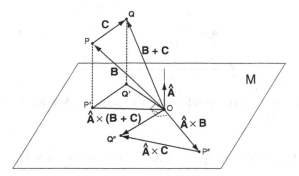

Fig. 1.14. Distributive law of cross product $\mathbf{A} \times (\mathbf{B} + \mathbf{C}) = \mathbf{A} \times \mathbf{B} + \mathbf{A} \times \mathbf{C}$

Let the triangle formed by the vectors $\mathbf{B}$, $\mathbf{C}$, and $\mathbf{B} + \mathbf{C}$ be arbitrarily oriented with respect to the vector $\mathbf{A}$ as shown in Fig. 1.14. Its projection on the plane M perpendicular to $\mathbf{A}$ is the triangle $OP'Q'$. Turn this triangle $90°$ around $\mathbf{A}$, we obtain another triangle $OP''Q''$. The three sides of the triangle $OP''Q''$ are $\widehat{\mathbf{A}} \times \mathbf{B}$, $\widehat{\mathbf{A}} \times \mathbf{C}$, and $\widehat{\mathbf{A}} \times (\mathbf{B} + \mathbf{C})$, where $\widehat{\mathbf{A}}$ is the unit vector along the direction of $\mathbf{A}$. It follows from the rule of vector addition that

$$\widehat{\mathbf{A}} \times (\mathbf{B} + \mathbf{C}) = \widehat{\mathbf{A}} \times \mathbf{B} + \widehat{\mathbf{A}} \times \mathbf{C}.$$

Multiplying both sides by the magnitude A, we obtain (1.22).

With the distributive law and (1.20) and (1.21), we can easily express the cross product $\mathbf{A} \times \mathbf{B}$ in terms of the components of $\mathbf{A}$ and $\mathbf{B}$:

$$\begin{aligned}
\mathbf{A} \times \mathbf{B} &= (A_x\mathbf{i} + A_y\mathbf{j} + A_z\mathbf{k}) \times (B_x\mathbf{i} + B_y\mathbf{j} + B_z\mathbf{k}) \\
&= A_xB_x\mathbf{i} \times \mathbf{i} + A_xB_y\mathbf{i} \times \mathbf{j} + A_xB_z\mathbf{i} \times \mathbf{k} \\
&\quad + A_yB_x\mathbf{j} \times \mathbf{i} + A_yB_y\mathbf{j} \times \mathbf{j} + A_yB_z\mathbf{j} \times \mathbf{k} \\
&\quad + A_zB_x\mathbf{k} \times \mathbf{i} + A_zB_y\mathbf{k} \times \mathbf{j} + A_zB_z\mathbf{k} \times \mathbf{k} \\
&= (A_yB_z - A_zB_y)\mathbf{i} + (A_zB_x - A_xB_z)\mathbf{j} + (A_xB_y - A_yB_x)\mathbf{k}. \quad (1.23)
\end{aligned}$$

This cumbersome equation can be more neatly expressed as the determinant

$$\mathbf{A} \times \mathbf{B} = \begin{vmatrix} \mathbf{i} & \mathbf{j} & \mathbf{k} \\ A_x & A_y & A_z \\ B_x & B_y & B_z \end{vmatrix}, \qquad (1.24)$$

with the understanding that it is to be expanded about its first row. The determinant form is not only easier to remember but also more convenient to use.

The cross product has a useful geometrical interpretation. Figure 1.15 shows a parallelogram having $\mathbf{A}$ and $\mathbf{B}$ as co-terminal edges. The area of this parallelogram is equal to the base A times the height h. But $h = B \sin\theta$, so

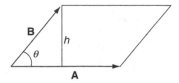

Fig. 1.15. The area of the parallelogram formed by **A** and **B** is equal to the magnitude of **A** × **B**

$$\text{Parallelogram Area} = Ah = AB \sin \theta = |\mathbf{A} \times \mathbf{B}| . \qquad (1.25)$$

Thus the magnitude of **A** × **B** is equal to the area of the parallelogram formed by **A** and **B**, its direction is normal to the plane of this parallelogram. This suggests that area may be treated as a vector quantity.

Since the area of the triangle formed by **A** and **B** as co-terminal edges is clearly half of the area of the parallelogram, so we also have

$$\text{Triangle Area} = \frac{1}{2} |\mathbf{A} \times \mathbf{B}| . \qquad (1.26)$$

Example 1.2.10. **The law of sine.** With the triangle in Fig. 1.6, show that

$$\frac{\sin \theta}{C} = \frac{\sin \alpha}{A} = \frac{\sin \beta}{B} .$$

Solution 1.2.10. The area of the triangle is equal to $\frac{1}{2} |\mathbf{A} \times \mathbf{B}| = \frac{1}{2} AB \sin \theta$. The same area is also given by $\frac{1}{2} |\mathbf{A} \times \mathbf{C}| = \frac{1}{2} AC \sin \beta$. Therefore,

$$AB \sin \theta = AC \sin \beta .$$

It follows $\dfrac{\sin \theta}{C} = \dfrac{\sin \beta}{B}$. Similarly, $\dfrac{\sin \theta}{C} = \dfrac{\sin \alpha}{A}$. Hence

$$\frac{\sin \theta}{C} = \frac{\sin \alpha}{A} = \frac{\sin \beta}{B} .$$

Lagrange Identity

The magnitude of $|\mathbf{A} \times \mathbf{B}|$ can be expressed in terms of $\mathbf{A}, \mathbf{B}$, and $\mathbf{A} \cdot \mathbf{B}$ through the equation

$$|\mathbf{A} \times \mathbf{B}|^2 = A^2 B^2 - (\mathbf{A} \cdot \mathbf{B})^2 , \qquad (1.27)$$

known as the *Lagrange identity*. This relation follows from the fact

$$|\mathbf{A} \times \mathbf{B}|^2 = (AB \sin\theta)^2 = A^2 B^2 (1 - \cos^2\theta)$$
$$= A^2 B^2 - A^2 B^2 \cos^2\theta = A^2 B^2 - (\mathbf{A} \cdot \mathbf{B})^2 .$$

This relation can also be shown by the components of the vectors. It follows from (1.24) that

$$|\mathbf{A} \times \mathbf{B}|^2 = (A_y B_z - A_z B_y)^2 + (A_z B_x - A_x B_z)^2 + (A_x B_y - A_y B_x)^2 \quad (1.28)$$

and

$$A^2 B^2 - (\mathbf{A} \cdot \mathbf{B})^2 = (A_x^2 + A_y^2 + A_z^2)(B_x^2 + B_y^2 + B_z^2) - (A_x B_x + A_y B_y + A_z B_z)^2 .$$
$$(1.29)$$

Multiplying out the right-hand sides of these two equations, we see that they are identical term by term.

1.2.7 Triple Products

Scalar Triple Product

The combination $(\mathbf{A} \times \mathbf{B}) \cdot \mathbf{C}$ is known as the triple scalar product. $\mathbf{A} \times \mathbf{B}$ is a vector. The dot product of this vector with the vector $\mathbf{C}$ gives a scalar. The triple scalar product has a direct geometrical interpretation. The three vectors can be used to define a parallelopiped as shown in Fig. 1.16. The magnitude of $\mathbf{A} \times \mathbf{B}$ is the area of the parallelogram base and its direction is normal (perpendicular) to the base. The projection of $\mathbf{C}$ onto the unit normal of the base is the height h of the parallelopiped. Therefore, $(\mathbf{A} \times \mathbf{B}) \cdot \mathbf{C}$ is equal to the area of the base times the height which is the volume of the parallelopiped:

$$\text{Parallelopiped Volume} = Area \times h = |\mathbf{A} \times \mathbf{B}| h = (\mathbf{A} \times \mathbf{B}) \cdot \mathbf{C}.$$

The volume of a tetrahedron is equal to one-third of the height times the area of the triangular base. Thus the volume of the tetrahedron formed by the

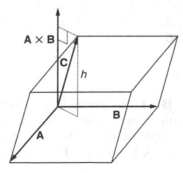

Fig. 1.16. The volume of the parallelopiped is equal to the triple scalar product of its edges as vectors

vectors $\mathbf{A}, \mathbf{B}$, and $\mathbf{C}$ as concurrent edges is equal to one-sixth of the scalar triple product of these three vectors:

$$\text{Tetrahedron Volume} = \frac{1}{3}h \times \frac{1}{2}|\mathbf{A} \times \mathbf{B}| = \frac{1}{6}(\mathbf{A} \times \mathbf{B}) \cdot \mathbf{C}.$$

In calculating the volume of the parallelopiped we can consider just as well $\mathbf{B} \times \mathbf{C}$ or $\mathbf{C} \times \mathbf{A}$ as the base. Since the volume is the same regardless of which side we choose as the base, we see that

$$(\mathbf{A} \times \mathbf{B}) \cdot \mathbf{C} = \mathbf{A} \cdot (\mathbf{B} \times \mathbf{C}) = (\mathbf{C} \times \mathbf{A}) \cdot \mathbf{B}. \tag{1.30}$$

The parentheses in this equation are often omitted, since the cross product must be performed first. If the dot product were performed first, the expression would become a scalar crossed into a vector, which is an undefined and meaningless operation. Without the parentheses, $\mathbf{A} \times \mathbf{B} \cdot \mathbf{C} = \mathbf{A} \cdot \mathbf{B} \times \mathbf{C}$, we see that in any scalar triple product, *the dot and the cross can be interchanged* without altering the value of the product. This is an easy way to remember this relation.

It is clear that if h is reduced to zero, the volume will become zero also. Therefore if $\mathbf{C}$ is in the same plane as $\mathbf{A}$ and $\mathbf{B}$, the scalar triple product $(\mathbf{A} \times \mathbf{B}) \cdot \mathbf{C}$ vanishes. In particular

$$(\mathbf{A} \times \mathbf{B}) \cdot \mathbf{A} = (\mathbf{A} \times \mathbf{B}) \cdot \mathbf{B} = 0. \tag{1.31}$$

A convenient expression in terms of components for the triple scalar product is provided by the determinant

$$\mathbf{A} \cdot (\mathbf{B} \times \mathbf{C}) = (A_x\mathbf{i} + A_y\mathbf{j} + A_z\mathbf{k}) \cdot \begin{vmatrix} \mathbf{i} & \mathbf{j} & \mathbf{k} \\ B_x & B_y & B_z \\ C_x & C_y & C_z \end{vmatrix} = \begin{vmatrix} A_x & A_y & A_z \\ B_x & B_y & B_z \\ C_x & C_y & C_z \end{vmatrix}. \tag{1.32}$$

The rules for interchanging rows of a determinant provide another verification of (1.30)

$$(\mathbf{A} \times \mathbf{B}) \cdot \mathbf{C} = \begin{vmatrix} C_x & C_y & C_z \\ A_x & A_y & A_z \\ B_x & B_y & B_z \end{vmatrix} = \begin{vmatrix} A_x & A_y & A_z \\ B_x & B_y & B_z \\ C_x & C_y & C_z \end{vmatrix} = \begin{vmatrix} B_x & B_y & B_z \\ C_x & C_y & C_z \\ A_x & A_y & A_z \end{vmatrix}.$$

Vector Triple Product

The triple product $\mathbf{A} \times (\mathbf{B} \times \mathbf{C})$ is a meaningful operation, because $\mathbf{B} \times \mathbf{C}$ is a vector, and can form cross product with $\mathbf{A}$ to give another vector (hence the name vector triple product). In this case, the parentheses are necessary, because $\mathbf{A} \times (\mathbf{B} \times \mathbf{C})$ and $(\mathbf{A} \times \mathbf{B}) \times \mathbf{C}$ are two different vectors. For example,

$$\mathbf{i} \times (\mathbf{i} \times \mathbf{j}) = \mathbf{i} \times \mathbf{k} = -\mathbf{j} \quad \text{and} \quad (\mathbf{i} \times \mathbf{i}) \times \mathbf{j} = 0 \times \mathbf{j} = 0.$$

The relation

$$\mathbf{A} \times (\mathbf{B} \times \mathbf{C}) = (\mathbf{A} \cdot \mathbf{C})\mathbf{B} - (\mathbf{A} \cdot \mathbf{B})\mathbf{C} \qquad (1.33)$$

is a very important identity. Because of its frequent use in a variety of problems, this relation should be memorized. This relation (sometimes known as ACB–ABC rule) can be verified by the direct but tedious method of expanding both sides into their cartesian components. A vector equation is, of course, independent of any particular coordinate system. Therefore, it might be more instructive to prove (1.33) without coordinate components.

Let $(\mathbf{B} \times \mathbf{C}) = \mathbf{D}$, hence $\mathbf{D}$ is perpendicular to the plane of $\mathbf{B}$ and $\mathbf{C}$. Now the vector $\mathbf{A} \times (\mathbf{B} \times \mathbf{C}) = \mathbf{A} \times \mathbf{D}$ is perpendicular to $\mathbf{D}$, therefore it is in the plane of $\mathbf{B}$ and $\mathbf{C}$. Thus we can write

$$\mathbf{A} \times (\mathbf{B} \times \mathbf{C}) = \alpha\mathbf{B} + \beta\mathbf{C}, \qquad (1.34)$$

where α and β are scalar constants. Furthermore, $\mathbf{A} \times (\mathbf{B} \times \mathbf{C})$ is also perpendicular to $\mathbf{A}$. So, the dot product of $\mathbf{A}$ with this vector must be zero:

$$\mathbf{A} \cdot [\mathbf{A} \times (\mathbf{B} \times \mathbf{C})] = \alpha\mathbf{A} \cdot \mathbf{B} + \beta\mathbf{A} \cdot \mathbf{C} = 0.$$

It follows that

$$\beta = -\alpha \frac{\mathbf{A} \cdot \mathbf{B}}{\mathbf{A} \cdot \mathbf{C}}$$

and (1.34) becomes

$$\mathbf{A} \times (\mathbf{B} \times \mathbf{C}) = \frac{\alpha}{\mathbf{A} \cdot \mathbf{C}}[(\mathbf{A} \cdot \mathbf{C})\mathbf{B} - (\mathbf{A} \cdot \mathbf{B})\mathbf{C}]. \qquad (1.35)$$

This equation is valid for any set of vectors. For the special case $\mathbf{B} = \mathbf{A}$, this equation reduces to

$$\mathbf{A} \times (\mathbf{A} \times \mathbf{C}) = \frac{\alpha}{\mathbf{A} \cdot \mathbf{C}}[(\mathbf{A} \cdot \mathbf{C})\mathbf{A} - (\mathbf{A} \cdot \mathbf{A})\mathbf{C}]. \qquad (1.36)$$

Take the dot product with $\mathbf{C}$, we have

$$\mathbf{C} \cdot [\mathbf{A} \times (\mathbf{A} \times \mathbf{C})] = \frac{\alpha}{\mathbf{A} \cdot \mathbf{C}}[(\mathbf{A} \cdot \mathbf{C})^2 - A^2 C^2]. \qquad (1.37)$$

Recall the property of the scalar triple product $\mathbf{C} \cdot (\mathbf{A} \times \mathbf{D}) = (\mathbf{C} \times \mathbf{A}) \cdot \mathbf{D}$, with $\mathbf{D} = (\mathbf{A} \times \mathbf{C})$ the left-hand side of the last equation becomes

$$\mathbf{C} \cdot [\mathbf{A} \times (\mathbf{A} \times \mathbf{C})] = (\mathbf{C} \times \mathbf{A}) \cdot (\mathbf{A} \times \mathbf{C}) = -|\mathbf{A} \times \mathbf{C}|^2.$$

Using the Lagrange identity (1.27) to express $|\mathbf{A} \times \mathbf{C}|^2$, we have

$$\mathbf{C} \cdot [\mathbf{A} \times (\mathbf{A} \times \mathbf{C})] = -[(A^2 C^2 - (\mathbf{A} \cdot \mathbf{C})^2]. \qquad (1.38)$$

Comparing (1.37) and (1.38) we see that

$$\frac{\alpha}{\mathbf{A} \cdot \mathbf{C}} = 1,$$

and (1.35) reduces to the ACB–ABC rule of (1.33).

All higher vector products can be simplified by repeated application of scalar and vector triple products.

Example 1.2.11. Use the scalar triple product to prove the distributive law of cross product: $\mathbf{A} \times (\mathbf{B} + \mathbf{C}) = \mathbf{A} \times \mathbf{B} + \mathbf{A} \times \mathbf{C}$.

Solution 1.2.11. First take a dot product $\mathbf{D} \cdot \mathbf{A} \times (\mathbf{B} + \mathbf{C})$ with an arbitrary vector $\mathbf{D}$, then regard $(\mathbf{B} + \mathbf{C})$ as one vector:

$$\mathbf{D} \cdot \mathbf{A} \times (\mathbf{B} + \mathbf{C}) = \mathbf{D} \times \mathbf{A} \cdot (\mathbf{B} + \mathbf{C}) = \mathbf{D} \times \mathbf{A} \cdot \mathbf{B} + \mathbf{D} \times \mathbf{A} \cdot \mathbf{C}$$
$$= \mathbf{D} \cdot \mathbf{A} \times \mathbf{B} + \mathbf{D} \cdot \mathbf{A} \times \mathbf{C} = \mathbf{D} \cdot [\mathbf{A} \times \mathbf{B} + \mathbf{A} \times \mathbf{C}].$$

(The first step is evident because dot and cross can be interchanged in the scalar triple product; in the second step we regard $\mathbf{D} \times \mathbf{A}$ as one vector and use the distributive law of the dot product; in the third step we interchange dot and cross again; in the last step we use again the distributive law of the dot product to factor out $\mathbf{D}$.) Since $\mathbf{D}$ can be any vector, it follows that $\mathbf{A} \times (\mathbf{B} + \mathbf{C}) = \mathbf{A} \times \mathbf{B} + \mathbf{A} \times \mathbf{C}$.

Example 1.2.12. Prove the general form of the Lagrange identity:

$$(\mathbf{A} \times \mathbf{B}) \cdot (\mathbf{C} \times \mathbf{D}) = (\mathbf{A} \cdot \mathbf{C})(\mathbf{B} \cdot \mathbf{D}) - (\mathbf{A} \cdot \mathbf{D})(\mathbf{B} \cdot \mathbf{C}).$$

Solution 1.2.12. First regard $(\mathbf{C} \times \mathbf{D})$ as one vector and interchange the cross and the dot in the scalar triple product $(\mathbf{A} \times \mathbf{B}) \cdot (\mathbf{C} \times \mathbf{D})$:

$$(\mathbf{A} \times \mathbf{B}) \cdot (\mathbf{C} \times \mathbf{D}) = \mathbf{A} \times \mathbf{B} \cdot (\mathbf{C} \times \mathbf{D}) = \mathbf{A} \cdot \mathbf{B} \times (\mathbf{C} \times \mathbf{D}).$$

Then we expand the vector triple product $\mathbf{B} \times (\mathbf{C} \times \mathbf{D})$ in

$$\mathbf{A} \cdot \mathbf{B} \times (\mathbf{C} \times \mathbf{D}) = \mathbf{A} \cdot [(\mathbf{B} \cdot \mathbf{D})\mathbf{C} - (\mathbf{B} \cdot \mathbf{C})\mathbf{D}].$$

Since $(\mathbf{B} \cdot \mathbf{D})$ and $(\mathbf{B} \cdot \mathbf{C})$ are scalars, the distributive law of dot product gives

$$\mathbf{A} \cdot [(\mathbf{B} \cdot \mathbf{D})\mathbf{C} - (\mathbf{B} \cdot \mathbf{C})\mathbf{D}] = (\mathbf{A} \cdot \mathbf{C})(\mathbf{B} \cdot \mathbf{D}) - (\mathbf{A} \cdot \mathbf{D})(\mathbf{B} \cdot \mathbf{C}).$$

Example 1.2.13. The dot and cross products of $\mathbf{u}$ with $\mathbf{A}$ are given by

$$\mathbf{A} \cdot \mathbf{u} = C; \qquad\qquad \mathbf{A} \times \mathbf{u} = \mathbf{B}.$$

Express $\mathbf{u}$ in terms of $\mathbf{A}, \mathbf{B}$, and C.

Solution 1.2.13.

$$\mathbf{A} \times (\mathbf{A} \times \mathbf{u}) = (\mathbf{A} \cdot \mathbf{u})\,\mathbf{A} - (\mathbf{A} \cdot \mathbf{A})\,\mathbf{u} = C\mathbf{A} - A^2\mathbf{u},$$
$$\mathbf{A} \times (\mathbf{A} \times \mathbf{u}) = \mathbf{A} \times \mathbf{B}.$$
$$C\mathbf{A} - A^2\mathbf{u} = A \times B$$
$$\mathbf{u} = \frac{1}{A^2}[C\mathbf{A} - \mathbf{A} \times \mathbf{B}].$$

Example 1.2.14. The force **F** experienced by the charge q moving with velocity **V** in the magnetic field **B** is given by the Lorentz force equation

$$\mathbf{F} = q\left(\mathbf{V} \times \mathbf{B}\right).$$

In three separate experiments, it was found

$$\mathbf{V} = \mathbf{i}, \qquad \mathbf{F}/q = 2\mathbf{k} - 4\mathbf{j},$$
$$\mathbf{V} = \mathbf{j}, \qquad \mathbf{F}/q = 4\mathbf{i} - \mathbf{k},$$
$$\mathbf{V} = \mathbf{k}, \qquad \mathbf{F}/q = \mathbf{j} - 2\mathbf{i}.$$

From these results determine the magnetic field **B**.

Solution 1.2.14. These results can be expressed as

$$\mathbf{i} \times \mathbf{B} = 2\mathbf{k} - 4\mathbf{j} \quad (1); \quad \mathbf{j} \times \mathbf{B} = 4\mathbf{i} - \mathbf{k} \quad (2); \quad \mathbf{k} \times \mathbf{B} = \mathbf{j} - 2\mathbf{i} \quad (3).$$

From (1)

$$\mathbf{i} \times (\mathbf{i} \times \mathbf{B}) = \mathbf{i} \times (2\mathbf{k} - 4\mathbf{j}) = -2\mathbf{j} - 4\mathbf{k},$$
$$\mathbf{i} \times (\mathbf{i} \times \mathbf{B}) = (\mathbf{i} \cdot \mathbf{B})\mathbf{i} - (\mathbf{i} \cdot \mathbf{i})\mathbf{B} = B_x\mathbf{i} - \mathbf{B};$$

therefore,

$$B_x\mathbf{i} - \mathbf{B} = -2\mathbf{j} - 4\mathbf{k} \quad or \quad \mathbf{B} = B_x\mathbf{i} + 2\mathbf{j} + 4\mathbf{k}.$$

From (2)

$$\mathbf{k} \cdot (\mathbf{j} \times \mathbf{B}) = \mathbf{k} \cdot (4\mathbf{i} - \mathbf{k}) = -1,$$
$$\mathbf{k} \cdot (\mathbf{j} \times \mathbf{B}) = (\mathbf{k} \times \mathbf{j}) \cdot \mathbf{B} = -\mathbf{i} \cdot \mathbf{B}.$$

Thus,

$$-\mathbf{i} \cdot \mathbf{B} = -1, \quad or \quad B_x = 1.$$

The final result is obtained just from these two conditions

$$\mathbf{B} = \mathbf{i} + 2\mathbf{j} + 4\mathbf{k}.$$

We can use the third condition as a consistency check

$$\mathbf{k} \times \mathbf{B} = \mathbf{k} \times (\mathbf{i} + 2\mathbf{j} + 4\mathbf{k}) = \mathbf{j} - 2\mathbf{i},$$

which is in agreement with (3).

Example 1.2.15. **Reciprocal vectors.** If **a**, **b**, **c** are three noncoplanar vectors,

$$\mathbf{a}' = \frac{\mathbf{b} \times \mathbf{c}}{\mathbf{a} \cdot \mathbf{b} \times \mathbf{c}}, \quad \mathbf{b}' = \frac{\mathbf{c} \times \mathbf{a}}{\mathbf{a} \cdot \mathbf{b} \times \mathbf{c}}, \quad \mathbf{c}' = \frac{\mathbf{a} \times \mathbf{b}}{\mathbf{a} \cdot \mathbf{b} \times \mathbf{c}}$$

are known as the reciprocal vectors. Show that any vector **r** can be expressed as

$$\mathbf{r} = (\mathbf{r} \cdot \mathbf{a}')\,\mathbf{a} + (\mathbf{r} \cdot \mathbf{b}')\,\mathbf{b} + (\mathbf{r} \cdot \mathbf{c}')\,\mathbf{c}.$$

Solution 1.2.15. Method I. Consider the vector product $(\mathbf{r} \times \mathbf{a}) \times (\mathbf{b} \times \mathbf{c})$. First, regard $(\mathbf{r} \times \mathbf{a})$ as one vector and expand

$$(\mathbf{r} \times \mathbf{a}) \times (\mathbf{b} \times \mathbf{c}) = [(\mathbf{r} \times \mathbf{a}) \cdot \mathbf{c}]\,\mathbf{b} - [(\mathbf{r} \times \mathbf{a}) \cdot \mathbf{b}]\,\mathbf{c}.$$

Then, regard $(\mathbf{b} \times \mathbf{c})$ as one vector and expand

$$(\mathbf{r} \times \mathbf{a}) \times (\mathbf{b} \times \mathbf{c}) = [(\mathbf{b} \times \mathbf{c}) \cdot \mathbf{r}]\,\mathbf{a} - [(\mathbf{b} \times \mathbf{c}) \cdot \mathbf{a}]\,\mathbf{r}.$$

Therefore,

$$[(\mathbf{r} \times \mathbf{a}) \cdot \mathbf{c}]\,\mathbf{b} - [(\mathbf{r} \times \mathbf{a}) \cdot \mathbf{b}]\,\mathbf{c} = [(\mathbf{b} \times \mathbf{c}) \cdot \mathbf{r}]\,\mathbf{a} - [(\mathbf{b} \times \mathbf{c}) \cdot \mathbf{a}]\,\mathbf{r}$$

or

$$[(\mathbf{b} \times \mathbf{c}) \cdot \mathbf{a}]\,\mathbf{r} = [(\mathbf{b} \times \mathbf{c}) \cdot \mathbf{r}]\,\mathbf{a} - [(\mathbf{r} \times \mathbf{a}) \cdot \mathbf{c}]\,\mathbf{b} + [(\mathbf{r} \times \mathbf{a}) \cdot \mathbf{b}]\,\mathbf{c}.$$

Since

$$-(\mathbf{r} \times \mathbf{a}) \cdot \mathbf{c} = -\mathbf{r} \cdot (\mathbf{a} \times \mathbf{c}) = \mathbf{r} \cdot (\mathbf{c} \times \mathbf{a}),$$
$$(\mathbf{r} \times \mathbf{a}) \cdot \mathbf{b} = \mathbf{r} \cdot (\mathbf{a} \times \mathbf{b}),$$

it follows that

$$\mathbf{r} = \frac{\mathbf{r} \cdot (\mathbf{b} \times \mathbf{c})}{(\mathbf{b} \times \mathbf{c}) \cdot \mathbf{a}}\,\mathbf{a} + \frac{\mathbf{r} \cdot (\mathbf{c} \times \mathbf{a})}{(\mathbf{b} \times \mathbf{c}) \cdot \mathbf{a}}\,\mathbf{b} + \frac{\mathbf{r} \cdot (\mathbf{a} \times \mathbf{b})}{(\mathbf{b} \times \mathbf{c}) \cdot \mathbf{a}}\,\mathbf{c}$$
$$= (\mathbf{r} \cdot \mathbf{a}')\,\mathbf{a} + (\mathbf{r} \cdot \mathbf{b}')\,\mathbf{b} + (\mathbf{r} \cdot \mathbf{c}')\,\mathbf{c}.$$

Method II. Let

$$\mathbf{r} = q_1 \mathbf{a} + q_2 \mathbf{b} + q_3 \mathbf{c}.$$

$$\mathbf{r} \cdot (\mathbf{b} \times \mathbf{c}) = q_1 \mathbf{a} \cdot (\mathbf{b} \times \mathbf{c}) + q_2 \mathbf{b} \cdot (\mathbf{b} \times \mathbf{c}) + q_3 \mathbf{c} \cdot (\mathbf{b} \times \mathbf{c}).$$

Since $(\mathbf{b} \times \mathbf{c})$ is perpendicular to **b** and perpendicular to **c**, therefore

$$\mathbf{b} \cdot (\mathbf{b} \times \mathbf{c}) = 0, \quad \mathbf{c} \cdot (\mathbf{b} \times \mathbf{c}) = 0.$$

Thus

$$q_1 = \frac{\mathbf{r} \cdot (\mathbf{b} \times \mathbf{c})}{\mathbf{a} \cdot (\mathbf{b} \times \mathbf{c})} = \mathbf{r} \cdot \mathbf{a}'.$$

Similarly,

$$q_2 = \frac{\mathbf{r} \cdot (\mathbf{c} \times \mathbf{a})}{\mathbf{b} \cdot (\mathbf{c} \times \mathbf{a})} = \frac{\mathbf{r} \cdot (\mathbf{c} \times \mathbf{a})}{\mathbf{a} \cdot (\mathbf{b} \times \mathbf{c})} = \mathbf{r} \cdot \mathbf{b}',$$

$$q_3 = \frac{\mathbf{r} \cdot (\mathbf{a} \times \mathbf{b})}{\mathbf{c} \cdot (\mathbf{a} \times \mathbf{b})} = \mathbf{r} \cdot \mathbf{c}'.$$

It follows that

$$\mathbf{r} = (\mathbf{r} \cdot \mathbf{a}')\,\mathbf{a} + (\mathbf{r} \cdot \mathbf{b}')\,\mathbf{b} + (\mathbf{r} \cdot \mathbf{c}')\,\mathbf{c}.$$

1.3 Lines and Planes

Much of analytic geometry can be simplified by the use of vectors. In analytic geometry, a point is a set of three coordinates (x, y, z). All points in space can be defined by the *position vector* $\mathbf{r}(x, y, z)$ (or just $\mathbf{r}$)

$$\mathbf{r} = x\mathbf{i} + y\mathbf{j} + z\mathbf{k}, \tag{1.39}$$

drawn from the origin to the point (x, y, z). To specify a particular point (x_0, y_0, z_0), we use the notation $\mathbf{r}_0(x_0, y_0, z_0)$

$$\mathbf{r}_0 = x_0\mathbf{i} + y_0\mathbf{j} + z_0\mathbf{k}. \tag{1.40}$$

With these notations, we can define lines and planes in space.

1.3.1 Straight Lines

There are several ways to specify a straight line in space. Let us first consider a line through a given point (x_0, y_0, z_0) in the direction of a known vector $\mathbf{v} = a\mathbf{i} + b\mathbf{j} + c\mathbf{k}$. If $\mathbf{r}(x, y, z)$ is any other point on the line, the vector $\mathbf{r} - \mathbf{r}_0$ is parallel to $\mathbf{v}$. Thus, we can write the equation of a straight line as

$$\mathbf{r} - \mathbf{r}_0 = t\mathbf{v}, \tag{1.41}$$

where t is any real number. This equation is called the parametric form of a straight line. It is infinitely long and fixed in space, as shown in Fig. 1.17. It cannot be moved parallel to itself as a free vector. This equation in the form of its components

$$(x - x_0)\mathbf{i} + (y - y_0)\mathbf{j} + (z - z_0)\mathbf{k} = ta\mathbf{i} + tb\mathbf{j} + tc\mathbf{k} \tag{1.42}$$

represents three equations

$$(x - x_0) = ta, \quad (y - y_0) = tb, \quad (z - z_0) = tc. \tag{1.43}$$

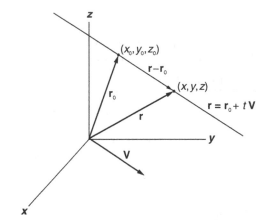

Fig. 1.17. A straight line in the parametric form

Now if a, b, c are not zero, we can solve for t in each of the three equations. The solutions must be equal to each other, since they are all equal to the same t.

$$\frac{x - x_0}{a} = \frac{y - y_0}{b} = \frac{z - z_0}{c}. \tag{1.44}$$

This is called the symmetric form of the equation of a line. If $\mathbf{v}$ is a normalized unit vector, then a, b, c are the direction cosines of the line.

If c happens to be zero, then (1.43) should be written as

$$\frac{x - x_0}{a} = \frac{y - y_0}{b}; \quad z = z_0. \tag{1.45}$$

The equation $z = z_0$ means that the line lies in the plane perpendicular to the z-axis, and the slope of the line is $\frac{b}{a}$. If both b and c are zero, then clearly the line is the intersection of the planes $y = y_0$ and $z = z_0$.

The parametric equation (1.41) has a useful interpretation when the parameter t means time. Consider a particle moving along this straight line. The equation $\mathbf{r} = \mathbf{r}_0 + t\mathbf{v}$ indicates when $t = 0$, the particle is at $\mathbf{r}_0$. As time goes on, the particle is moving with a constant velocity $\mathbf{v}$, or $\dfrac{d\mathbf{r}}{dt} = \mathbf{v}$.

Perpendicular Distance Between Two Skew Lines

Two lines which are not parallel and which do not meet are said to be skew lines. To find the perpendicular distance between them is a difficult problem in analytical geometry. With vectors, it is relatively easy.

Let the equations of two such lines be

$$\mathbf{r} = \mathbf{r}_1 + t\mathbf{v}_1, \tag{1.46}$$

$$\mathbf{r} = \mathbf{r}_2 + t'\mathbf{v}_2. \tag{1.47}$$

Let a on line 1 and b on line 2 be the end points of the common perpendicular on these two lines. We shall suppose that the position vector $\mathbf{r}_a$ from origin to a is given by (1.46) with $t = t_1$, and the position vector $\mathbf{r}_b$, by (1.47) with $t' = t_2$. Accordingly

$$\mathbf{r}_a = \mathbf{r}_1 + t_1 \mathbf{v}_1, \tag{1.48}$$
$$\mathbf{r}_b = \mathbf{r}_2 + t_2 \mathbf{v}_2. \tag{1.49}$$

Since $\mathbf{r}_b - \mathbf{r}_a$ is perpendicular to both $\mathbf{v}_1$ and $\mathbf{v}_2$, it must be in the direction of $(\mathbf{v}_1 \times \mathbf{v}_2)$. If d is the length of $\mathbf{r}_b - \mathbf{r}_a$, then

$$\mathbf{r}_b - \mathbf{r}_a = \frac{\mathbf{v}_1 \times \mathbf{v}_2}{|\mathbf{v}_1 \times \mathbf{v}_2|} d.$$

Since $\mathbf{r}_b - \mathbf{r}_a = \mathbf{r}_2 - \mathbf{r}_1 + t_2 \mathbf{v}_2 - t_1 \mathbf{v}_1$,

$$\mathbf{r}_2 - \mathbf{r}_1 + t_2 \mathbf{v}_2 - t_1 \mathbf{v}_1 = \frac{\mathbf{v}_1 \times \mathbf{v}_2}{|\mathbf{v}_1 \times \mathbf{v}_2|} d. \tag{1.50}$$

Then take the dot product with $\mathbf{v}_1 \times \mathbf{v}_2$ on both sides of this equation. Since $\mathbf{v}_1 \times \mathbf{v}_2 \cdot \mathbf{v}_1 = \mathbf{v}_1 \times \mathbf{v}_2 \cdot \mathbf{v}_2 = 0$, the equation becomes

$$(\mathbf{r}_2 - \mathbf{r}_1) \cdot (\mathbf{v}_1 \times \mathbf{v}_2) = |\mathbf{v}_1 \times \mathbf{v}_2| d;$$

therefore,

$$d = \frac{(\mathbf{r}_2 - \mathbf{r}_1) \cdot (\mathbf{v}_1 \times \mathbf{v}_2)}{|\mathbf{v}_1 \times \mathbf{v}_2|}. \tag{1.51}$$

This must be the perpendicular distance between the two lines. Clearly, if $d = 0$, the two lines meet. Therefore the condition for the two lines to meet is

$$(\mathbf{r}_2 - \mathbf{r}_1) \cdot (\mathbf{v}_1 \times \mathbf{v}_2) = 0. \tag{1.52}$$

To determine the coordinates of a and b, take the dot product of (1.50) first with $\mathbf{v}_1$, then with $\mathbf{v}_2$

$$(\mathbf{r}_2 - \mathbf{r}_1) \cdot \mathbf{v}_1 + t_2 \mathbf{v}_2 \cdot \mathbf{v}_1 - t_1 \mathbf{v}_1 \cdot \mathbf{v}_1 = 0, \tag{1.53}$$
$$(\mathbf{r}_2 - \mathbf{r}_1) \cdot \mathbf{v}_2 + t_2 \mathbf{v}_2 \cdot \mathbf{v}_2 - t_1 \mathbf{v}_1 \cdot \mathbf{v}_2 = 0. \tag{1.54}$$

These two equations can be solved for t_1 and t_2. With them, $\mathbf{r}_a$ and $\mathbf{r}_b$ can be found from (1.48) and (1.49).

Example 1.3.1. Find the coordinates of the end points a and b of the common perpendicular to the following two lines

$$\mathbf{r} = 9\mathbf{j} + 2\mathbf{k} + t(3\mathbf{i} - \mathbf{j} + \mathbf{k}),$$
$$\mathbf{r} = -6\mathbf{i} - 5\mathbf{j} + 10\mathbf{k} + t'(-3\mathbf{i} + 2\mathbf{j} + 4\mathbf{k}).$$

Solution 1.3.1. The first line passes through the point $r_1(0, 9, 2)$ in the direction of $v_1 = 3i - j + k$. The second line passes the point $r_2(-6, -5, 10)$ in the direction of $v_2 = -3i + 2j + 4k$. From (1.53) and (1.54),

$$(r_2 - r_1) \cdot v_1 + t_2 v_2 \cdot v_1 - t_1 v_1 \cdot v_1 = 4 - 7t_2 - 11t_1 = 0,$$
$$(r_2 - r_1) \cdot v_2 + t_2 v_2 \cdot v_2 - t_1 v_1 \cdot v_2 = 22 + 29t_2 + 7t_1 = 0.$$

The solution of these two equations is

$$t_1 = 1, \quad t_2 = -1.$$

Therefore, by (1.48) and (1.49),

$$r_a = r_1 + t_1 v_1 = 3i + 8j + 3k,$$
$$r_b = r_2 + t_2 v_2 = -3i - 7j + 6k.$$

Example 1.3.2. Find the perpendicular distance between the two lines of the previous example, and an equation for the perpendicular line.

Solution 1.3.2. The perpendicular distance d is simply

$$d = |r_a - r_b| = |6i + 15j - 3k| = 3\sqrt{30}.$$

It can be readily verified that this is the same as given by (1.51). The perpendicular line can be represented by the equation

$$r = r_a + t(r_a - r_b) = (3 + 6t)\,i + (8 + 15t)\,j + (3 - 3t)\,k,$$

or equivalently

$$\frac{x - 3}{6} = \frac{y - 8}{15} = \frac{z - 3}{-3}.$$

This line can also be represented by

$$r = r_b + s\,(r_a - r_b) = (-3 + 6s)\,i + (-7 + 15s)\,j + (6 - 3s)\,k,$$

or

$$\frac{x + 3}{6} = \frac{y + 7}{15} = \frac{z - 6}{-3}.$$

Example 1.3.3. Find (a) the perpendicular distance of the point $(5, 4, 2)$ from the line

$$\frac{x + 1}{2} = \frac{y - 3}{3} = \frac{z - 1}{-1},$$

and (b) also the coordinates of the point where the perpendicular meets the line, and (c) an equation for the line of the perpendicular.

Solution 1.3.3. The line passes the point $\mathbf{r}_0 = -\mathbf{i} + 3\mathbf{j} + \mathbf{k}$ and is in the direction of $\mathbf{v} = 2\mathbf{i} + 3\mathbf{j} - \mathbf{k}$. The parametric form of the line is

$$\mathbf{r} = \mathbf{r}_0 + t\mathbf{v} = -\mathbf{i} + 3\mathbf{j} + \mathbf{k} + t\left(2\mathbf{i} + 3\mathbf{j} - \mathbf{k}\right).$$

Let the position vector to $(5, 4, 2)$ be

$$\mathbf{r}_1 = 5\mathbf{i} + 4\mathbf{j} + 2\mathbf{k}.$$

The distance d from the point $\mathbf{r}_1(5, 4, 2)$ to the line is the cross product of $(\mathbf{r}_1 - \mathbf{r}_0)$ with the unit vector in the $\mathbf{v}$ direction

$$d = \left|(\mathbf{r}_1 - \mathbf{r}_0) \times \frac{\mathbf{v}}{v}\right| = \left|(6\mathbf{i} + \mathbf{j} + \mathbf{k}) \times \frac{2\mathbf{i} + 3\mathbf{j} - \mathbf{k}}{\sqrt{4 + 9 + 1}}\right| = 2\sqrt{6}.$$

Let p be the point where the perpendicular meets the line. Since p is on the given line, the position vector to p must satisfy the equation of the given line with a specific t. Let that specific t be t_1,

$$\mathbf{r}_p = -\mathbf{i} + 3\mathbf{j} + \mathbf{k} + t_1\left(2\mathbf{i} + 3\mathbf{j} - \mathbf{k}\right).$$

Since $(\mathbf{r}_p - \mathbf{r}_0)$ is perpendicular to $\mathbf{v}$, their dot product must be zero

$$(\mathbf{r}_1 - \mathbf{r}_p) \cdot \mathbf{v} = \left[6\mathbf{i} + \mathbf{j} + \mathbf{k} - t_1\left(2\mathbf{i} + 3\mathbf{j} - \mathbf{k}\right)\right] \cdot \left(2\mathbf{i} + 3\mathbf{j} - \mathbf{k}\right) = 0.$$

This gives $t_1 = 1$. It follows that

$$\mathbf{r}_p = -\mathbf{i} + 3\mathbf{j} + \mathbf{k} + 1\left(2\mathbf{i} + 3\mathbf{j} - \mathbf{k}\right) = \mathbf{i} + 6\mathbf{j}.$$

In other words, the coordinates of the foot of the perpendicular is $(1, 6, 0)$. The equation of the perpendicular can be obtained from the fact that it passes $\mathbf{r}_1$ (or $\mathbf{r}_p$) and is in the direction of the vector from $\mathbf{r}_p$ to $\mathbf{r}_1$. So the equation can be written as

$$\mathbf{r} = \mathbf{r}_1 + t\left(\mathbf{r}_1 - \mathbf{r}_p\right) = 5\mathbf{i} + 4\mathbf{j} + 2\mathbf{k} + t(4\mathbf{i} - 2\mathbf{j} + 2\mathbf{k})$$

or

$$\frac{x - 5}{4} = \frac{y - 4}{-2} = \frac{z - 2}{2}.$$

1.3.2 Planes in Space

A set of parallel planes in space can be determined by a vector normal (perpendicular) to these planes. A particular plane can be specified by an additional condition, such as the perpendicular distance between the origin and the plane, or a given point that lies on the plane.

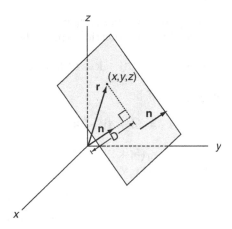

Fig. 1.18. A plane in space. The position vector **r** from the origin to any point on the plane must satisfy the equation $\mathbf{r} \cdot \mathbf{n} = D$ where **n** is the unit normal to the plane and D is the perpendicular distance between the origin and the plane

Suppose the unit normal to the plane is known to be

$$\mathbf{n} = A\mathbf{i} + B\mathbf{j} + C\mathbf{k} \tag{1.55}$$

and the distance between this plane and the origin is D as shown in Fig. 1.18. If (x, y, z) is a point (any point) on the plane, then it is clear from the figure that the projection of $\mathbf{r}(x, y, z)$ on **n** must satisfy the equation

$$\mathbf{r} \cdot \mathbf{n} = D. \tag{1.56}$$

Multiplying out its components, we have the familiar equation of a plane

$$Ax + By + Cz = D. \tag{1.57}$$

This equation will not be changed if both sides are multiplied by the same constant. The result represents, of course, the same plane. However, it is to be emphasized that if the right-hand side D is interpreted as the distance between the plane and the origin, then the coefficients A, B, C on the left-hand side must satisfy the condition $A^2 + B^2 + C^2 = 1$, since they should be the direction cosine of the unit normal.

The plane can also be uniquely specified if in addition to the unit normal **n**, a point (x_0, y_0, z_0) that lies on the plane is known. In this case, the vector $\mathbf{r} - \mathbf{r_0}$ must be perpendicular to **n**. That means the dot product with **n** must vanish,

$$(\mathbf{r} - \mathbf{r_0}) \cdot \mathbf{n} = 0. \tag{1.58}$$

Multiplying out the components, we can write this equation as

$$Ax + By + Cz = Ax_0 + By_0 + Cz_0. \tag{1.59}$$

This is in the same form of (1.57). Clearly the distance between this plane and the origin is

$$D = Ax_0 + By_0 + Cz_0. \tag{1.60}$$

In general, to find distances between points and lines or planes, it is far simpler to use vectors as compared with calculations in analytic geometry without vectors.

Example 1.3.4. Find the perpendicular distance from the point $(1, 2, 3)$ to the plane described by the equation $3x - 2y + 5z = 10$.

Solution 1.3.4. The unit normal to the plane is

$$\mathbf{n} = \frac{3}{\sqrt{9+4+25}}\mathbf{i} - \frac{2}{\sqrt{9+4+25}}\mathbf{j} + \frac{5}{\sqrt{9+4+25}}\mathbf{k}.$$

The distance from the origin to the plane is

$$D = \frac{10}{\sqrt{9+4+25}} = \frac{10}{\sqrt{38}}.$$

The length of the projection of $\mathbf{r}_1 = \mathbf{i} + 2\mathbf{j} + 3\mathbf{k}$ on $\mathbf{n}$ is

$$\ell = \mathbf{r}_1 \cdot \mathbf{n} = \frac{3}{\sqrt{38}} - \frac{4}{\sqrt{38}} + \frac{15}{\sqrt{38}} = \frac{14}{\sqrt{38}}.$$

The distance from $(1, 2, 3)$ to the plane is therefore

$$d = |\ell - D| = \frac{14}{\sqrt{38}} - \frac{10}{\sqrt{38}} = \frac{4}{\sqrt{38}}.$$

Another way to find the solution is to note that the required distance is equal to the projection on $\mathbf{n}$ of any vector joining the given point with a point on the plane. Note that

$$\mathbf{r}_0 = D\mathbf{n} = \frac{30}{38}\mathbf{i} - \frac{20}{38}\mathbf{j} + \frac{50}{38}\mathbf{k}$$

is the position vector of the foot of the perpendicular from the origin to the plane. Therefore

$$d = |(\mathbf{r}_1 - \mathbf{r}_0) \cdot \mathbf{n}| = \frac{4}{\sqrt{38}}.$$

Example 1.3.5. Find the coordinates of the foot of the perpendicular from the point $(1, 2, 3)$ to the plane of the last example.

Solution 1.3.5. Let the position vector from the origin to the foot of the perpendicular be $\mathbf{r}_p$. The vector $\mathbf{r}_1 - \mathbf{r}_p$ is perpendicular to the plane, therefore it is parallel to the unit normal vector $\mathbf{n}$ of the plane,

$$\mathbf{r}_1 - \mathbf{r}_p = k\mathbf{n}.$$

It follows that $|\mathbf{r}_1 - \mathbf{r}_p| = k$. Since $|\mathbf{r}_1 - \mathbf{r}_p| = d$, so $k = d$. Thus,

$$\mathbf{r}_p = \mathbf{r}_1 - d\mathbf{n} = \mathbf{i} + 2\mathbf{j} + 3\mathbf{k} - \frac{4}{\sqrt{38}}\left(\frac{3}{\sqrt{38}}\mathbf{i} - \frac{2}{\sqrt{38}}\mathbf{j} + \frac{5}{\sqrt{9+4+25}}\mathbf{k}\right).$$

Hence, the coordinates of the foot of the perpendicular are $\left(\frac{26}{38}, \frac{84}{38}, \frac{94}{38}\right)$.

Example 1.3.6. A plane intersects the x, y, and z axes, respectively, at $(a, 0, 0)$, $(0, b, 0)$, and $(0, 0, c)$ (Fig. 1.19). Find (a) a unit normal to this plane, (b) the perpendicular distance between the origin and this plane, (c) the equation for this plane.

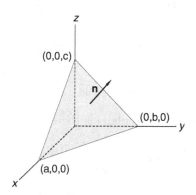

Fig. 1.19. The plane $bcx + acy + abz = abc$ cuts the three axes at $(a, 0, 0)$, $(0, a, 0)$, $(0, 0, a)$, respectively

Solution 1.3.6. Let $\mathbf{r}_1 = a\mathbf{i}$, $\mathbf{r}_2 = b\mathbf{j}$, $\mathbf{r}_3 = c\mathbf{k}$. The vector from $(a, 0, 0)$ to $(0, b, 0)$ is $\mathbf{r}_2 - \mathbf{r}_1 = b\mathbf{j} - a\mathbf{i}$, and the vector from $(a, 0, 0)$ to $(0, 0, c)$ is $\mathbf{r}_3 - \mathbf{r}_1 = c\mathbf{k} - a\mathbf{i}$. The unit normal to this plane must be in the same direction as the cross product of these two vectors:

$$\mathbf{n} = \frac{(b\mathbf{j} - a\mathbf{i}) \times (c\mathbf{k} - a\mathbf{i})}{|(b\mathbf{j} - a\mathbf{i}) \times (c\mathbf{k} - a\mathbf{i})|},$$

$$(b\mathbf{j} - a\mathbf{i}) \times (c\mathbf{k} - a\mathbf{i}) = \begin{vmatrix} \mathbf{i} & \mathbf{j} & \mathbf{k} \\ -a & b & 0 \\ -a & 0 & c \end{vmatrix} = bc\mathbf{i} + ac\mathbf{j} + ab\mathbf{k},$$

$$\mathbf{n} = \frac{bc\mathbf{i} + ac\mathbf{j} + ab\mathbf{k}}{\left((bc)^2 + (ac)^2 + (ab)^2\right)^{1/2}}.$$

If the perpendicular distance from the origin to the plane is D, then

$$D = \mathbf{r}_1 \cdot \mathbf{n} = \mathbf{r}_2 \cdot \mathbf{n} = \mathbf{r}_3 \cdot \mathbf{n},$$

$$D = \frac{abc}{\left((bc)^2 + (ac)^2 + (ab)^2\right)^{1/2}}.$$

In general, the position vector $\mathbf{r} = x\mathbf{i} + y\mathbf{j} + z\mathbf{k}$ to any point (x, y, z) on the plane must satisfy the equation

$$\mathbf{r} \cdot \mathbf{n} = D,$$

$$\frac{xbc + yac + zab}{\left((bc)^2 + (ac)^2 + (ab)^2\right)^{1/2}} = \frac{abc}{\left((bc)^2 + (ac)^2 + (ab)^2\right)^{1/2}}.$$

Therefore the equation of this plane can be written as

$$bcx + acy + abz = abc$$

or as

$$\frac{x}{a} + \frac{y}{b} + \frac{z}{c} = 1.$$

Another way to find an equation for the plane is to note that the scalar triple product of three coplanar vectors is equal to zero. If the position vector from the origin to any point (x, y, z) on the plane is $\mathbf{r} = x\mathbf{i} + y\mathbf{j} + z\mathbf{k}$, then the three vectors $\mathbf{r} - a\mathbf{i}$, $b\mathbf{j} - a\mathbf{i}$ and $c\mathbf{k} - a\mathbf{i}$ are in the same plane. Therefore,

$$(\mathbf{r} - a\mathbf{i}) \cdot (b\mathbf{j} - a\mathbf{i}) \times (c\mathbf{k} - a\mathbf{i}) = \begin{vmatrix} x - a & y & z \\ -a & b & 0 \\ -a & 0 & c \end{vmatrix}$$

$$= bc\,(x - a) + acy + abz = 0$$

or

$$bcx + acy + abz = abc.$$

Exercises

1. If the vectors $\mathbf{A} = 2\mathbf{i} + 3\mathbf{k}$ and $\mathbf{B} = \mathbf{i} - \mathbf{k}$, find $|\mathbf{A}|$, $|\mathbf{B}|$, $\mathbf{A} + \mathbf{B}$, $\mathbf{A} - \mathbf{B}$, and $\mathbf{A} \cdot \mathbf{B}$. What is the angle between the vectors $\mathbf{A}$ and $\mathbf{B}$?
   Ans. $\sqrt{13}$, $\sqrt{2}$, $3\mathbf{i} + 2\mathbf{k}$, $\mathbf{i} + 4\mathbf{k}$, -1, $101°$.

2. For what value of c are the vectors $c\mathbf{i} + \mathbf{j} + \mathbf{k}$ and $-\mathbf{i} + 2\mathbf{k}$ perpendicular?
   Ans. 2.

3. If $\mathbf{A} = \mathbf{i} + 2\mathbf{j} + 2\mathbf{k}$ and $\mathbf{B} = -6\mathbf{i} + 2\mathbf{j} + 3\mathbf{k}$, find the projection of $\mathbf{A}$ on $\mathbf{B}$, and the projection of $\mathbf{B}$ on $\mathbf{A}$.
   Ans. $4/7$, $4/3$.

4. Show that the vectors $\mathbf{A} = 3\mathbf{i} - 2\mathbf{j} + \mathbf{k}$, $\mathbf{B} = \mathbf{i} - 3\mathbf{j} + 5\mathbf{k}$, and $\mathbf{C} = 2\mathbf{i} + \mathbf{j} - 4\mathbf{k}$ form a right triangle.

5. Use vectors to prove that the line joining the midpoints of two sides of any triangle is parallel to the third side and half its length.

6. Use vectors to show that for any triangle, the medians (the three lines drawn from each vertex to the midpoint of the opposite side) all pass the same point. The point is at two-thirds of the way of the median from the vertex.

7. If $\mathbf{A} = 2\mathbf{i} - 3\mathbf{j} - \mathbf{k}$ and $\mathbf{B} = \mathbf{i} + 4\mathbf{j} - 2\mathbf{k}$, find $\mathbf{A} \times \mathbf{B}$ and $\mathbf{B} \times \mathbf{A}$.
   Ans. $10\mathbf{i} + 3\mathbf{j} + 11\mathbf{k}$, $-10\mathbf{i} - 3\mathbf{j} - 11\mathbf{k}$.

8. Find the area of a parallelogram having diagonals $\mathbf{A} = 3\mathbf{i} + \mathbf{j} - 2\mathbf{k}$ and $\mathbf{B} = \mathbf{i} - 3\mathbf{j} + 4\mathbf{k}$.
   Ans. $5\sqrt{3}$.

9. Evaluate $(2\mathbf{i} - 3\mathbf{j}) \cdot [(\mathbf{i} + \mathbf{j} - \mathbf{k}) \times (3\mathbf{j} - \mathbf{k})]$.
   Ans. 4.

10. Find the volume of the parallelepied whose edges are represented by $\mathbf{A} = 2\mathbf{i} - 3\mathbf{j} + 4\mathbf{k}$, $\mathbf{B} = \mathbf{i} + 2\mathbf{j} - \mathbf{k}$, and $\mathbf{C} = 3\mathbf{i} - \mathbf{j} + 2\mathbf{k}$.
    Ans. 7.

11. Find the constant a such that the vectors $2\mathbf{i} - \mathbf{j} + \mathbf{k}$, $\mathbf{i} + 2\mathbf{j} - 3\mathbf{k}$ and $3\mathbf{i} + a\mathbf{j} + 5\mathbf{k}$ are coplanar.
    Ans. $a = -4$.

12. Show that (a) $(\mathbf{b} \times \mathbf{c}) \times (\mathbf{c} \times \mathbf{a}) = \mathbf{c}(\mathbf{a} \cdot \mathbf{b} \times \mathbf{c})$;
    (b) $(\mathbf{a} \times \mathbf{b}) \cdot (\mathbf{b} \times \mathbf{c}) \times (\mathbf{c} \times \mathbf{a}) = (\mathbf{a} \cdot \mathbf{b} \times \mathbf{c})^2$.
    Hint: to prove (a) first regard $(\mathbf{b} \times \mathbf{c})$ as one vector, then note $\mathbf{b} \times \mathbf{c} \cdot \mathbf{c} = 0$

13. If $\mathbf{a}, \mathbf{b}, \mathbf{c}$ are non-coplanar (so $\mathbf{a} \cdot \mathbf{b} \times \mathbf{c} \neq 0$), and

$$\mathbf{a}' = \frac{\mathbf{b} \times \mathbf{c}}{\mathbf{a} \cdot \mathbf{b} \times \mathbf{c}}, \quad \mathbf{b}' = \frac{\mathbf{c} \times \mathbf{a}}{\mathbf{a} \cdot \mathbf{b} \times \mathbf{c}}, \quad \mathbf{c}' = \frac{\mathbf{a} \times \mathbf{b}}{\mathbf{a} \cdot \mathbf{b} \times \mathbf{c}},$$

show that
   (a) $\mathbf{a}' \cdot \mathbf{a} = \mathbf{b}' \cdot \mathbf{b} = \mathbf{c}' \cdot \mathbf{c} = 1$,
   (b) $\mathbf{a}' \cdot \mathbf{b} = \mathbf{a}' \cdot \mathbf{c} = 0, \quad \mathbf{b}' \cdot \mathbf{a} = \mathbf{b}' \cdot \mathbf{c} = 0, \quad \mathbf{c}' \cdot \mathbf{a} = \mathbf{c}' \cdot \mathbf{b} = 0$,
   (c) if $\mathbf{a} \cdot \mathbf{b} \times \mathbf{c} = V$ then $\mathbf{a}' \cdot \mathbf{b}' \times \mathbf{c}' = 1/V$.

14. Find the perpendicular distance from the point $(-1, 0, 1)$ to the line $\mathbf{r} = 3\mathbf{i} + 2\mathbf{j} + 3\mathbf{k} + t\,(\mathbf{i} + 2\mathbf{j} + 3\mathbf{k})$.
    Ans. $\sqrt{10}$.

15. Find the coordinates of the foot of the perpendicular from the point $(1, 2, 1)$ to the line joining the origin to the point $(2, 2, 5)$.
    Ans. $(2/3, 2/3, 5/3)$.

16. Find the length and equation of the line which is the common perpendicular to the two lines

$$\frac{x - 4}{2} = \frac{y + 2}{1} = \frac{z - 3}{-1}, \quad \frac{x + 7}{3} = \frac{y + 2}{2} = \frac{z - 1}{1}.$$

    Ans. $\sqrt{35}$, $\dfrac{x - 2}{3} = \dfrac{y + 3}{-5} = \dfrac{z - 4}{1}$.

17. Find the distance from $(-2, 4, 5)$ to the plane $2x + 6y - 3z = 10$.
    Ans. $5/7$

18. Find the equation of the plane that is perpendicular to the vector $\mathbf{i} + \mathbf{j} - \mathbf{k}$ and passes through the point $(1, 2, 1)$.
    Ans. $x + y - z = 2$.

19. Find an equation for the plane determined by the points $(2, -1, 1)$, $(3, 2, -1)$, and $(-1, 3, 2)$.
    Ans. $11x + 5y + 13z = 30$.

2

Vector Calculus

So far we have been discussing constant vectors, but the most interesting applications of vectors involve vector functions. The simplest example is a position vector that depends on time. Such a vector can be differentiated with respect to time. The first and second derivatives are simply the velocity and acceleration of the particle whose position is given by the position vector. In this case, the coordinates of the tip of the position vector are functions of time.

Even more interesting are quantities which depend on the position in space. Such quantities are said to form fields. The word "field" has the connotation that the space has some physical properties. For example, the electrical field created by a static charge is that space surrounding the charge which has now been given a certain property, known as the electrical field. Every point in this field is associated with an electric field vector whose magnitude and direction depend on the location of the point. This electric field vector will manifest itself when another charge is brought to that point. Mathematically a vector field is simply a vector function, each of its three components depends on the coordinates of the point. The field may also dependent on time, such as the electric field in the electromagnetic wave.

There are scalar fields, by which we mean that the field is characterized at each point by a single number. Of course the number may change in time, but usually we are talking about the field at a given instant. For example, temperature at different point in space is different, so the temperature is a function of the position. Thus temperature is a scalar field. It is possible to derive one kind of field from another. For example, the directional derivatives of a scalar field lead to a vector field, known as the gradient.

With vector differential calculus, we develop a set of precise terms, such as gradient, divergence, and curl, to describe the rate of change of vector functions with respect to the spatial coordinates. With vector integral calculus, we establish relationships between line, surface, and volume integrals through the theorems of Gauss and Stokes. These are important attributes of vector fields, in terms of which many fundamental laws of physics are expressed.

In this chapter, we shall assume that the vector functions are continuous and differentiable, and the region of interests is simply connected unless otherwise specified. However, this does not mean that singularities and multiple connected regions are not of our concern. They have important implications in physical problems. We will more carefully define and discuss these terms at appropriate places.

2.1 The Time Derivative

Differentiating a vector function is a simple extension of differentiating scalar quantities. If the vector $\mathbf{A}$ depends on time t only, then the derivative of $\mathbf{A}$ with respect to t is defined as

$$\frac{d\mathbf{A}}{dt} = \lim_{\Delta t \to 0} \frac{\mathbf{A}(t + \Delta t) - \mathbf{A}(t)}{\Delta t} = \lim_{\Delta t \to 0} \frac{\Delta \mathbf{A}}{\Delta t}. \tag{2.1}$$

From this definition it follows that the sums and products involving vector quantities can be differentiated as in ordinary calculus; that is

$$\frac{d}{dt}(\mathbf{A} + \mathbf{B}) = \frac{d\mathbf{A}}{dt} + \frac{d\mathbf{B}}{dt}, \tag{2.2}$$

$$\frac{d}{dt}(\mathbf{A} \cdot \mathbf{B}) = \mathbf{A}\frac{d\mathbf{B}}{dt} + \frac{d\mathbf{A}}{dt} \cdot \mathbf{B}, \tag{2.3}$$

$$\frac{d}{dt}(\mathbf{A} \times \mathbf{B}) = \mathbf{A} \times \frac{d\mathbf{B}}{dt} + \frac{d\mathbf{A}}{dt} \times \mathbf{B}. \tag{2.4}$$

Since $\Delta \mathbf{A}$ has components $\Delta A_x, \Delta A_y$, and ΔA_z,

$$\frac{d\mathbf{A}}{dt} = \lim_{\Delta t \to 0} \frac{\Delta A_x \mathbf{i} + \Delta A_y \mathbf{j} + \Delta A_z \mathbf{k}}{\Delta t} = \frac{dA_x}{dt}\mathbf{i} + \frac{dA_y}{dt}\mathbf{j} + \frac{dA_z}{dt}\mathbf{k}. \tag{2.5}$$

The time derivatives of a vector is thus equal to the vector sum of the time derivative of its components.

2.1.1 Velocity and Acceleration

Of particular importance is the case where $\mathbf{A}$ is the position vector $\mathbf{r}$,

$$\mathbf{r}(t) = x(t)\,\mathbf{i} + y(t)\,\mathbf{j} + z(t)\,\mathbf{k}. \tag{2.6}$$

If t changes, the tip of $\mathbf{r}$ traces out a space curve as shown in Fig. 2.1. If a particle is moving along this space curve, then $d\mathbf{r}/dt$ is clearly the velocity $\mathbf{v}$ of the particle along this trajectory

$$\mathbf{v} = \frac{d\mathbf{r}}{dt} = \lim_{\Delta t \to 0} \frac{\Delta \mathbf{r}}{\Delta t} = \lim_{\Delta t \to 0} \frac{\Delta x \mathbf{i} + \Delta y \mathbf{j} + \Delta z \mathbf{k}}{\Delta t} = \frac{dx}{dt}\mathbf{i} + \frac{dy}{dt}\mathbf{j} + \frac{dz}{dt}\mathbf{k} = v_x \mathbf{i} + v_y \mathbf{j} + v_z \mathbf{k}. \tag{2.7}$$

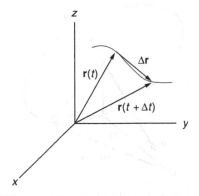

Fig. 2.1. The tip of **r** traces out the trajectory of a particle moving in space, $\Delta \mathbf{r}$ is independent of the origin

It is important to note that the direction of $\Delta \mathbf{r}$ is unrelated to the direction of **r**. In other words the velocity is independent of the origin chosen. Similarly, the acceleration is defined as the rate of change of velocity

$$\mathbf{a} = \frac{d\mathbf{v}}{dt} = \frac{dv_x}{dt}\mathbf{i} + \frac{dv_y}{dt}\mathbf{j} + \frac{dv_z}{dt}\mathbf{k} = \frac{d^2x}{dt^2}\mathbf{i} + \frac{d^2y}{dt^2}\mathbf{j} + \frac{d^2z}{dt^2}\mathbf{k} = \frac{d^2\mathbf{r}}{dt^2}. \tag{2.8}$$

The acceleration is also independent of the origin.

Notation of differentiation with respect to time. A convenient and widely used notation (Newton's notation) is that a single dot above a symbol denotes the first time derivative and two dots denote the second time derivative, and so on. Thus

$$\mathbf{v} = \frac{d\mathbf{r}}{dt} = \dot{\mathbf{r}} = \dot{x}\mathbf{i} + \dot{y}\mathbf{j} + \dot{z}\mathbf{k}, \tag{2.9}$$

$$\mathbf{a} = \dot{\mathbf{v}} = \ddot{\mathbf{r}} = \ddot{x}\mathbf{i} + \ddot{y}\mathbf{j} + \ddot{z}\mathbf{k}. \tag{2.10}$$

2.1.2 Angular Velocity Vector

For a particle moving around a circle, shown in Fig. 2.2, the rate of change of the angular position is called angular speed ω:

$$\omega = \lim_{\Delta t \to 0} \frac{\Delta \theta}{\Delta t} = \frac{d\theta}{dt} = \dot{\theta}. \tag{2.11}$$

The velocity **v** of the particle is, by definition,

$$\mathbf{v} = \frac{d\mathbf{r}}{dt} = \dot{\mathbf{r}}, \tag{2.12}$$

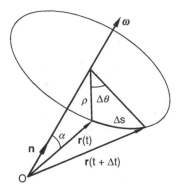

Fig. 2.2. Angular velocity vector $\boldsymbol{\omega}$. The velocity $\mathbf{v}$ of a particle moving around a circle is given by $\mathbf{v} = \boldsymbol{\omega} \times \mathbf{r}$

where $\mathbf{r}$ is the position vector drawn from the origin to the position of the particle. The magnitude of the velocity is given by

$$v = |\mathbf{v}| = \lim_{\Delta t \to 0} \frac{\Delta s}{\Delta t} = \lim_{\Delta t \to 0} \frac{\varrho \Delta \theta}{\Delta t} = \varrho \omega, \tag{2.13}$$

where ϱ is the radius of the circle. The direction of the velocity is, of course, tangent to the circle.

Now, let $\mathbf{n}$ be the unit vector drawn from the origin to the center of the circle, pointing in the positive direction of advance of a right-hand screw when turned in the same sense as the rotation of the particle. Since $|\mathbf{n} \times \mathbf{r}| = r \sin \alpha = \varrho$, as shown in Fig. 2.2, the magnitude of the velocity can be written as

$$v = \varrho \omega = |\mathbf{n} \times \mathbf{r}| \, \omega. \tag{2.14}$$

If we define the angular velocity vector $\boldsymbol{\omega}$ as

$$\boldsymbol{\omega} = \omega \mathbf{n}, \tag{2.15}$$

then we can write the velocity $\mathbf{v}$ as

$$\mathbf{v} = \dot{\mathbf{r}} = \boldsymbol{\omega} \times \mathbf{r}. \tag{2.16}$$

Recalling the definition of cross product of two vectors, one can easily see that both direction and magnitude of the velocity are given by this equation.

A particle moving in space, even though not in a circle, may always be considered at a given instant to be moving in a circular path. Even a straight line can be considered as a circle with infinite radius. The path, which the particle describes during an infinitesimal time interval δt, may be represented as an infinitesimal arc of a circle. Therefore, at any moment, an instantaneous angular velocity vector can be defined to describe the general motion. The instantaneous velocity is then given by (2.16).

Example 2.1.1. Show that the linear momentum, defined as $\mathbf{p} = m\dot{\mathbf{r}}$, always lies in a fixed plane in a central force field. (A central force field means that the force $\mathbf{F}$ is in the radial direction, such as gravitational and electrostatic forces, in other words $\mathbf{F}$ is parallel to $\mathbf{r}$.)

Solution 2.1.1. Let us form the angular momentum $\mathbf{L}$

$$\mathbf{L} = \mathbf{r} \times \mathbf{p} = \mathbf{r} \times m\dot{\mathbf{r}}.$$

Differentiating with respect to time, we have

$$\dot{\mathbf{L}} = \dot{\mathbf{r}} \times \mathbf{p} + \mathbf{r} \times \dot{\mathbf{p}}.$$

Now $\dot{\mathbf{r}} \times \mathbf{p} = \dot{\mathbf{r}} \times m\dot{\mathbf{r}} = 0$ and $\mathbf{r} \times \dot{\mathbf{p}} = \mathbf{r} \times m\ddot{\mathbf{r}}$. According to Newton's second law $m\ddot{\mathbf{r}} = \mathbf{F}$ and $\mathbf{F}$ is parallel to $\mathbf{r}$, therefore $\mathbf{r} \times \mathbf{F} = \mathbf{r} \times m\ddot{\mathbf{r}} = 0$. Thus, $\dot{\mathbf{L}} = 0$. In other words, $\mathbf{L}$ is a constant vector. Furthermore, $\mathbf{L}$ is perpendicular to $\mathbf{p}$, since $\mathbf{r} \times \mathbf{p}$ is perpendicular to $\mathbf{p}$. Therefore $\mathbf{p}$ must always lie in the plane perpendicular to the constant vector $\mathbf{L}$.

Example 2.1.2. Suppose a particle is rotating around the z-axis with a constant angular velocity ω as shown in Fig. 2.3. Find the velocity and acceleration of the particle.

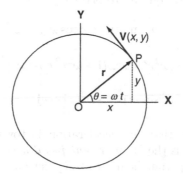

Fig. 2.3. Particle rotating around z-axis with a constant angular velocity ω

Solution 2.1.2. *Method I.* Since the particle is moving in a circular path and z is not changing in this motion, we will consider only the x and y components. The angular velocity vector is in the $\mathbf{k}$ (unit vector along the z axis) direction, $\boldsymbol{\omega} = \omega\mathbf{k}$, and the position vector $\mathbf{r}$ drawn from the origin to the point (x, y) is perpendicular to $\mathbf{k}$. Therefore

$$\mathbf{v} = \boldsymbol{\omega} \times \mathbf{r}.$$

The direction of $\mathbf{v}$ is perpendicular to $\mathbf{k}$ and perpendicular to $\mathbf{r}$, that is, in the tangential direction of the circle. The magnitude of the velocity is

$$v = \omega r \sin(\pi/2) = \omega r.$$

Explicitly

$$\mathbf{v} = \boldsymbol{\omega} \times \mathbf{r} = \begin{vmatrix} \mathbf{i} & \mathbf{j} & \mathbf{k} \\ 0 & 0 & \omega \\ x & y & 0 \end{vmatrix} = -\omega y \mathbf{i} + \omega x \mathbf{j}.$$

The acceleration is given by

$$\mathbf{a} = \frac{d\mathbf{v}}{dt} = \frac{d}{dt}(\boldsymbol{\omega} \times \mathbf{r}) = \dot{\boldsymbol{\omega}} \times \mathbf{r} + \boldsymbol{\omega} \times \dot{\mathbf{r}}.$$

Since $\boldsymbol{\omega}$ is a constant, $\dot{\boldsymbol{\omega}} = \mathbf{0}$. Moreover, $\dot{\mathbf{r}} = \mathbf{v} = \boldsymbol{\omega} \times \mathbf{r}$. Thus

$$\mathbf{a} = \boldsymbol{\omega} \times (\boldsymbol{\omega} \times \mathbf{r}) = \omega^2 \mathbf{k} \times (\mathbf{k} \times \mathbf{r}).$$

Hence, $\mathbf{a}$ is in the $-\mathbf{r}$ direction and its magnitude is equal to $\omega^2 r$.

Method II. The position vector can be explicitly written as

$$\mathbf{r} = x(t)\mathbf{i} + y(t)\mathbf{j} = r\cos\omega t\,\mathbf{i} + r\sin\omega t\,\mathbf{j}. \tag{2.17}$$

The velocity and acceleration are, respectively,

$$\mathbf{v} = \dot{\mathbf{r}} = \dot{x}\mathbf{i} + \dot{y}\mathbf{j} = -\omega r\sin\omega t\,\mathbf{i} + \omega r\cos\omega t\,\mathbf{j} = \omega(-y\mathbf{i} + x\mathbf{j}), \tag{2.18}$$

$$v = (\mathbf{v}\cdot\mathbf{v})^{1/2} = \left(\omega^2 r^2 \sin^2\omega t\,\mathbf{i} + \omega^2 r^2 \cos^2\omega t\right)^{1/2} = \omega r \tag{2.19}$$

$$\mathbf{a} = \dot{\mathbf{v}} = -\omega^2 \cos\omega t\,\mathbf{i} - \omega^2 \sin\omega t\,\mathbf{j} = -\omega^2(x\mathbf{i} + y\mathbf{j}) = -\omega^2 \mathbf{r} = -\frac{v^2}{r^2}\mathbf{r}. \tag{2.20}$$

We see immediately that the acceleration is toward the center with a magnitude of $\omega^2 r$. This is the familiar *centripetal acceleration.*

Also the velocity is perpendicular to the position vector since

$$\mathbf{v}\cdot\mathbf{r} = \omega(-y\mathbf{i} + x\mathbf{j})\cdot(x\mathbf{i} + y\mathbf{j}) = 0.$$

In this example, the magnitude of $\mathbf{r}$ is a constant, we have explicitly shown that the velocity is perpendicular to $\mathbf{r}$. This fact is also a consequence of the following general theorem.

If the magnitude of a vector is not changing, the vector is always orthogonal (perpendicular) to its derivative.

This follows from the fact that if $\mathbf{r} \cdot \mathbf{r} = r_0^2$ and r_0 is a constant, then

$$\frac{d}{dt}(\mathbf{r} \cdot \mathbf{r}) = \frac{d}{dt}r_0^2 = 0,$$

$$\frac{d}{dt}(\mathbf{r} \cdot \mathbf{r}) = \frac{d\mathbf{r}}{dt} \cdot \mathbf{r} + \mathbf{r} \cdot \frac{d\mathbf{r}}{dt} = 2\frac{d\mathbf{r}}{dt} \cdot \mathbf{r} = 0.$$

When the dot product of two vectors is zero, the two vectors are perpendicular to each other.

This can also be seen from geometry. In Fig. 2.1, if $\mathbf{r}(t)$ and $\mathbf{r}(t + \Delta t)$ have the same length, then $\Delta\mathbf{r}$ is the base of an isosceles triangle. When $\Delta t \to 0$, the angle between $\mathbf{r}(t)$ and $\mathbf{r}(t + \Delta t)$ also goes to zero. In that case, the two base angles approach $90°$, which means $\Delta\mathbf{r}$ is perpendicular to $\mathbf{r}$. Since Δt is a scalar, the direction of $\Delta\mathbf{r}/\Delta t$ is determined by $\Delta\mathbf{r}$. Therefore $d\mathbf{r}/dt$ is perpendicular to $\mathbf{r}$.

This theorem is limited neither to the position vector, nor to the time derivative. For example, if vector $\mathbf{A}$ is a function of the arc distance s measured from some fixed point, as long as the magnitude of $\mathbf{A}$ is a constant, it can be shown in the same way that $d\mathbf{A}/ds$ is always perpendicular to $\mathbf{A}$. This theorem is of considerable importance and should always be kept in mind.

Velocity Vector Field. Sometimes the name velocity (or acceleration) vector field is used to mean that at every point (x, y, z) there is a velocity vector whose magnitude and direction depend on where the point is. In other words, the velocity is a vector function which has three components. Each component can be a function of (x, y, z). For example, consider a rotating body. The velocity of the material of the body at any point is a vector which is a function of position. In general, a vector function may also explicitly dependent on time t. For example, in a continuum, such as a fluid, the velocity of the particles in the continuum is a vector field which is not only a function of position but may also of time. To find the acceleration, we can use the chain rule:

$$\mathbf{a} = \frac{d\mathbf{v}}{dt} = \frac{\partial\mathbf{v}}{\partial x}\frac{dx}{dt} + \frac{\partial\mathbf{v}}{\partial y}\frac{dy}{dt} + \frac{\partial\mathbf{v}}{\partial z}\frac{dz}{dt} + \frac{\partial\mathbf{v}}{\partial t}. \tag{2.21}$$

Since

$$\frac{dx}{dt} = v_x, \quad \frac{dy}{dt} = v_y, \quad \frac{dz}{dt} = v_z,$$

It follows that

$$\mathbf{a} = v_x\frac{\partial\mathbf{v}}{\partial x} + v_y\frac{\partial\mathbf{v}}{\partial y} + v_z\frac{\partial\mathbf{v}}{\partial z} + \frac{\partial\mathbf{v}}{\partial t}. \tag{2.22}$$

Therefore the acceleration may also be a vector field.

Example 2.1.3. A body is rotating around the z-axis with an angular velocity ω, find the velocity of the particles in the body as a function of the position, and use (2.22) to find the acceleration of these particles.

Solution 2.1.3. The angular velocity vector is $\boldsymbol{\omega} = \omega\mathbf{k}$, and the velocity of any point in the body is

$$\mathbf{v} = \boldsymbol{\omega} \times \mathbf{r} = \begin{vmatrix} \mathbf{i} & \mathbf{j} & \mathbf{k} \\ 0 & 0 & \omega \\ x & y & z \end{vmatrix} = -\omega y\mathbf{i} + \omega x\mathbf{j}.$$

Thus the components of the velocity vector are

$$v_x(x,y,z) = -\omega y, \quad v_y(x,y,z) = \omega x, \quad v_z(x,y,z) = 0,$$

and

$$\frac{\partial \mathbf{v}}{\partial x} = \omega\mathbf{j}, \quad \frac{\partial \mathbf{v}}{\partial y} = -\omega\mathbf{i}, \quad \frac{\partial \mathbf{v}}{\partial z} = \mathbf{0}, \quad \frac{\partial \mathbf{v}}{\partial t} = \mathbf{0}.$$

Hence according to (2.22)

$$\mathbf{a} = (-\omega y)\omega\mathbf{j} + (\omega x)(-\omega\mathbf{i}) = -\omega^2(x\mathbf{i} + y\mathbf{j}).$$

If we define $\boldsymbol{\varrho} = x\mathbf{i} + y\mathbf{j}$, then the magnitude of $\boldsymbol{\varrho}$ is the perpendicular distance between the particle and the rotating axis. Thus

$$\mathbf{a} = -\omega^2\boldsymbol{\varrho},$$

which shows that every particle has a centripetal acceleration $\omega^2\rho$, as expected.

2.2 Differentiation in Noninertial Reference Systems

The acceleration $\mathbf{a}$ in Newton's equation $\mathbf{F} = m\mathbf{a}$ is to be measured in an inertial reference system. An inertial reference system is either a coordinate system fixed in space, or a system moving with a constant velocity relative to the fixed system. A coordinate system fixed on the earth is not an inertial system because the earth is rotating.

The derivatives of a vector in a noninertial system are, of course, different from those in a fixed system. To find the relationships between them, let us first consider a moving coordinate system that has the same origin as a fixed system. Intuition tells us that, in this case, the only possible relative motion between the coordinate systems is a rotation. To transform the derivatives from one system to the other, we need to take this rotation into account.

Let us denote the quantities associated with the moving system by a prime. The position vector of a particle expressed in terms of the basis vector of the fixed system is

$$\mathbf{r} = x\mathbf{i} + y\mathbf{j} + z\mathbf{k}. \tag{2.23}$$

The same position vector expressed in the moving coordinate system whose origin coincides with that of the fixed system becomes

$$\mathbf{r} = x'\mathbf{i}' + y'\mathbf{j}' + z'\mathbf{k}',\tag{2.24}$$

where $\mathbf{i}'$, $\mathbf{j}'$, $\mathbf{k}'$ are the basis vectors of the moving system.

The velocity $\mathbf{v}$ is by definition the time derivative of the position vector in the fixed system

$$\mathbf{v} = \frac{d\mathbf{r}}{dt} = \frac{dx}{dt}\mathbf{i} + \frac{dy}{dt}\mathbf{j} + \frac{dz}{dt}\mathbf{k}.\tag{2.25}$$

If we express the time derivative of $\mathbf{r}$ in the moving system, then with (2.24) we have

$$\mathbf{v} = \frac{d\mathbf{r}}{dt} = \frac{dx'}{dt}\mathbf{i}' + \frac{dy'}{dt}\mathbf{j}' + \frac{dz'}{dt}\mathbf{k}' + x'\frac{d\mathbf{i}'}{dt} + y'\frac{d\mathbf{j}'}{dt} + z'\frac{d\mathbf{k}'}{dt},\tag{2.26}$$

since $\mathbf{i}'$, $\mathbf{j}'$, $\mathbf{k}'$ are fixed in the moving system, so they are not constant in time. Clearly, the velocity seen in the moving system is

$$\mathbf{v}' = \frac{dx'}{dt}\mathbf{i}' + \frac{dy'}{dt}\mathbf{j}' + \frac{dz'}{dt}\mathbf{k}' = \frac{D\mathbf{r}}{Dt}.\tag{2.27}$$

This equation also defines the operation D/Dt, which simply means the time derivative in the moving system. The notation

$$\frac{D\mathbf{r}}{Dt} = \dot{\mathbf{r}}\tag{2.28}$$

is also often used. As mentioned earlier, a dot on top of a symbol means the time derivative. In addition, it usually means the time derivative in the moving system. Note that while the position vector has the same appearance in both the fixed and the moving system as seen in (2.23) and (2.24), the velocity vector, or any other derivative, has more terms in the moving system than in the fixed system as seen in (2.26) and (2.25) . The three derivatives $(dx'/dt, dy'/dt, dz'/dt)$ are not the components of the velocity vector $\mathbf{v}$ in the moving system, they only appear to be the velocity components to a stationary observer in the moving system. The velocity vector $\mathbf{v}$ expressed in the moving system is given by (2.26), which can be written as

$$\mathbf{v} = \frac{D\mathbf{r}}{Dt} + x'\frac{d\mathbf{i}'}{dt} + y'\frac{d\mathbf{j}'}{dt} + z'\frac{d\mathbf{k}'}{dt}.\tag{2.29}$$

Since $\mathbf{i}'$, $\mathbf{j}'$, $\mathbf{k}'$ are unit vectors, their magnitudes are not changing. Therefore their derivatives must be perpendicular to themselves. For example, $d\mathbf{i}'/dt$ is perpendicular to $\mathbf{i}'$, and lies in the plane of $\mathbf{j}'$ and $\mathbf{k}'$. Thus we can write

$$\frac{d\mathbf{i}'}{dt} = c\mathbf{j}' - b\mathbf{k}',\tag{2.30}$$

where c and $-b$ are two constants. (The reason for choosing these particular symbols for the coefficients of the linear combination is for convenience, as will be clear in a moment). Similarly

$$\frac{d\mathbf{j}'}{dt} = a\mathbf{k}' - f\mathbf{i}', \tag{2.31}$$

$$\frac{d\mathbf{k}'}{dt} = e\mathbf{i}' - d\mathbf{j}'. \tag{2.32}$$

But $\mathbf{i}' = \mathbf{j}' \times \mathbf{k}'$, so

$$\frac{d\mathbf{i}'}{dt} = \frac{d\mathbf{j}'}{dt} \times \mathbf{k}' + \mathbf{j}' \times \frac{d\mathbf{k}'}{dt} = (a\mathbf{k}' - f\mathbf{i}') \times \mathbf{k}' + \mathbf{j}' \times (e\mathbf{i}' - d\mathbf{j}') = f\mathbf{j}' - e\mathbf{k}'. \tag{2.33}$$

Comparing (2.30) and (2.33), we see that $f = c$ and $e = b$. Similarly, from $\mathbf{j}' = \mathbf{k}' \times \mathbf{i}'$ one can show that

$$\frac{d\mathbf{j}'}{dt} = d\mathbf{k}' - c\mathbf{i}'. \tag{2.34}$$

It is clear from (2.31) and (2.34) that $d = a$ and $c = f$.

It follows that

$$\begin{aligned}
x'\frac{d\mathbf{i}'}{dt} + y'\frac{d\mathbf{j}'}{dt} + z'\frac{d\mathbf{k}'}{dt} &= x'\left(c\mathbf{j}' - b\mathbf{k}'\right) + y'\left(a\mathbf{k}' - c\mathbf{i}'\right) + z'\left(b\mathbf{i}' - a\mathbf{j}'\right) \\
&= \mathbf{i}'\left(bz' - cy'\right) + \mathbf{j}'\left(cx' - az'\right) + \mathbf{k}'\left(ay' - bx'\right) \\
&= \begin{vmatrix} \mathbf{i}' & \mathbf{j}' & \mathbf{k}' \\ a & b & c \\ x' & y' & z' \end{vmatrix}.
\end{aligned} \tag{2.35}$$

If we define

$$\boldsymbol{\omega} = a\mathbf{i}' + b\mathbf{j}' + c\mathbf{k}', \tag{2.36}$$

with $\mathbf{r}$ given by (2.24), we can write (2.35) as

$$x'\frac{d\mathbf{i}'}{dt} + y'\frac{d\mathbf{j}'}{dt} + z'\frac{d\mathbf{k}'}{dt} = \boldsymbol{\omega} \times \mathbf{r}. \tag{2.37}$$

The meaning of $\boldsymbol{\omega} \times \mathbf{r}$ is exactly the same as in (2.16). We have thus demonstrated explicitly that the most general relative motion of two coordinate systems having a common origin is a rotation with an instantaneous angular velocity $\boldsymbol{\omega}$. Furthermore, (2.29) becomes

$$\mathbf{v} = \mathbf{v}' + \boldsymbol{\omega} \times \mathbf{r}. \tag{2.38}$$

Often this equation is written in the form

$$\frac{d\mathbf{r}}{dt} = \left(\frac{D}{Dt} + \boldsymbol{\omega} \times\right)\mathbf{r} = \dot{\mathbf{r}} + \boldsymbol{\omega} \times \mathbf{r}, \tag{2.39}$$

with the understanding that the time derivative on the left-hand side is in the fixed system, and on the right-hand side, all quantities are to be expressed in the rotating system.

This analysis is not limited to the position vector. For any vector $\mathbf{A}$, we can follow exactly the same procedure and show that

$$\boxed{\frac{d\mathbf{A}}{dt} = \frac{D}{Dt}\mathbf{A} + \boldsymbol{\omega} \times \mathbf{A}},$$ (2.40)

where

$$\frac{D}{Dt}\mathbf{A} = \dot{\mathbf{A}} = \dot{A}'_x \mathbf{i}' + \dot{A}'_y \mathbf{j}' + \dot{A}'_z \mathbf{k}',$$

$$\boldsymbol{\omega} \times \mathbf{A} = A'_x \frac{d\mathbf{i}'}{dt} + A'_y \frac{d\mathbf{j}'}{dt} + A'_z \frac{d\mathbf{k}'}{dt}.$$

Example 2.2.1. Show that the time derivative of the angular velocity vector is the same in either the fixed or the rotating system.

Solution 2.2.1. Since

$$\frac{d\boldsymbol{\omega}}{dt} = \frac{D}{Dt}\boldsymbol{\omega} + \boldsymbol{\omega} \times \boldsymbol{\omega},$$

but $\boldsymbol{\omega} \times \boldsymbol{\omega} = \mathbf{0}$, therefore the time derivative in the rotating system $\dot{\boldsymbol{\omega}}$ is the same time derivative in the fixed system.

Example 2.2.2. If the rotating system and the fixed system have the same origin, express the acceleration $\mathbf{a}$ in the fixed system in terms of $\mathbf{a}'$, $\mathbf{v}'$, $\boldsymbol{\omega}$, $\dot{\boldsymbol{\omega}}$ of the rotating system.

Solution 2.2.2. By definition $\mathbf{a} = d\mathbf{v}/dt$. So by (2.40)

$$\mathbf{a} = \frac{d\mathbf{v}}{dt} = \frac{D}{Dt}\mathbf{v} + \boldsymbol{\omega} \times \mathbf{v}.$$

Since

$$\mathbf{v} = \frac{d\mathbf{r}}{dt} = \frac{D}{Dt}\mathbf{r} + \boldsymbol{\omega} \times \mathbf{r} = \dot{\mathbf{r}} + \boldsymbol{\omega} \times \mathbf{r},$$

$$\frac{D}{Dt}\mathbf{v} = \frac{D}{Dt}(\dot{\mathbf{r}} + \boldsymbol{\omega} \times \mathbf{r}) = \ddot{\mathbf{r}} + \dot{\boldsymbol{\omega}} \times \mathbf{r} + \boldsymbol{\omega} \times \dot{\mathbf{r}},$$

$$\boldsymbol{\omega} \times \mathbf{v} = \boldsymbol{\omega} \times (\dot{\mathbf{r}} + \boldsymbol{\omega} \times \mathbf{r}) = \boldsymbol{\omega} \times \dot{\mathbf{r}} + \boldsymbol{\omega} \times (\boldsymbol{\omega} \times \mathbf{r}).$$

Therefore,

$$\mathbf{a} = \frac{D}{Dt}\mathbf{v} + \boldsymbol{\omega} \times \mathbf{v} = \ddot{\mathbf{r}} + \dot{\boldsymbol{\omega}} \times \mathbf{r} + \boldsymbol{\omega} \times \dot{\mathbf{r}} + \boldsymbol{\omega} \times \dot{\mathbf{r}} + \boldsymbol{\omega} \times (\boldsymbol{\omega} \times \mathbf{r})$$

$$= \ddot{\mathbf{r}} + \dot{\boldsymbol{\omega}} \times \mathbf{r} + 2\boldsymbol{\omega} \times \dot{\mathbf{r}} + \boldsymbol{\omega} \times (\boldsymbol{\omega} \times \mathbf{r}).$$

Since $\ddot{\mathbf{r}} = \mathbf{a}'$ and $\dot{\mathbf{r}} = \mathbf{v}'$,

$$\mathbf{a} = \mathbf{a}' + \dot{\boldsymbol{\omega}} \times \mathbf{r} + 2\boldsymbol{\omega} \times \mathbf{v}' + \boldsymbol{\omega} \times (\boldsymbol{\omega} \times \mathbf{r}).$$

In general, the primed system may have both translational and rotational motion. This can be thought as a translation followed by a rotation. It is clear from Fig. 2.4 that $\mathbf{r} = \mathbf{r}' + \mathbf{r}_0$, so

$$\mathbf{v} = \frac{d\mathbf{r}}{dt} = \frac{d\mathbf{r}'}{dt} + \frac{d\mathbf{r}_0}{dt}$$

$$= \left(\frac{D\mathbf{r}'}{Dt} + \boldsymbol{\omega} \times \mathbf{r}' \right) + \frac{d\mathbf{r}_0}{dt}. \qquad (2.41)$$

The translational velocity of the coordinates is simply $\mathbf{v}_0 = d\mathbf{r}_0/dt$. Similarly the linear acceleration of the coordinates is $\mathbf{a}_0 = d^2\mathbf{r}_0/dt^2$. Therefore, the acceleration of the particle is given by

$$\mathbf{a} = \frac{d\mathbf{v}}{dt} = \frac{d}{dt}\left(\frac{D\mathbf{r}'}{Dt} + \boldsymbol{\omega} \times \mathbf{r}' \right) + \frac{d^2\mathbf{r}_0}{dt^2}$$

$$= \frac{D}{Dt}\left(\frac{D\mathbf{r}'}{Dt} + \boldsymbol{\omega} \times \mathbf{r}' \right) + \boldsymbol{\omega} \times \left(\frac{D\mathbf{r}'}{Dt} + \boldsymbol{\omega} \times \mathbf{r}' \right) + \mathbf{a}_0. \qquad (2.42)$$

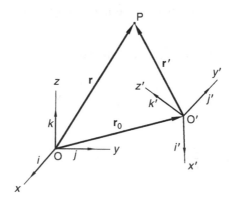

Fig. 2.4. Geometry of the coordinate systems. The primed system is a noninertial reference frame which has both translational and rotational motion relative to the fixed system

Therefore the general equations for the transformation from a fixed system to a moving system are

$$\mathbf{v} = \mathbf{v}' + \boldsymbol{\omega} \times \mathbf{r}' + \mathbf{v}_0, \tag{2.43}$$

$$\mathbf{a} = \mathbf{a}' + \dot{\boldsymbol{\omega}} \times \mathbf{r}' + 2\boldsymbol{\omega} \times \mathbf{v}' + \boldsymbol{\omega} \times (\boldsymbol{\omega} \times \mathbf{r}') + \mathbf{a}_0. \tag{2.44}$$

The term $\dot{\boldsymbol{\omega}} \times \mathbf{r}'$ is known as the *transverse acceleration* and the term $2\boldsymbol{\omega} \times \mathbf{v}'$ is called the *coriolis acceleration*. The *centripetal acceleration* $\boldsymbol{\omega} \times (\boldsymbol{\omega} \times \mathbf{r})$ is always directed toward the center.

2.3 Theory of Space Curve

Suppose we have a particle moving on a space curve as shown in Fig. 2.5. At certain time, the particle is at some point P. In a time interval Δt the particle moves to another point P' along the path. The arc distance between P and P' is Δs. Let $\mathbf{t}$ be the unit vector in the direction of the tangent of the curve at P. The velocity of the particle is of course in the direction of $\mathbf{t}$,

$$\mathbf{v} = \frac{d\mathbf{r}}{dt} = v\mathbf{t}, \tag{2.45}$$

where v is the magnitude of the velocity, which is given by

$$v = \lim_{\Delta t \to 0} \frac{\Delta s}{\Delta t} = \frac{ds}{dt}. \tag{2.46}$$

The point P can also be specified by s, the distance along the curve measured from a fixed point to P. Then by chain rule,

$$\mathbf{v} = \frac{d\mathbf{r}}{dt} = \frac{ds}{dt}\frac{d\mathbf{r}}{ds} = v\frac{d\mathbf{r}}{ds}. \tag{2.47}$$

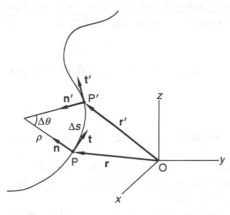

Fig. 2.5. Space curve. The tangent vector $\mathbf{t}$ and the normal vector $\mathbf{n}$ determine the osculating plane which may turn in space

Comparing (2.45) and (2.47), we have

$$\frac{d\mathbf{r}}{ds} = \mathbf{t}, \tag{2.48}$$

hardly a surprising result. Clearly, as $\Delta s \to 0$, $\left|\dfrac{\Delta \mathbf{r}}{\Delta s}\right| = 1$.

Now, $\mathbf{t}$ is a unit vector, which means its magnitude is not changing, therefore its derivative must be perpendicular to itself. Let $\mathbf{n}$ be a unit vector perpendicular to $\mathbf{t}$, so we can write

$$\frac{d\mathbf{t}}{ds} = \kappa \mathbf{n}, \tag{2.49}$$

where κ is the magnitude of $d\mathbf{t}/ds$ and is called the curvature. The vector $\mathbf{n}$ is known as the normal vector and is perpendicular to $\mathbf{t}$. The reciprocal of the curvature $\varrho = 1/\kappa$ is known as the radius of the curvature. Equation (2.49) defines both κ and $\mathbf{n}$, and tells us how fast the unit tangent vector changes direction as we move along the curve.

The acceleration of the particle is

$$\mathbf{a} = \frac{d\mathbf{v}}{dt} = \frac{d}{dt}(v\mathbf{t}) = \dot{v}\mathbf{t} + v\dot{\mathbf{t}},$$

where

$$\dot{\mathbf{t}} = \frac{d\mathbf{t}}{dt} = \frac{ds}{dt}\frac{d\mathbf{t}}{ds} = v\frac{d\mathbf{t}}{ds} = v\kappa\mathbf{n}. \tag{2.50}$$

Therefore the acceleration can be written as

$$\mathbf{a} = \dot{v}\mathbf{t} + v^2\kappa\mathbf{n} = \dot{v}\mathbf{t} + \frac{v^2}{\varrho}\mathbf{n}. \tag{2.51}$$

The tangential component of $\mathbf{a}$ corresponds to the change in the magnitude of v, and the normal component of $\mathbf{a}$ corresponds to the change in direction of $\mathbf{v}$. The normal component is the familiar centripetal acceleration.

For the circular motion in example 2.1.2, $\mathbf{r} = r\cos\omega t\mathbf{i} + r\sin\omega t\mathbf{j}$ and $v = \omega r$,

$$\mathbf{t} = \frac{1}{v}\mathbf{v} = \frac{1}{\omega r}(-r\omega\sin\omega t\mathbf{i} + r\omega\cos\omega t\mathbf{j}) = -\sin\omega t\mathbf{i} + \omega\cos\omega t\mathbf{j}.$$

Since by (2.50),

$$\frac{d\mathbf{t}}{ds} = \frac{1}{v}\frac{d\mathbf{t}}{dt} = \frac{1}{\omega r}(-\omega\cos\omega t\mathbf{i} - \omega\sin\omega t\mathbf{j}) = -\frac{1}{r^2}(r\cos\omega t\mathbf{i} + r\sin\omega t\mathbf{j}) = -\frac{1}{r^2}\mathbf{r}.$$

But by definition, $d\mathbf{t}/ds = \kappa\mathbf{n}$. So, in this case, $\mathbf{n} = -\mathbf{r}/r$, $\kappa = 1/r$, and $\varrho = 1/\kappa = r$. In other words, the radius of curvature in a circular motion is equal to the radius of the circle.

In a small region, we can approximate Δs by the arc of a circle. The radius of this circle is the radius of curvature of the curve as shown in Fig. 2.5.

The motion may not be confined in a plane, although both the velocity and acceleration lie in the plane of **t** and **n**, known as the *osculating plane*. In general, there is another degree of freedom for the motion, namely the arc as a whole may turn. In other words, the osculating plane is not necessarily fixed in space. We need another factor to compute the derivatives of the acceleration.

Let us define a third vector, known as *binormal vector*,

$$\mathbf{b} = \mathbf{t} \times \mathbf{n}. \tag{2.52}$$

Since both **t** and **n** are unit vectors and they are perpendicular to each other, therefore **b** is also a unit vector and is perpendicular to both **t** and **n**. It follows from the definition that, in a right-hand system,

$$\mathbf{b} \times \mathbf{t} = \mathbf{n}, \quad \mathbf{b} \times \mathbf{n} = -\mathbf{t}, \quad \mathbf{n} \times \mathbf{t} = -\mathbf{b}. \tag{2.53}$$

All vectors associated with the curve at the point P can be written as a linear combination of $\mathbf{t}, \mathbf{n}$, and $\mathbf{b}$ which form a basis at P. Now we evaluate $d\mathbf{b}/ds$ and $d\mathbf{n}/ds$.

Since **b** is perpendicular to **t**, so $\mathbf{b} \cdot \mathbf{t} = 0$. Afer differentiating we have

$$\frac{d}{ds}(\mathbf{b} \cdot \mathbf{t}) = \frac{d\mathbf{b}}{ds} \cdot \mathbf{t} + \mathbf{b} \cdot \frac{d\mathbf{t}}{ds} = \frac{d\mathbf{b}}{ds} \cdot \mathbf{t} + \mathbf{b} \cdot \kappa\mathbf{n} = 0. \tag{2.54}$$

Hence, $d\mathbf{b}/ds \cdot \mathbf{t} = -\kappa\mathbf{b} \cdot \mathbf{n}$. Since **b** is perpendicular to **n**, $\mathbf{b} \cdot \mathbf{n} = 0$. Thus, $d\mathbf{b}/ds \cdot \mathbf{t} = 0$, which means $d\mathbf{b}/ds$ is perpendicular to **t**. On the other hand, since **b** is a unit vector, so $d\mathbf{b}/ds$ is perpendicular **b**. Therefore $d\mathbf{b}/ds$ must be in the direction of **n**. Let

$$d\mathbf{b}/ds = \gamma\mathbf{n}, \tag{2.55}$$

where γ, by definition, is the magnitude of $d\mathbf{b}/ds$ and is called the *torsion of the curve*.

To obtain $d\mathbf{n}/ds$, we use (2.53),

$$\frac{d\mathbf{n}}{ds} = \frac{d}{ds}(\mathbf{b} \times \mathbf{t}) = \frac{d\mathbf{b}}{ds} \times \mathbf{t} + \mathbf{b} \times \frac{d\mathbf{t}}{ds} = \gamma\mathbf{n} \times \mathbf{t} + \mathbf{b} \times \kappa\mathbf{n} = -\gamma\mathbf{b} - \kappa\mathbf{t}. \tag{2.56}$$

The set of equations

$$\boxed{\frac{d\mathbf{t}}{ds} = \kappa\mathbf{n}, \quad \frac{d\mathbf{n}}{ds} = -(\gamma\mathbf{b} + \kappa\mathbf{t}), \quad \frac{d\mathbf{b}}{ds} = \gamma\mathbf{n}} \tag{2.57}$$

are the famous *Frenet–Serret formulas*. They are fundamental equations in differential geometry.

Example 2.3.1. Find the arc length s of the curve $\mathbf{r}(t) = a\cos t\mathbf{i} + a\sin t\mathbf{j}$ between $t = 0$ and $t = T$. Express $\mathbf{r}$ as a function of s.

Solution 2.3.1. Since $ds/dt = v$ and $v = (\mathbf{v} \cdot \mathbf{v})^{1/2} = (\dot{\mathbf{r}} \cdot \dot{\mathbf{r}})^{1/2}$,

$$ds = v\,dt = (\dot{\mathbf{r}} \cdot \dot{\mathbf{r}})^{1/2}dt.$$

This seemingly trivial formula is actually very useful in a variety of problems.
    In the present case

$$ds = [(-a\sin t\mathbf{i} + a\cos t\mathbf{j}) \cdot (-a\sin t\mathbf{i} + a\cos t\mathbf{j})]^{1/2}\,dt = a\,dt,$$

$$s = \int_0^T a\,dt = aT.$$

In general $s = at$ and $t = s/a$, thus

$$\mathbf{r}(s) = a\cos\frac{s}{a}\mathbf{i} + a\sin\frac{s}{a}\mathbf{j}.$$

Example 2.3.2. A circular helix is given by $\mathbf{r} = a\cos t\mathbf{i} + a\sin t\mathbf{j} + bt\mathbf{k}$, calculate $\mathbf{t}, \mathbf{n}, \mathbf{b}$ and κ, ρ, γ for this curve.

Solution 2.3.2.

$$\mathbf{v} = \dot{\mathbf{r}} = -a\sin t\mathbf{i} + a\cos t\mathbf{j} + b\mathbf{k},$$

$$v = (\mathbf{v} \cdot \mathbf{v})^{1/2} = [(a^2\sin^2 t + a^2\cos^2 t) + b^2]^{1/2} = (a^2 + b^2)^{1/2},$$

$$\mathbf{t} = \frac{1}{v}\mathbf{v} = \frac{1}{(a^2 + b^2)^{1/2}}(-a\sin t\mathbf{i} + a\cos t\mathbf{j} + b\mathbf{k}).$$

Since

$$\frac{d\mathbf{t}}{dt} = \frac{ds}{dt}\frac{d\mathbf{t}}{ds} = v\frac{d\mathbf{t}}{ds},$$

$$\frac{d\mathbf{t}}{ds} = \frac{1}{v}\frac{d\mathbf{t}}{dt} = \frac{1}{(a^2 + b^2)}(-a\cos t\mathbf{i} - a\sin t\mathbf{j}) = \kappa\mathbf{n},$$

$$\kappa^2 = (\kappa\mathbf{n} \cdot \kappa\mathbf{n}) = \frac{1}{(a^2 + b^2)^2}(a^2\cos^2 t + a^2\sin^2 t) = \frac{a^2}{(a^2 + b^2)^2},$$

$$\kappa = \frac{a}{(a^2 + b^2)}, \qquad \rho = \frac{1}{\kappa} = \frac{a^2 + b^2}{a},$$

$$\mathbf{n} = \frac{1}{\kappa}\frac{d\mathbf{t}}{ds} = -\cos t\mathbf{i} - \sin t\mathbf{j}.$$

$$\mathbf{b} = \mathbf{t} \times \mathbf{n} = \frac{1}{(a^2 + b^2)^{1/2}} \begin{vmatrix} \mathbf{i} & \mathbf{j} & \mathbf{k} \\ -a\sin t & a\cos t & b \\ -\cos t & -\sin t & 0 \end{vmatrix}$$

$$= \frac{1}{(a^2 + b^2)^{1/2}} [b \sin t\mathbf{i} - b\cos t\mathbf{j} + (a\sin^2 t + a\cos^2 t)\mathbf{k}]$$

$$= \frac{1}{(a^2 + b^2)^{1/2}} [b \sin t\mathbf{i} - b\cos t\mathbf{j} + a\mathbf{k}].$$

Use chain rule again,

$$\frac{d\mathbf{b}}{dt} = \frac{ds}{dt}\frac{d\mathbf{b}}{ds} = v\frac{d\mathbf{b}}{ds},$$

so

$$\frac{d\mathbf{b}}{ds} = \frac{1}{v}\frac{d\mathbf{b}}{dt} = \frac{b}{(a^2 + b^2)}[\cot t\mathbf{i} + \sin t\mathbf{j}] = \gamma\mathbf{n},$$

$$\gamma = -\frac{b}{(a^2 + b^2)}.$$

2.4 The Gradient Operator

The application of vector methods to physical problems most frequently takes the form of differential operations. We have discussed the rate of change with respect to time which allows us to define velocity and acceleration vectors for the motion of a particle. Now we begin a more systematic study of the rate of change with respect to the spatial coordinates. The most important differential operator is the gradient.

2.4.1 The Gradient of a Scalar Function

Before we discuss the gradient, let us briefly review the notation of derivative and partial derivative in calculus

$$\frac{df(x)}{dx} = \lim_{\Delta x \to 0} \frac{f(x + \Delta x) - f(x)}{\Delta x} = \lim_{\Delta x \to 0} \frac{\Delta f}{\Delta x}.$$

If $\Delta x \to 0$ is implicitly understood, we can multiply both sides by Δx and write

$$\Delta f = f(x + \Delta x) - f(x) = \frac{df}{dx}\Delta x. \tag{2.58}$$

This equation can also be derived from the Taylor expansion of $f(x + \Delta x)$ around x:

$$f(x + \Delta x) = f(x) + \frac{df}{dx}\Delta x + \frac{1}{2}\frac{d^2 f}{dx^2}(\Delta x)^2 + \cdots.$$

Equation (2.58) is obtained if one drops terms of $(\Delta x)^n$ with $n \geq 2$ as $\Delta x \to 0$.

Similarly in terms of partial derivatives

$$f(x + \Delta x, y) - f(x, y) = \frac{\partial f}{\partial x} \Delta x,$$

or

$$f(x, y + \Delta y, z + \Delta z) - f(x, y, z + \Delta z) = \frac{\partial f}{\partial y} \Delta y.$$

Let the difference of the scalar function φ between two nearby points $(x + \Delta x, y + \Delta y, z + \Delta z)$ and (x, y, z) be $\Delta \varphi$:

$$\Delta \varphi = \varphi \left(x + \Delta x, y + \Delta y, z + \Delta z \right) - \varphi \left(x, y, z \right).$$

This equation can be written as

$$\begin{aligned} \Delta \varphi = {} & \varphi \left(x + \Delta x, y + \Delta y, z + \Delta z \right) \\ & - [\varphi \left(x, y + \Delta y, z + \Delta z \right) - \varphi(x, y + \Delta y, z + \Delta z)] \\ & - [\varphi \left(x, y, z + \Delta z \right) - \varphi \left(x, y, z + \Delta z \right)] - \varphi \left(x, y, z \right), \end{aligned}$$

since the quantities in the two brackets cancel out. Removing the brackets, we have

$$\begin{aligned} \Delta \varphi = {} & \varphi \left(x + \Delta x, y + \Delta y, z + \Delta z \right) - \varphi \left(x, y + \Delta y, z + \Delta z \right) \\ & + \varphi(x, y + \Delta y, z + \Delta z) - \varphi \left(x, y, z + \Delta z \right) \\ & + \varphi \left(x, y, z + \Delta z \right) - \varphi \left(x, y, z \right). \end{aligned}$$

With the definition of partial derivative, the above equation can be written as

$$\Delta \varphi = \frac{\partial \varphi}{\partial x} \Delta x + \frac{\partial \varphi}{\partial y} \Delta y + \frac{\partial \varphi}{\partial z} \Delta z. \tag{2.59}$$

The displacement vector from (x, y, z) to $(x + \Delta x, y + \Delta y, z + \Delta z)$ is, of course,

$$\Delta \mathbf{r} = \Delta x \mathbf{i} + \Delta y \mathbf{j} + \Delta z \mathbf{k}.$$

One can readily verify that

$$\left(\mathbf{i} \frac{\partial \varphi}{\partial x} + \mathbf{j} \frac{\partial \varphi}{\partial y} + \mathbf{k} \frac{\partial \varphi}{\partial z} \right) \cdot (\Delta x \mathbf{i} + \Delta y \mathbf{j} + \Delta z \mathbf{k}) = \frac{\partial \varphi}{\partial x} \Delta x + \frac{\partial \varphi}{\partial y} \Delta y + \frac{\partial \varphi}{\partial z} \Delta z.$$

Thus,

$$\Delta \varphi = \left(\mathbf{i} \frac{\partial \varphi}{\partial x} + \mathbf{j} \frac{\partial \varphi}{\partial y} + \mathbf{k} \frac{\partial \varphi}{\partial z} \right) \cdot \Delta \mathbf{r}. \tag{2.60}$$

The vector in the parenthesis is called the *gradient* of φ, and is usually written as grad φ or $\nabla \varphi$,

$$\nabla \varphi = \mathbf{i} \frac{\partial \varphi}{\partial x} + \mathbf{j} \frac{\partial \varphi}{\partial y} + \mathbf{k} \frac{\partial \varphi}{\partial z}. \tag{2.61}$$

Since φ is an arbitrary scalar function, it is convenient to define the differential operation in terms of the *gradient operator* ∇ (sometimes known as del or del operator)

$$\nabla = \mathbf{i}\frac{\partial}{\partial x} + \mathbf{j}\frac{\partial}{\partial y} + \mathbf{k}\frac{\partial}{\partial z}. \tag{2.62}$$

This is a vector operator and obeys the same convention as the derivative notation. If a function is placed on the left-hand side of it, $\varphi\nabla$ is still an operator and by itself means nothing. What is to be differentiated must be placed on the right of ∇. When it operates on a scalar function, it turns $\nabla\varphi$ into a vector with definite magnitude and direction. It also has a definite physical meaning.

Example 2.4.1. Show that $\nabla r = \hat{\mathbf{r}}$ and $\nabla f(r) = \hat{\mathbf{r}} df/dr$, where $\hat{\mathbf{r}}$ is a unit vector along the position vector $\mathbf{r} = x\mathbf{i} + y\mathbf{j} + z\mathbf{k}$ and r is the magnitude of $\mathbf{r}$.

Solution 2.4.1.

$$\nabla r = \left(\mathbf{i}\frac{\partial}{\partial x} + \mathbf{j}\frac{\partial}{\partial y} + \mathbf{k}\frac{\partial}{\partial z} \right) r,$$

$$\mathbf{i}\frac{\partial r}{\partial x} = \mathbf{i}\frac{\partial}{\partial x}\left(x^2 + y^2 + z^2 \right)^{1/2} = \frac{\mathbf{i}x}{(x^2 + y^2 + z^2)^{1/2}} = \frac{\mathbf{i}x}{r}, \quad \text{etc.}$$

$$\nabla r = \frac{\mathbf{i}x}{r} + \frac{\mathbf{j}y}{r} + \frac{\mathbf{k}z}{r} = \frac{x\mathbf{i} + y\mathbf{j} + z\mathbf{k}}{r} = \frac{\mathbf{r}}{r} = \hat{\mathbf{r}}.$$

$$\nabla f(r) = \mathbf{i}\frac{\partial f}{\partial x} + \mathbf{j}\frac{\partial f}{\partial y} + \mathbf{k}\frac{\partial f}{\partial z},$$

$$\mathbf{i}\frac{\partial f}{\partial x} = \mathbf{i}\frac{df}{dr}\frac{\partial r}{\partial x} = \mathbf{i}\frac{df}{dr}\frac{x}{r}, \quad \text{etc.}$$

$$\nabla f(r) = \mathbf{i}\frac{df}{dr}\frac{x}{r} + \mathbf{j}\frac{df}{dr}\frac{y}{r} + \mathbf{k}\frac{df}{dr}\frac{z}{r} = \frac{x\mathbf{i} + y\mathbf{j} + z\mathbf{k}}{r}\frac{df}{dr} = \hat{\mathbf{r}}\frac{df}{dr}.$$

Example 2.4.2. Show that $(\mathbf{A} \cdot \nabla)\mathbf{r} = \mathbf{A}$.

Solution 2.4.2.

$$(\mathbf{A} \cdot \nabla)\mathbf{r} = \left[(A_x\mathbf{i} + A_y\mathbf{j} + A_z\mathbf{k}) \cdot \left(\mathbf{i}\frac{\partial}{\partial x} + \mathbf{j}\frac{\partial}{\partial y} + \mathbf{k}\frac{\partial}{\partial z} \right) \right]\mathbf{r}$$

$$= \left(A_x\frac{\partial}{\partial x} + A_y\frac{\partial}{\partial y} + A_z\frac{\partial}{\partial z} \right)(x\mathbf{i} + y\mathbf{j} + z\mathbf{k})$$

$$= A_x\mathbf{i} + A_y\mathbf{j} + A_z\mathbf{k} = \mathbf{A}.$$

2.4.2 Geometrical Interpretation of Gradient

To see the physical meaning of $\nabla\varphi$, let us substitute (2.61) into (2.60)

$$\Delta\varphi = \nabla\varphi \cdot \Delta\mathbf{r}.$$

Taking the limit as $\Delta\mathbf{r}$ approaches zero yields the differential form of this equation:

$$d\varphi = \boldsymbol{\nabla}\varphi \cdot d\mathbf{r} = \frac{\partial\varphi}{\partial x}dx + \frac{\partial\varphi}{\partial y}dy + \frac{\partial\varphi}{\partial z}dz. \qquad (2.63)$$

Now, $\varphi(x, y, z) = C$ represents a surface in space. For example, $\varphi(x, y, z) = x + y + z$ and $\varphi = C$ represents a family of parallel planes, as discussed in the previous chapter. Different values of C simply mean the different perpendicular distances between the plane and the origin. Another example, $\varphi(x, y, z) = x^2 + y^2 + z^2 = 4$ is the surface of a sphere of radius 2. Changing 4 to 9 simply means another sphere of radius 3.

If the two near-by points lie on the same surface $\varphi = C$, then clearly $d\varphi = 0$, since $\varphi(x + dx, y + dy, z + dz) = \varphi(x, y, z) = C$. In this case $d\mathbf{r}$ is, of course, a vector on this surface and

$$d\varphi = \boldsymbol{\nabla}\varphi \cdot d\mathbf{r} = 0. \qquad (2.64)$$

Since the dot product of $\boldsymbol{\nabla}\varphi$ and $d\mathbf{r}$ is equal to zero, $\boldsymbol{\nabla}\varphi$ must be perpendicular to $d\mathbf{r}$. Therefore $\boldsymbol{\nabla}\varphi$ is normal (perpencicular) to the surface $\varphi = C$ as shown in Fig. 2.6.

We can look at it in another way. Let the unit vector in the direction $d\mathbf{r}$ be $\mathbf{d}$ and the magnitude of $d\mathbf{r}$ be dr, then the dot product of (2.63) can be written as

$$d\varphi = \boldsymbol{\nabla}\varphi \cdot \mathbf{d}\, dr,$$

or

$$\frac{d\varphi}{dr} = \boldsymbol{\nabla}\varphi \cdot \mathbf{d}. \qquad (2.65)$$

This means that the rate of change of φ in the direction of $\mathbf{d}$ is equal to $\boldsymbol{\nabla}\varphi \cdot \mathbf{d}$. Furthermore, since

$$\boldsymbol{\nabla}\varphi \cdot \mathbf{d} = |\boldsymbol{\nabla}\varphi| \cos\theta,$$

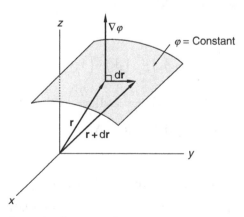

Fig. 2.6. Gradient of a scalar function. $\boldsymbol{\nabla}\varphi$ is a vector normal to the surface of $\varphi = $ constant

where θ is the angle between $d\mathbf{r}$ and $\nabla\varphi$, the maximum rate of change occurs at $\theta = 0$. This means that if $d\mathbf{r}$ and $\nabla\varphi$ are in the same direction, the change of φ is the largest. Therefore the meaning of $\nabla\varphi$ can be summarized as follows:

The vector $\nabla\varphi$ is in the direction of the steepest increase in φ and the magnitude of the vector $\nabla\varphi$ is equal to the rate of increase in that direction.

Example 2.4.3. Find the unit normal to the surface described by $\varphi(x, y, z) = 2x^2 + 4yz - 5z^2 = -10$ at $(3, -1, 2)$.

Solution 2.4.3. First check the point $(3, -1, 2)$ is indeed on the surface: $2(3)^2 + 4(-1)2 - 5(2)^2 = -10$. Recall the unit normal to the surface at any point is $\mathbf{n} = \nabla\varphi/|\nabla\varphi|$.

$$\nabla\varphi = \left(\mathbf{i}\frac{\partial}{\partial x} + \mathbf{j}\frac{\partial}{\partial y} + \mathbf{k}\frac{\partial}{\partial z}\right)\left(2x^2 + 4yz - 5z^2\right) = 4x\mathbf{i} + 4z\mathbf{j} + (4y - 10z)\,\mathbf{k}.$$

$$\mathbf{n} = \left[\frac{\nabla\varphi}{|\nabla\varphi|}\right]_{3,-1,2} = \frac{12\mathbf{i} + 8\mathbf{j} - 24\mathbf{k}}{(12^2 + 8^2 + 24^2)^{1/2}} = \frac{3\mathbf{i} + 2\mathbf{j} - 6\mathbf{k}}{\sqrt{46}}.$$

Example 2.4.4. Find the maximum rate of increase for the surface $\varphi(x, y, z) = 100 + xyz$ at the point $(1, 3, 2)$. In which direction is the maximum rate of increase?

Solution 2.4.4. The maximum rate of increase is $|\nabla\varphi|_{1,3,2}$.

$$\nabla\varphi = \left(\mathbf{i}\frac{\partial}{\partial x} + \mathbf{j}\frac{\partial}{\partial y} + \mathbf{k}\frac{\partial}{\partial z}\right)(100 + xyz) = yz\mathbf{i} + xz\mathbf{j} + xy\mathbf{k},$$

$$|\nabla\varphi|_{1,3,2} = |6\mathbf{i} + 2\mathbf{j} + 3\mathbf{k}| = (36 + 4 + 9)^{1/2} = 9.$$

The direction of the maximum increase is given by

$$\nabla\varphi|_{1,3,2} = 6\mathbf{i} + 2\mathbf{j} + 3\mathbf{k}.$$

Example 2.4.5. Find the rate of increase for the surface $\varphi(x, y, z) = xy^2 + yz^3$ at the point $(2, -1, 1)$ in the direction of $\mathbf{i} + 2\mathbf{j} + 2\mathbf{k}$.

Solution 2.4.5.

$$\nabla\varphi = \left(\mathbf{i}\frac{\partial}{\partial x} + \mathbf{j}\frac{\partial}{\partial y} + \mathbf{k}\frac{\partial}{\partial z}\right)(xy^2 + yz^3) = y^2\mathbf{i} + (2xy + z^3)\,\mathbf{j} + 3yz^2\mathbf{k},$$

$$\nabla\varphi_{2,-1,1} = \mathbf{i} - 3\mathbf{j} - 3\mathbf{k}.$$

The unit vector along $\mathbf{i} + 2\mathbf{j} + 2\mathbf{k}$ is

$$\mathbf{n} = \frac{\mathbf{i} + 2\mathbf{j} + 2\mathbf{k}}{\sqrt{1 + 4 + 4}} = \frac{1}{3}(\mathbf{i} + 2\mathbf{j} + 2\mathbf{k}).$$

The rate of increase is

$$\frac{d\varphi}{dr} = \nabla\varphi \cdot \mathbf{n} = (\mathbf{i} - 3\mathbf{j} - 3\mathbf{k}) \cdot \frac{1}{3}(\mathbf{i} + 2\mathbf{j} + 2\mathbf{k}) = -\frac{11}{3}.$$

Example 2.4.6. Find the equation of the tangent plane to the surface described by $\varphi(x, y, z) = 2xz^2 - 3xy - 4x = 7$ at the point $(1, -1, 2)$.

Solution 2.4.6. If $\mathbf{r}_0$ is a vector from the origin to the point $(1, -1, 2)$ and $\mathbf{r}$ is a vector to any point in the tangent plane, then $\mathbf{r} - \mathbf{r}_0$ lies in the tangent plane. The tangent plane at $(1, -1, 2)$ is normal to the gradient at that point, so we have

$$\nabla\varphi|_{1,-1,2} \cdot (\mathbf{r} - \mathbf{r}_0) = 0.$$

$$\nabla\varphi|_{1,-1,2} = \left[\left(2z^2 - 3y - 4\right)\mathbf{i} - 3x\mathbf{j} - 4xz\mathbf{k}\right]_{1,-1,2} = 7\mathbf{i} - 3\mathbf{j} + 8\mathbf{k}.$$

Therefore the tangent plane is given by the equation

$$(7\mathbf{i} - 3\mathbf{j} + 8\mathbf{k}) \cdot [(x - 1)\mathbf{i} + (y + 1)\mathbf{j} + (z - 2)\mathbf{k}] = 0,$$
$$7(x - 1) - 3(y + 1) + 8(z - 2) = 0,$$
$$7x - 3y + 8z = 26.$$

2.4.3 Line Integral of a Gradient Vector

Line integrals occur frequently in physical sciences. The most familiar is probably the work done by a force $\mathbf{F}$ between A and B along some path Γ:

$$\text{Work}(A \to B) = \int_{A,\Gamma}^{B} \mathbf{F} \cdot d\mathbf{r},$$

where

$$d\mathbf{r} = \mathbf{i}\,dx + \mathbf{j}\,dy + \mathbf{k}\,dz$$

is the differential displacement vector from (x, y, z) to $(x + dx, y + dy, z + dz)$. Sometimes $d\mathbf{l}$ is used in place of $d\mathbf{r}$ to emphasize that the differential displacement vector is along a certain path for the line integral. We will not use this convention here.

For any vector field $\mathbf{A}(x, y, z)$, the line integral

$$\int_{A,\Gamma}^{B} \mathbf{A} \cdot d\mathbf{r} = \int_{A,\Gamma}^{B} (A_x\,dx + A_y\,dy + A_z\,dz) \tag{2.66}$$

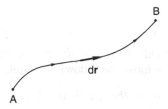

Fig. 2.7. Path of a line integral. The differential displacement dr is along a specified curve in space between A and B

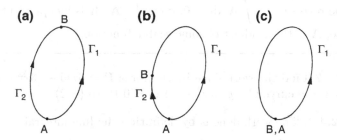

Fig. 2.8. Path independence of the line integral of $\boldsymbol{\nabla}\varphi \cdot d\mathbf{r}$. (**a**) The value of the integral from A to B along Γ_1 is the same as along Γ_2. (**b**) They are still the same even though Γ_2 is much shorter. (**c**) As Γ_2 shrinks to zero, the integral along Γ_2 vanishes. The integral along Γ_1, which becomes a loop integral, must also be zero

is the sum of contributions $\mathbf{A} \cdot d\mathbf{r}$ for each differential displacement $d\mathbf{r}$ along the curve Γ in space from A to B as shown in Fig. 2.7. The line integral is also called *path integral* because it is carried out along a specific path Γ.

In general, the result depends on the path taken between A and B. However, if $\mathbf{A} = \boldsymbol{\nabla}\varphi$ for some scalar function φ, the integral is independent of the path.

$$\int_{A,\Gamma}^{B} \mathbf{A} \cdot d\mathbf{r} = \int_{A,\Gamma}^{B} \boldsymbol{\nabla}\varphi \cdot d\mathbf{r} = \int_{A}^{B} d\varphi = \varphi(B) - \varphi(A), \tag{2.67}$$

where we have used (2.63) to convert $\boldsymbol{\nabla}\varphi \cdot d\mathbf{r}$ to the total differential $d\varphi$. Since the result depends only on the position of the two end points, the integral is independent of path. In this case, the integral from A to B in Fig. 2.8 gives the same value whether it is carried out along Γ_1 or along Γ_2. This remains to be true as we bring the two points A and B closer, no matter how short Γ_2 becomes. When B is brought to the same place as A, the line integral over Γ_2 obviously vanishes because the length of Γ_2 is equal to zero. So the line integral over Γ_1 must also be zero. The line integral over Γ_1 is an integral around a closed loop:

$$\boxed{\oint \boldsymbol{\nabla}\varphi \cdot d\mathbf{r} = 0.} \tag{2.68}$$

The symbol $\oint$ means the integration is over a closed loop. The line integral around the closed loop is called the *circulation* of the vector field $\mathbf{A}$ around the closed loop Γ. Thus we have the following result:

> *When a vector field $\mathbf{A}$ is the gradient of a scalar function φ, the circulation of $\mathbf{A}$ around any closed curve is zero.*

Sometimes this is called the *fundamental theorem of gradient*. The argument can be reversed. If $\oint \mathbf{A} \cdot \mathrm{d}\mathbf{r} = 0$, then $\int_A^B \mathbf{A} \cdot \mathrm{d}\mathbf{r}$ is independent of path. In that case, $\mathbf{A}$ is the gradient of some scalar function φ.

Example 2.4.7. Find the work done by the force $\mathbf{F} = 6xy\mathbf{i} + \left(3x^2 - 3y^2\right)\mathbf{j}$ in a plane along the curve $C: y = x^2 - x$ from $(0,0)$ to $(2,2)$.

Solution 2.4.7. The work done is by definition the line integral

$$W = \int_C \mathbf{F} \cdot \mathrm{d}\mathbf{r} = \int_C \left[6xy\mathbf{i} + \left(3x^2 - 3y^2\right)\mathbf{j}\right] \cdot (\mathbf{i}\,\mathrm{d}x + \mathbf{j}\,\mathrm{d}y + \mathbf{k}\,\mathrm{d}z)$$

$$= \int_C \left[6xy\,\mathrm{d}x + (3x^2 - 3y^2)\,\mathrm{d}y\right].$$

There are more than one way to carry out this integration along curve C. *Method I.* Change all variables into x.

$$y = x^2 - x, \quad \mathrm{d}y = (2x - 1)\mathrm{d}x,$$

$$W = \int_C \left[6xy\,\mathrm{d}x + \left(3x^2 - 3y^2\right)\mathrm{d}y\right]$$

$$= \int_0^2 \{6x\left(x^2 - x\right)\mathrm{d}x + [3x^2 - 3(x^2 - x)^2](2x - 1)\}\,\mathrm{d}x$$

$$= \int_0^2 \{-6x^5 + 15x^4 - 6x^2\}\mathrm{d}x = \left[-x^6 + 3x^5 - 2x^3\right]_0^2 = 16.$$

Method II. The curve C can be considered as the trajectory described by the tip of the position vector $\mathbf{r}(t) = x\,(t)\,\mathbf{i} + y\,(t)\,\mathbf{j}$ with t as a parameter. It can be readily verified that with $x = t$ and $y = t^2 - t$, the curve $y = x^2 - x$ is traced out. Therefore the curve C is given by

$$\mathbf{r}(t) = x\,(t)\,\mathbf{i} + y\,(t)\,\mathbf{j} = t\mathbf{i} + \left(t^2 - t\right)\mathbf{j}.$$

The point $(0,0)$ corresponds to $t = 0$, and $(2,2)$ corresponds to $t = 2$. Now we can change all variables into t.

$$\mathbf{F}\left(x,y,z\right)\cdot d\mathbf{r}=\mathbf{F}\left(x\left(t\right),y\left(t\right),z\left(t\right)\right)\cdot\frac{d\mathbf{r}}{dt}dt,$$

$$\frac{d\mathbf{r}}{dt}=\mathbf{i}+\left(2t-1\right)\mathbf{j},$$

$$\mathbf{F}=6xy\mathbf{i}+\left(3x^2-3y^2\right)\mathbf{j}=6t\left(t^2-t\right)\mathbf{i}+\left[3t^2-3\left(x^2-x\right)^2\right]\mathbf{j},$$

$$\mathbf{F}\cdot d\mathbf{r}=\{6t\left(t^2-t\right)+\left[3t^2-3\left(x^2-x\right)^2\right]\left(2t-1\right)\}dt,$$

$$W=\int_C\mathbf{F}\cdot d\mathbf{r}=\int_0^2\{6t\left(t^2-t\right)+\left[3t^2-3\left(x^2-x\right)^2\right]\left(2t-1\right)\}dt=16.$$

Example 2.4.8. Calculate the line integral of the last example from the point $(0,0)$ to the point (x_1,y_1) along the path which runs straight from $(0,0)$ to $(x_1,0)$ and thence to (x_1,y_1). Make a similar calculation for the path which runs along the other two sides of the rectangle, via the point $(0,y_1)$. If $(x_1,y_1)=(2,2)$, what is the value of the integral?

Solution 2.4.8.

$$I_1(x_1,y_1)=\int_{C_1}\mathbf{F}\cdot d\mathbf{r}=\int_{C_1}\left[6xy\,dx+\left(3x^2-3y^2\right)dy\right],$$

$$C_1\;:\;(0,0)\to(x_1,0)\to(x_1,y_1).$$

From $(0,0)\to(x_1,0):\;\;y=0,\;dy=0,$

$$\int_{0,0}^{x_1,0}\left[6xy\,dx+\left(3x^2-3y^2\right)dy\right]=0.$$

From $(x_1,0)\to(x_1,y_1):\;\;x=x_1,\;dx=0,$

$$\int_{x_1,0}^{x_1,y_1}\left[6xy\,dx+\left(3x^2-3y^2\right)dy\right]=\int_0^{y_1}(3x_1^2-3y^2)dy$$

$$=\left[3x_1^2y-y^3\right]_0^{y_1}=3x_1^2y_1-y_1^3,$$

$$I_1(x_1,y_1)=\left[\int_{0,0}^{x_1,0}+\int_{x_1,0}^{x_1,y_1}\right]\left[6xy\,dx+\left(3x^2-3y^2\right)dy\right]=3x_1^2y_1-y_1^3,$$

$$I_2(x_1,y_1)=\int_{C_2}\left[6xy\,dx+\left(3x^2-3y^2\right)dy\right],$$

$$C_2\;:\;(0,0)\to(0,y_1)\to(x_1,y_1).$$

From $(0,0)\to(0,y_1):\;\;x=0,\;dx=0,$

$$\int_{0,0}^{0,y_1}\left[6xy\,dx+\left(3x^2-3y^2\right)dy\right]=\int_0^{y_1}(-3y^2)dy=[-y^3]_0^{y_1}=-y_1^3.$$

From $(0, y_1) \rightarrow (x_1, y_1):$ $y = y_1,\ dy = 0,$

$$\int_{0,y_1}^{x_1,y_1} \left[6xy\ dx + \left(3x^2 - 3y^2\right) dy \right] = \int_0^{x_1} 6xy_1\ dx = 3x_1^2 y_1,$$

$$I_2(x_1, y_1) = \left[\int_{0,0}^{0,y_1} + \int_{0,y_1}^{x_1,y_1} \right] \left[6xy\ dx + \left(3x^2 - 3y^2\right) dy \right] = -y_1^3 + 3x_1^2 y_1.$$

Clearly $I_1(x_1, y_1) = I_2(x_1, y_1)$, and $I_1(2,2) = 3\,(2)^2\,2 - (2)^3 = 16.$

Example 2.4.9. From the last two examples, it is clear that the line integral $\int_C \mathbf{F} \cdot d\mathbf{r}$ with $\mathbf{F} = 6xy\mathbf{i} + \left(3x^2 - 3y^2\right)\mathbf{j}$ depends only on the end points and is independent of the path of integration, therefore $\mathbf{F} = \nabla\varphi$. Find $\varphi(x, y)$ and show that $\int_{0,0}^{2,2} \mathbf{F} \cdot d\mathbf{r} = \varphi(2,2) - \varphi(0,0)$.

Solution 2.4.9.

$$\nabla\varphi = \mathbf{i}\frac{\partial\varphi}{\partial x} + \mathbf{j}\frac{\partial\varphi}{\partial y} = 6xy\mathbf{i} + \left(3x^2 - 3y^2\right)\mathbf{j} = \mathbf{F},$$

$$\frac{\partial\varphi}{\partial x} = 6xy \implies \varphi = 3x^2 y + f(y),$$

$$\frac{\partial\varphi}{\partial y} = 3x^2 - 3y^2 = 3x^2 + \frac{df(y)}{dy},$$

$$\frac{df(y)}{dy} = -3y^2 \implies f(y) = -y^3 + k\ (k \text{ is a constant}).$$

Thus,

$$\varphi(x, y) = 3x^2 y - y^3 + k.$$

$$\int_{0,0}^{2,2} \mathbf{F} \cdot d\mathbf{r} = \int_{0,0}^{2,2} \nabla\varphi \cdot d\mathbf{r} = \varphi(2,2) - \varphi(0,0) = 16 + k - k = 16.$$

Note that

$$\int_{0,0}^{x_1,y_1} \mathbf{F} \cdot d\mathbf{r} = \varphi(x_1, y_1) - \varphi(0,0) = 3x_1^2 y_1 - y_1^3,$$

in agreement with the result of the last example.

Example 2.4.10. Find the line integral $\int_{0,0}^{2,1} \mathbf{F} \cdot d\mathbf{r}$ with $\mathbf{F} = xy\mathbf{i} - y^2\mathbf{j}$ along the path (a) $y = (1/2)x$, (b) $y = (1/4)x^2$, (c) from $(0,0)$ straight up to $(0,1)$ and then along a horizontal line to $(2,1)$.

Solution 2.4.10. $\int_{0,0}^{2,1} \mathbf{F} \cdot d\mathbf{r} = \int_{0,0}^{2,1} \left(xy\, dx - y^2\, dy \right)$ along

$$(a)\ y = \frac{1}{2}x, \quad \text{so} \quad dy = \frac{1}{2}dx,$$

$$\int_{0,0}^{2,1} \left(xy\, dx - y^2 dy \right) = \int_0^2 \left(\frac{1}{2}x^2 dx - \frac{1}{8}x^2 dx \right) = \left[\frac{3}{8} \cdot \frac{1}{3}x^3 \right]_0^2 = 1.$$

$$(b)\ y = \frac{1}{4}x^2, \quad \text{so} \quad dy = \frac{1}{2}x\, dx,$$

$$\int_{0,0}^{2,1} \left(xy\, dx - y^2\, dy \right) = \int_0^2 \left(\frac{1}{4}x^3\, dx - \frac{1}{32}x^5\, dx \right) = \left[\frac{1}{16}x^4 - \frac{1}{32 \cdot 6}x^6 \right]_0^2 = \frac{2}{3}.$$

(c) From $(0,0)$ straight up to $(0,1)$: $x = 0$ so $dx = 0$;
then from $(0,1)$ along a horizontal line to $(2,1)$: $y = 1$ and $dy = 0$,

$$\int_{0,0}^{2,1} \left(xy\, dx - y^2\, dy \right) = \int_{0,0}^{0,1} \left(xy\, dx - y^2\, dy \right) + \int_{0,1}^{2,1} \left(xy\, dx - y^2\, dy \right)$$

$$= -\int_0^1 y^2\, dy + \int_0^2 x\, dx = -\frac{1}{3} + 2 = \frac{5}{3}.$$

In general the line integral $\int_C \mathbf{F} \cdot d\mathbf{r}$ depends on the path of integration as shown in the last example. However, if $\mathbf{F} = \nabla\varphi$, the line integral is independent of the path of integration. We are going to discuss the condition under which $\mathbf{F}$ can be expressed as the gradient of a scalar function.

2.5 The Divergence of a Vector

Just as we can operate with ∇ on a scalar field, we can also operate with ∇ on a vector field $\mathbf{A}$ by taking the dot product. With their components, this operation gives

$$\nabla \cdot \mathbf{A} = \left(\mathbf{i}\frac{\partial}{\partial x} + \mathbf{j}\frac{\partial}{\partial y} + \mathbf{k}\frac{\partial}{\partial z} \right) \cdot (\mathbf{i}A_x + \mathbf{j}A_y + \mathbf{k}A_z)$$

$$= \frac{\partial A_x}{\partial x} + \frac{\partial A_y}{\partial y} + \frac{\partial A_z}{\partial z}. \tag{2.69}$$

Just as the dot product of two vectors is a scalar, $\nabla \cdot \mathbf{A}$ is also a scalar. This sum, called the *divergence* of $\mathbf{A}$ (or div $\mathbf{A}$), is a special combination of derivatives.

Example 2.5.1. Show that $\nabla \cdot \mathbf{r} = 3$ and $\nabla \cdot \mathbf{r} f(r) = 3f(r) + r(\mathrm{d}f/\mathrm{d}r)$.

Solution 2.5.1.

$$\nabla \cdot \mathbf{r} = \left(\mathbf{i}\frac{\partial}{\partial x} + \mathbf{j}\frac{\partial}{\partial y} + \mathbf{k}\frac{\partial}{\partial z}\right) \cdot (\mathbf{i}x + \mathbf{j}y + \mathbf{k}z)$$

$$= \frac{\partial x}{\partial x} + \frac{\partial y}{\partial y} + \frac{\partial z}{\partial z} = 3.$$

$$\nabla \cdot \mathbf{r}f(r) = \left(\mathbf{i}\frac{\partial}{\partial x} + \mathbf{j}\frac{\partial}{\partial y} + \mathbf{k}\frac{\partial}{\partial z}\right) \cdot (\mathbf{i}xf(r) + \mathbf{j}yf(r) + \mathbf{k}zf(r))$$

$$= \frac{\partial}{\partial x}[xf(r)] + \frac{\partial}{\partial y}[yf(r)] + \frac{\partial}{\partial z}[zf(r)]$$

$$= f(r) + x\frac{\partial f}{\partial x} + f(r) + y\frac{\partial f}{\partial y} + f(r) + z\frac{\partial f}{\partial z}$$

$$= 3f(r) + x\frac{\mathrm{d}f}{\mathrm{d}r}\frac{\partial r}{\partial x} + y\frac{\mathrm{d}f}{\mathrm{d}r}\frac{\partial r}{\partial y} + z\frac{\mathrm{d}f}{\mathrm{d}r}\frac{\partial r}{\partial z}.$$

$$\frac{\partial r}{\partial x} = \frac{\partial}{\partial x}\left(x^2 + y^2 + z^2\right)^{1/2} = \frac{1}{2}\frac{2x}{\left(x^2 + y^2 + z^2\right)^{1/2}} = \frac{x}{r};$$

$$\frac{\partial r}{\partial y} = \frac{y}{r}; \qquad \frac{\partial r}{\partial z} = \frac{z}{r}.$$

$$\nabla \cdot \mathbf{r}f(r) = 3f(r) + \frac{x^2}{r}\frac{\mathrm{d}f}{\mathrm{d}r} + \frac{y^2}{r}\frac{\mathrm{d}f}{\mathrm{d}r} + \frac{z^2}{r}\frac{\mathrm{d}f}{\mathrm{d}r}$$

$$= 3f(r) + \frac{x^2 + y^2 + z^2}{r}\frac{\mathrm{d}f}{\mathrm{d}r} = 3f(r) + r\frac{\mathrm{d}f}{\mathrm{d}r}.$$

2.5.1 The Flux of a Vector Field

To gain some physical feeling for the divergence of a vector field, it is helpful to introduce the concept of *flux* (Latin for "flow"). Consider a fluid of density ϱ moving with velocity $\mathbf{v}$. We ask for the total mass of fluid which crosses an area Δa perpendicular to the direction of flow in a time Δt. As shown in Fig. 2.9a, all the fluid in the rectangular pipe of length $v\Delta t$ with the patch Δa as its base will cross Δa in the time interval Δt. The volume of this pipe is

(a)   **(b)**

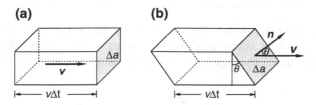

Fig. 2.9. Flux through the base. (a) Flux through $\Delta a = \rho v \Delta a$. (b) Flux through the tilted $\Delta a = \rho \mathbf{v} \cdot \mathbf{n}\Delta a$

$v\Delta t\Delta a$, and it contains a total mass $\varrho v\Delta t\Delta a$. Dividing Δt will give the mass across Δa per unit time, which by definition is the rate of the flow

$$\text{Rate of flow through } \Delta a = \varrho v\Delta a.$$

Now let us consider the case shown in Fig. 2.9b. In this case the area Δa is not perpendicular to the direction of the flow. The total mass which will flow through this tilted Δa in time Δt is just the density times the volume of this pipe with the slanted bases. That volume is $v\Delta t\Delta a \cos\theta$, where θ is the angle between the velocity vector $\mathbf{v}$ and $\mathbf{n}$, the unit normal to Δa. But $v\cos\theta = \mathbf{v}\cdot\mathbf{n}$. So, multiplying by ϱ and dividing Δt, we have

$$\text{Rate of flow through tilted } \Delta a = \varrho\mathbf{v}\cdot\mathbf{n}\Delta a.$$

To get the total flow through any surface S, first we divide the whole surface into little patches which are so small that over any one patch the surface is practically flat. Then we sum up the contributions from all the patches. As the patches become smaller and more numerous without limit, the sum becomes a surface integral. Accordingly,

$$\text{Total flow through } S = \iint_S \varrho\mathbf{v}\cdot\mathbf{n}\,da. \tag{2.70}$$

If we define $\mathbf{J} = \varrho\mathbf{v}$, (2.70) is known as the flux of $\mathbf{J}$ through the surface S

$$\text{Flux of } \mathbf{J} \text{ through } S = \iint_S \mathbf{J}\cdot\mathbf{n}\,da. \tag{2.71}$$

Originally it means the rate of the flow, the word flux is now generalized to mean the surface integral of the normal component of a vector. For example, the vector might be the electric field $\mathbf{E}$. Although electric field is not flowing in the sense in which fluid flows, we still say things like "the flux of $\mathbf{E}$ through a closed surface is equal to the total charge inside" to help us to visualize the electric field lines "flowing" out of the electric charges.

Example 2.5.2. Let $\mathbf{J} = \varrho v_0 \mathbf{k}$ where ϱ is the density of the fluid and $v_0 \mathbf{k}$ is its velocity ($\mathbf{k}$ is the unit vector in the z direction). Calculate the flux of $\mathbf{J}$ (the flow rate of the fluid) through a hemispherical surface of radius b.

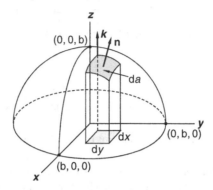

Fig. 2.10. Surface element on a hemisphere. The projected area on the xy plane is $dx\, dy = \cos\theta\, da$ where θ is the angle between the tangent plane at da and the xy plane

Solution 2.5.2. The equation of the spherical surface is $\varphi(x, y, z) = x^2 + y^2 + z^2 = b^2$. Therefore the unit normal to the surface is

$$\mathbf{n} = \frac{\nabla\varphi}{|\nabla\varphi|} = \frac{2x\mathbf{i} + 2y\mathbf{j} + 2z\mathbf{k}}{(4x^2 + 4y^2 + 4z^2)^{1/2}} = \frac{x\mathbf{i} + y\mathbf{j} + z\mathbf{k}}{b}.$$

Let the flux of $\mathbf{J}$ through the hemisphere be Φ, which is given by

$$\Phi = \iint_S \mathbf{J} \cdot \mathbf{n}\, da = \iint_S \varrho v_0 \mathbf{k} \cdot \mathbf{n}\, da,$$

where da is an element of the surface area on the hemisphere as shown in Fig. 2.10. This surface area projects onto $dx\, dy$ in the xy plane. Let θ be the acute angle between da (actually the tangent plane at da) and the xy plane. Then we have $dx\, dy = \cos\theta\, da$. The integral becomes

$$\Phi = \iint_S \varrho v_0 \mathbf{k} \cdot \mathbf{n}\, da = \iint_S \varrho v_0 \mathbf{k} \cdot \mathbf{n} \frac{1}{\cos\theta} dx\, dy,$$

where the limit on x and y must be such that we integrate over the projected area in the xy plane which is inside a circle of radius b. The angle between two planes is the same as the angle between the normals to the planes. Since $\mathbf{n}$ is the unit normal to da and $\mathbf{k}$ is the unit normal to xy plane, the angle θ is between $\mathbf{n}$ and $\mathbf{k}$. Thus, $\cos\theta = \mathbf{n} \cdot \mathbf{k}$. Therefore,

$$\Phi = \iint_S \varrho v_0 \mathbf{k} \cdot \mathbf{n} \frac{1}{\cos\theta} \mathrm{d}x\ \mathrm{d}y = \iint_S \varrho v_0\ \mathrm{d}x\ \mathrm{d}y = \varrho v_0 \pi b^2.$$

Note that this result is the same as the flux through the circular flat area in the xy plane. In fact, it is exactly the same as the flux through any surface whose boundary is the circle of radius b in the xy plane, since we obtained the result without using the explicit expression of $\mathbf{n}$.

2.5.2 Divergence Theorem

The *divergence theorem* is also known as *Gauss' theorem*. It relates the flux of a vector field through a closed surface S to the divergence of the vector field in the enclosed volume

$$\iint_{\text{closed surface } S} \mathbf{A} \cdot \mathbf{n}\ \mathrm{d}a = \iiint_{\text{vol. enclosed in } S} \boldsymbol{\nabla} \cdot \mathbf{A}\ \mathrm{d}V. \qquad (2.72)$$

The surface integral is over a closed surface as shown in Fig. 2.11. The unit normal vector $\mathbf{n}$ is pointing outward from the enclosed volume. The right-hand side of this equation is the integration of the divergence over the volume that is enclosed in the surface.

To prove this theorem, we cut the volume V up into a very large number of tiny (differential) cubes. The volume integral is the sum of the integrals over all the cubes.

Imagine we have a parallelepiped with six surfaces enclosing a volume V. We separate the volume into two cubes by a cut as in Fig. 2.12. Note that the sum of the flux through the six surfaces of the cube on the left and the flux through the six surfaces of the cube on the right is equal to the flux through the six surfaces of the original parallelepiped before it was cut. This is because

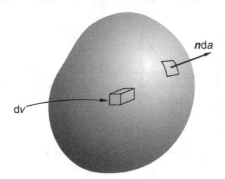

Fig. 2.11. The divergence theorem. The volume is enclosed by the surface. The integral of the divergence over the volume inside is equal to the flux through the outside surface

(a)   **(b)**

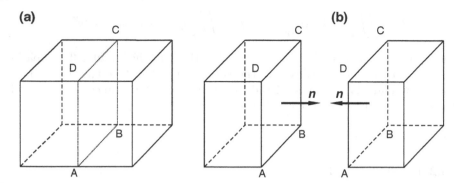

Fig. 2.12. The flux out of the touching sides of the two neighboring cubes cancel each other

the unit normal vectors on the touching sides of the two neighboring cubes are equal and opposite to each other. So the contributions to the flux for the two cubes from these two sides exactly cancel. Thus it must be generally true that the sum of the flux through the surfaces of all the cubes is equal to the integral over those surfaces that have no touching neighbors, i.e., over the original outside surface. So if we can prove the result for a small cube, we can prove it for any arbitrary volume.

Consider the flux of $\mathbf{A}$ through the surfaces of the small cube of volume $\Delta V = \Delta x \Delta y \Delta z$ shown in Fig. 2.13. The unit vector perpendicular to the surface ABCD is clearly $-\mathbf{j}$ $(\mathbf{n} = -\mathbf{j})$. The flux through this surface is therefore given by

$$\mathbf{A}(x,y,z) \cdot (-\mathbf{j})\Delta a = -A_y(x,y,z)\Delta x \Delta z.$$

The flux is defined as the outgoing "flow." The minus sign simply means the flux is flowing into the volume. Similarly, the unit normal to the surface EFGH is $\mathbf{j}$, and the flux through this surface is

$$\mathbf{A}(x, y + \Delta y, z) \cdot \mathbf{j}\Delta a = A_y(x, y + \Delta y, z)\Delta x \Delta z.$$

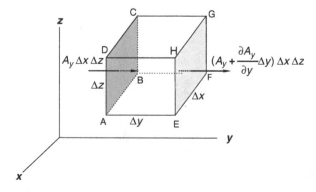

Fig. 2.13. The flux through the left and right face of a infinitesimal cube

Note that for every point (x, y, z) on ABCD, the corresponding point on EFGH is $(x, y + \Delta y, z)$. The net flux through these two surfaces is simply the sum of the two:

$$[A_y(x, y + \Delta y, z) - A_y(x, y, z)] \, \Delta x \Delta z = \frac{\partial A_y}{\partial y} \Delta y \Delta x \Delta z = \frac{\partial A_y}{\partial y} \Delta V. \quad (2.73)$$

By applying similar reasoning to the flux components in the two other directions, we find the total flux through all the surfaces of the cube is

$$\sum_{\text{cube}} \mathbf{A} \cdot \mathbf{n} \, da = \left(\frac{\partial A_x}{\partial x} + \frac{\partial A_y}{\partial y} + \frac{\partial A_z}{\partial z} \right) \Delta V = (\boldsymbol{\nabla} \cdot \mathbf{A}) \Delta V. \quad (2.74)$$

This shows that the outward flux from the surface of an infinitesimal cube is equal to the divergence of the vector multiplied by the volume of the cube. Thus the divergence has the following physical meaning:

The divergence of a vector $\mathbf{A}$ at a point is the total outward flux of $\mathbf{A}$ per unit volume in the neighborhood of that point.

For any finite volume, the total flux of $\mathbf{A}$ through the outside surface enclosing the volume is equal to the sum of the fluxes out of all the infinitesimal interior cubes, and the flux out of each cube is equal to the divergence of $\mathbf{A}$ times the volume of the cube. Therefore the integral of the normal component of a vector over any closed surface is equal to the integral of the divergence of the vector over the volume enclosed by the surface,

$$\oiint_S \mathbf{A} \cdot \mathbf{n} \, da = \iiint_V \boldsymbol{\nabla} \cdot \mathbf{A} \, dV. \quad (2.75)$$

The small circle on the double integral sign means the surface integral is over a closed surface. The volume integral is understood to be over the entire region inside the closed surface. This is the divergent theorem of (2.72), which is sometimes called the *fundamental theorem for divergence*.

A flow field $\mathbf{A}$ is said to be *solenoidal* if everywhere the divergence of $\mathbf{A}$ is equal to zero $(\boldsymbol{\nabla} \cdot \mathbf{A} = 0)$. An incompressible fluid must flow out of a given volume as rapidly as it flows in. The divergence of such a flow field must be zero, therefore the field is solenoidal.

On the other hand if $\mathbf{A}$ is such a field that at certain point $\boldsymbol{\nabla} \cdot \mathbf{A} \neq 0$, then there is a net outward flow from a small volume surrounding that point. Fluid must be "created" or "put in" at that point. If $\boldsymbol{\nabla} \cdot \mathbf{A}$ is negative, fluid must be "taken out" at that point. Therefore we come to the following conclusion.

The divergence of flow field at a point is a measure of the strength of the source (or sink) of the flow at that point.

Example 2.5.3. Verify the divergence theorem by evaluating both sides of (2.72) with $\mathbf{A} = x\mathbf{i} + y\mathbf{j} + z\mathbf{k}$ over a cylinder described by $x^2 + y^2 = 4$ and $0 \leq z \leq 4$.

Solution 2.5.3. Since $\nabla \cdot \mathbf{A} = \left(\mathbf{i}\dfrac{\partial}{\partial x} + \mathbf{j}\dfrac{\partial}{\partial y} + \mathbf{k}\dfrac{\partial}{\partial z} \right) \cdot (x\mathbf{i} + y\mathbf{j} + z\mathbf{k}) = 3$, the volume integral is

$$\iiint_V \nabla \cdot \mathbf{A} \, dV = 3 \iiint_V dV = 3(\pi 2^2)4 = 48\pi,$$

which is simply three times the volume of the cylinder. The surface of the cylinder consists of the top, bottom, and curved side surfaces. Therefore the surface integral can be divided into three parts

$$\oiint_S \mathbf{A} \cdot \mathbf{n} \, da = \iint_{\text{top}} \mathbf{A} \cdot \mathbf{n} \, da + \iint_{\text{bottom}} \mathbf{A} \cdot \mathbf{n} \, da + \iint_{\text{curved}} \mathbf{A} \cdot \mathbf{n} \, da.$$

For the top surface

$$\iint_{\text{top}} \mathbf{A} \cdot \mathbf{n} \, da = \iint_{\text{top}} (x\mathbf{i} + y\mathbf{j} + 4\mathbf{k}) \cdot \mathbf{k} \, da = 4 \iint_{\text{top}} da = 4\pi 2^2 = 16\pi.$$

For the bottom surface

$$\iint_{\text{bottom}} \mathbf{A} \cdot \mathbf{n} \, da = \iint_{\text{bottom}} (x\mathbf{i} + y\mathbf{j} + 0\mathbf{k}) \cdot (-\mathbf{k}) da = 0.$$

For the curved side surface, we must first find the unit normal $\mathbf{n}$. Since the surface is described by $\varphi(x, y) = x^2 + y^2 = 4$,

$$\mathbf{n} = \frac{\nabla \varphi}{|\nabla \varphi|} = \frac{2x\mathbf{i} + 2y\mathbf{j}}{(4x^2 + 4y^2)^{1/2}} = \frac{x\mathbf{i} + y\mathbf{j}}{2},$$

$$\mathbf{A} \cdot \mathbf{n} = (x\mathbf{i} + y\mathbf{j} + z\mathbf{k}) \cdot \frac{x\mathbf{i} + y\mathbf{j}}{2} = \frac{1}{2}(x^2 + y^2) = 2,$$

$$\iint_{\text{curved}} \mathbf{A} \cdot \mathbf{n} \, da = 2 \iint_{\text{curved}} da = 2(2\pi \cdot 2) 4 = 32\pi.$$

Therefore

$$\oiint_S \mathbf{A} \cdot \mathbf{n} \, da = 16\pi + 0 + 32\pi = 48\pi,$$

which is the same as the volume integral.

2.5.3 Continuity Equation

One of the most important applications of the divergence theorem is using it to express the conservation laws in differential forms. As an example, consider a fluid of density ϱ moving with velocity $\mathbf{v}$. According to (2.70), the rate at which the fluid flows out of a closed surface is

$$\text{Rate of outward flow through a closed surface} = \oiint_S \varrho \mathbf{v} \cdot \mathbf{n} \, da. \qquad (2.76)$$

Now because of the conservation of mass, this rate of out flow must be equal to the rate of decrease of the fluid inside the volume that is enclosed by the surface. Therefore

$$\oiint_S \varrho \mathbf{v} \cdot \mathbf{n} \, da = - \iiint_V \frac{\partial \varrho}{\partial t} dV. \qquad (2.77)$$

The negative sign accounts for the fact that the fluid inside is decreasing if the flow is outward. Using the divergence theorem

$$\oiint_S \varrho \mathbf{v} \cdot \mathbf{n} \, da = \iiint_V \boldsymbol{\nabla} \cdot (\varrho \mathbf{v}) \, dV, \qquad (2.78)$$

we have

$$\iiint_V \left[\boldsymbol{\nabla} \cdot (\varrho \mathbf{v}) + \frac{\partial \varrho}{\partial t} \right] dV = 0. \qquad (2.79)$$

Since the volume V in this equation, the integrand must equal to zero, or

$$\boxed{\boldsymbol{\nabla} \cdot (\varrho \mathbf{v}) + \frac{\partial \varrho}{\partial t} = 0.} \qquad (2.80)$$

This important equation, known as the continuity equation, relates the density and the velocity at the same point in a differential form. Many other conservation laws can be similarly expressed.

For an incompressible fluid, ϱ is not changing in time. In that case, the divergence of the velocity must be zero,

$$\boldsymbol{\nabla} \cdot \mathbf{v} = 0. \qquad (2.81)$$

Singularities in the Field. In deriving these integral theorems, we require the scalar and vector fields to be continuous and finite at every point. Often there are points, lines, or surfaces in space at which fields become discontinuous or even infinite. Examples are the electric fields produced by point, line, or surface charges. One way of dealing with this situation is to eliminate these volume elements, by appropriate surfaces, from the domain to which the theorems are to be applied. Another scheme is to "smear out" the discontinuous quantities, such as using charge densities, so that the fields are again well behaved. Still another powerful way is to use Dirac delta function. Sect. 2.10.2 is a specific example of these procedures.

2.6 The Curl of a Vector

The cross product of the gradient operator ∇ with vector $\mathbf{A}$ gives us another special combination of the derivatives of the components of $\mathbf{A}$

$$\nabla \times \mathbf{A} = \begin{vmatrix} \mathbf{i} & \mathbf{j} & \mathbf{k} \\ \dfrac{\partial}{\partial x} & \dfrac{\partial}{\partial y} & \dfrac{\partial}{\partial z} \\ A_x & A_y & A_z \end{vmatrix}$$

$$= \mathbf{i} \left(\frac{\partial A_z}{\partial y} - \frac{\partial A_y}{\partial z} \right) + \mathbf{j} \left(\frac{\partial A_x}{\partial z} - \frac{\partial A_z}{\partial x} \right) + \mathbf{k} \left(\frac{\partial A_y}{\partial x} - \frac{\partial A_x}{\partial y} \right). \tag{2.82}$$

It is a vector known as the *curl* of $\mathbf{A}$. The name curl suggests that it has something to do with rotation. In fact, in European texts the word rotation (or rot) is used in place of curl. In Example 2.1.3, we have considered the motion of a body rotating around the z-axis with angular velocity $\boldsymbol{\omega}$. The velocity of the particles in the body is $\mathbf{v} = -\omega y \mathbf{i} + \omega x \mathbf{j}$. The circular characteristic of this velocity field is manifested in the curl of the velocity

$$\nabla \times \mathbf{v} = \begin{vmatrix} \mathbf{i} & \mathbf{j} & \mathbf{k} \\ \dfrac{\partial}{\partial x} & \dfrac{\partial}{\partial y} & \dfrac{\partial}{\partial z} \\ -\omega y & \omega x & 0 \end{vmatrix} = 2\omega \mathbf{k}, \tag{2.83}$$

which shows that the curl of $\mathbf{v}$ is twice the angular velocity of the rotating body.

If this velocity field describes a fluid flow, curl $\mathbf{v}$ is called the *vorticity* vector of the fluid. It points in the direction around which a vortex motion takes place and is a measure of the the angular velocity of the flow. A small paddle wheel placed in the field will tend to rotate in regions where $\nabla \times \mathbf{v} \neq 0$. The paddle wheel will remain stationary in those regions where $\nabla \times \mathbf{v} = 0$. If the curl of a vector field is equal to zero everywhere, the field is called *irrotational*.

Example 2.6.1. Show that (a) $\nabla \times \mathbf{r} = 0$; (b) $\nabla \times \mathbf{r} f(r) = 0$ where $\mathbf{r}$ is the position vector.

Solution 2.6.1. (a) Since $\mathbf{r} = x\mathbf{i} + y\mathbf{j} + z\mathbf{k}$, so

$$\nabla \times \mathbf{r} = \begin{vmatrix} \mathbf{i} & \mathbf{j} & \mathbf{k} \\ \dfrac{\partial}{\partial x} & \dfrac{\partial}{\partial y} & \dfrac{\partial}{\partial z} \\ x & y & z \end{vmatrix} = 0.$$

(b)

$$\nabla \times \mathbf{r} f\left(r\right) = \begin{vmatrix} \mathbf{i} & \mathbf{j} & \mathbf{k} \\ \dfrac{\partial}{\partial x} & \dfrac{\partial}{\partial y} & \dfrac{\partial}{\partial z} \\ xf\left(r\right) & yf\left(r\right) & zf\left(r\right) \end{vmatrix} = \left\{\dfrac{\partial}{\partial y}\left[zf\left(r\right)\right] - \dfrac{\partial}{\partial z}\left[yf\left(r\right)\right]\right\}\mathbf{i}$$

$$+ \left\{\dfrac{\partial}{\partial z}\left[xf\left(r\right)\right] - \dfrac{\partial}{\partial x}\left[zf\left(r\right)\right]\right\}\mathbf{j} + \left\{\dfrac{\partial}{\partial x}\left[yf\left(r\right)\right] - \dfrac{\partial}{\partial y}\left[xf\left(r\right)\right]\right\}\mathbf{k}$$

$$= \left\{z\dfrac{\partial}{\partial y}f\left(r\right) - y\dfrac{\partial}{\partial z}f\left(r\right)\right\}\mathbf{i} + \left\{x\dfrac{\partial}{\partial z}f\left(r\right) - z\dfrac{\partial}{\partial x}f\left(r\right)\right\}\mathbf{j}$$

$$+ \left\{y\dfrac{\partial}{\partial x}f\left(r\right) - x\dfrac{\partial}{\partial y}f\left(r\right)\right\}\mathbf{k}.$$

Since

$$\dfrac{\partial}{\partial y}f\left(r\right) = \dfrac{df}{dr}\dfrac{\partial r}{\partial y} \text{ and } r = \left(x^2 + y^2 + z^2\right)^{1/2},$$

$$\dfrac{\partial}{\partial y}f\left(r\right) = \dfrac{df}{dr}\left(-\dfrac{y}{\left(x^2 + y^2 + z^2\right)^{1/2}}\right) = -\dfrac{df}{dr}\dfrac{y}{r},$$

$$\dfrac{\partial}{\partial x}f\left(r\right) = -\dfrac{df}{dr}\dfrac{x}{r}; \qquad \dfrac{\partial}{\partial z}f\left(r\right) = -\dfrac{df}{dr}\dfrac{z}{r}.$$

Therefore

$$\nabla \times \mathbf{r} f\left(r\right) = \dfrac{df}{dr}\left\{-z\dfrac{y}{r} + y\dfrac{z}{r}\right\}\mathbf{i} + \dfrac{df}{dr}\left\{-x\dfrac{z}{r} + z\dfrac{x}{r}\right\}\mathbf{j} + \dfrac{df}{dr}\left\{-y\dfrac{x}{r} + x\dfrac{y}{r}\right\}\mathbf{k} = \mathbf{0}.$$

2.6.1 Stokes' Theorem

Stokes' theorem relates the line integral of a vector function around a closed loop C to a surface integral of the curl of that vector over a surface S that spans the loop. The theorem states that

$$\int_{\text{closed loop } C} \mathbf{A} \cdot \mathbf{dr} = \iint_{\text{area bounded by } C} \left(\nabla \times \mathbf{A}\right) \cdot \mathbf{n}\, da, \qquad (2.84)$$

where $\mathbf{dr}$ is a directed line element along a closed curve C and S is any surface bounded by C. At any point on the surface, the unit normal vector $\mathbf{n}$ is perpendicular to the surface element da at that point as shown in Fig. 2.14. The sign of $\mathbf{n}$ is determined by the convention of the "right-hand rule." Curl the fingers of your right hand in the direction $\mathbf{dr}$, then your thumb points in the positive direction of $\mathbf{n}$. If the curve C lies in a plane, the simplest surface spans C is a flat surface. Now imagine the flat surface is a flexible membrane

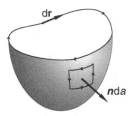

Fig. 2.14. Stokes' theorem. The integral of the curl over the surface is equal to the line integral around the closed boundary curve

which can continuously expand but remains attached to curve C. A sequence of curved surfaces is generated. Stokes theorem applies to all such surfaces. The positive direction of $\mathbf{n}$ for the flat surface is clear from the right-hand rule. As the surface expands, $\mathbf{n}$ moves along with it. For example, with the direction of $\mathbf{dr}$ shown in Fig. 2.14, the normal vector $\mathbf{n}$ of the nearly flat surface bounded by C is pointing "downward" according to the right-hand rule. When the surface is expanded into the final shape, $\mathbf{n}$ is turned to "outward" direction as shown in the figure.

The surface in Stokes' theorem must be two sided. A one-sided surface can be constructed by taking a long strip of paper, giving it a half twist, and joining the ends. If we tried to color one side of the surface we would find the whole thing colored. A belt of this shape is called a Moebius surface. Such a surface is not orientable since we cannot define the sense of the normal vector $\mathbf{n}$. Stokes' theorem applies only to the *orientable surface*, furthermore, the boundary curve of the surface must not cross itself.

To prove Stokes' theorem, we divide the surface into a large number of small rectangles. The surface integral on the right-hand side of (2.84) is of course just the sum of the surface integrals over all the small rectangles. If the line integrals around all the small rectangles are traced in the same direction, each interior line will be traced twice – once in each direction. Thus the line integrals of $\mathbf{A} \cdot \mathbf{dr}$ from all the interior lines will sum up to zero, since each term will appear twice with opposite sign. Therefore the sum of the line integrals around all the small rectangles will equal to the line integral around the boundary curve C, as shown in Fig. 2.15. So if we prove the result for a small rectangle, we will have proved it for any closed curve C.

Since the surface is to be composed of an infinitely large number of these infinitesimal rectangles, we may consider each to be a plane rectangle. Let us orient the coordinate axes so that one of these rectangles lies in the xy-plane, the sides will be of length Δx and Δy as shown in Fig. 2.16. Let the coordinates of the center of the loop be (x, y, z). We designate the corners of this rectangle by A, B, C, D. So, the line integral around this rectangle is

$$\oint_{\text{ABCD}} \mathbf{A} \cdot \mathbf{dr} = \int_{\text{AB}} \mathbf{A} \cdot (\mathbf{i} \ dx) + \int_{\text{BC}} \mathbf{A} \cdot (\mathbf{j} \ dy) + \int_{\text{CD}} \mathbf{A} \cdot (-\mathbf{i} \ dx) + \int_{\text{DA}} \mathbf{A} \cdot (-\mathbf{j} \ dy).$$
$$(2.85)$$

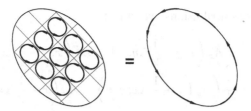

Fig. 2.15. Proof of Stokes' theorem. The surface is divided into differential surface elements. Circulation along interior lines cancel and the result is a circulation around the perimeter of the original surface

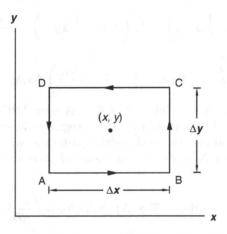

Fig. 2.16. The line integral of $\mathbf{A} \cdot d\mathbf{r}$ around the four sides of the infinitesimal square is equal to the surface integral of $\nabla \times \mathbf{A}$ over the area of this square

We use the symbol $\oint$ to mean the line integral is over a closed loop.

Now we may approximate the integral by the average value of the integrand multiplied by the length of the integration interval. The average value of $\mathbf{A}$ on the line AB may be taken to be the value of $\mathbf{A}$ at the midpoint of AB. The coordinates at the midpoint of AB is $\left(x, y - \frac{1}{2}\Delta y, z\right)$. Thus

$$\int_{AB} \mathbf{A} \cdot (\mathbf{i}\, dx) = \int_{AB} A_x\, dx = A_x\left(x, y - \frac{1}{2}\Delta y, z\right)\Delta x.$$

Similarly,

$$\int_{BC} \mathbf{A} \cdot (\mathbf{j}\, dy) = \int_{BC} A_y\, dy = A_y\left(x + \frac{1}{2}\Delta x, y, z\right)\Delta y,$$

$$\int_{CD} \mathbf{A} \cdot (-\mathbf{i}\, dx) = -\int_{CD} A_x\, dx = -A_x\left(x, y + \frac{1}{2}\Delta y, z\right)\Delta x,$$

$$\int_{DA} \mathbf{A} \cdot (-\mathbf{j}\, dy) = -\int_{DA} A_y\, dy = -A_y\left(x - \frac{1}{2}\Delta x, y, z\right)\Delta y.$$

Summing all these contributions, we obtain

$$\oint_{ABCD} \mathbf{A} \cdot d\mathbf{r} = \left(A_x \left(x, y - \frac{1}{2}\Delta y, z \right) - A_x \left(x, y + \frac{1}{2}\Delta y, z \right) \right) \Delta x$$
$$+ \left(A_y \left(x + \frac{1}{2}\Delta x, y, z \right) - A_y \left(x - \frac{1}{2}\Delta x, y, z \right) \right) \Delta y.$$

Since

$$A_y \left(x + \frac{1}{2}\Delta x, y, z \right) - A_y \left(x - \frac{1}{2}\Delta x, y, z \right) = \frac{\partial A_y}{\partial x}\Delta x,$$

$$A_x \left(x, y - \frac{1}{2}\Delta y, z \right) - A_x \left(x, y + \frac{1}{2}\Delta y, z \right) = -\frac{\partial A_x}{\partial y}\Delta y,$$

so we have

$$\oint_{ABCD} \mathbf{A} \cdot d\mathbf{r} = \left(\frac{\partial A_y}{\partial x} - \frac{\partial A_x}{\partial y} \right) \Delta x \Delta y. \tag{2.86}$$

Next consider the surface integral of $\boldsymbol{\nabla} \times \mathbf{A}$ over ABCD. In this case the unit normal $\mathbf{n}$ is just $\mathbf{k}$. Again we take the integral to be equal to the average value of the integrand over the area multiplied by the area of the integration. The average value of $\mathbf{A}$ is simply the value of $\mathbf{A}$ evaluated at the center. Therefore

$$\iint_{ABCD} (\boldsymbol{\nabla} \times \mathbf{A}) \cdot \mathbf{n}\, da = (\boldsymbol{\nabla} \times \mathbf{A}) \cdot \mathbf{k}\Delta x \Delta y = \left(\frac{\partial A_y}{\partial x} - \frac{\partial A_x}{\partial y} \right) \Delta x \Delta y, \tag{2.87}$$

which is the same as (2.86). This result can be interpreted as follows:

The component of $\boldsymbol{\nabla} \times \mathbf{A}$ in the direction of $\mathbf{n}$ at a point P is the circulation of $\mathbf{A}$ per unit area around P in the plane normal to $\mathbf{n}$.

The circulation of a vector field around any closed loop can now be easily related to the curl of that field. We fill the loop with a surface S and add the circulations around a set of infinitesimal squares covering this surface. Thus we have

$$\sum \oint_{ABCD} \mathbf{A} \cdot d\mathbf{r} = \sum \iint_{ABCD} (\boldsymbol{\nabla} \times \mathbf{A}) \cdot \mathbf{n}\, da, \tag{2.88}$$

which can be written as

$$\oint_C \mathbf{A} \cdot d\mathbf{r} = = \iint_S (\boldsymbol{\nabla} \times \mathbf{A}) \cdot \mathbf{n}\, da. \tag{2.89}$$

This is Stokes' theorem. Sometimes this theorem is referred to as the *fundamental theorem for curls*. Note that this theorem holds for any surface S as long as the boundary of the surface is the closed loop C.

Example 2.6.2. Verify Stokes' theorem by finding the circulation of the vector field $\mathbf{A} = 4y\mathbf{i} + x\mathbf{j} + 2z\mathbf{k}$ around a square of radius $2a$ in the xy plane, centered at the origin and the surface integral $\iint (\nabla \times \mathbf{A}) \cdot \mathbf{n} \, da$ over the surface of the square.

Solution 2.6.2. We compute the circulation by calculating the line integral around each of the four sides of the square. From $(a, -a, 0)$ to $(a, a, 0)$, $x = a$, $dx = 0$, $z = 0$:

$$I_1 = \int_{a,-a,0}^{a,a,0} \mathbf{A} \cdot \mathbf{dr} = \int_{a,-a,0}^{a,a,0} (4y\mathbf{i} + x\mathbf{j} + 2z\mathbf{k}) \cdot (\mathbf{i} \, dx + \mathbf{j} \, dy + \mathbf{k} \, dz)$$

$$= \int_{a,-a,0}^{a,a,0} (4y \, dx + x \, dy + 2z \, dz) = \int_{-a}^{a} a \, dy = 2a^2.$$

From $(a, a, 0)$ to $(-a, a, 0)$, $y = a$, $dy = 0$, $z = 0$:

$$I_2 = \int_{a,a,0}^{-a,a,0} \mathbf{A} \cdot \mathbf{dr} = \int_{a}^{-a} 4a \, dx = -8a^2.$$

From $(-a, a, 0)$ to $(-a, -a, 0)$, $x = -a$, $dx = 0$, $z = 0$:

$$I_3 = \int_{-a,a,0}^{-a,-a,0} \mathbf{A} \cdot \mathbf{dr} = \int_{a}^{-a} (-a) dy = 2a^2.$$

From $(-a, -a, 0)$ to $(a, -a, 0)$, $y = -a$, $dy = 0$, $z = 0$:

$$I_4 = \int_{-a,-a,0}^{a,-a,0} \mathbf{A} \cdot \mathbf{dr} = \int_{-a}^{a} 4(-a) dx = -8a^2.$$

Therefore the circulation is

$$\oint_C \mathbf{A} \cdot \mathbf{dr} = I_1 + I_2 + I_3 + I_4 = -12a^2.$$

Now we compute the surface integral. First $\mathbf{n} \, da = \mathbf{k} \, dx \, dy$ over the square and the curl of $\mathbf{A}$ is

$$\nabla \times \mathbf{A} = \begin{vmatrix} \mathbf{i} & \mathbf{j} & \mathbf{k} \\ \dfrac{\partial}{\partial x} & \dfrac{\partial}{\partial y} & \dfrac{\partial}{\partial z} \\ 4y & x & 2z \end{vmatrix} = -3\mathbf{k}.$$

Thus

$$\iint_S \nabla \times \mathbf{A} \cdot \mathbf{n} \, da = -3 \iint_S dx \, dy = -3(2a)^2 = -12a^2,$$

which is the same as the circulation, satisfying Stokes' theorem.

Example 2.6.3. Verify Stokes' theorem by evaluating both sides of (2.84) with $\mathbf{A} = 4y\mathbf{i} + x\mathbf{j} + 2z\mathbf{k}$. This time the surface is over the hemisphere described by $\varphi(x, y, z) = x^2 + y^2 + z^2 = 4$ and the loop C is given by the circle $x^2 + y^2 = 4$.

Solution 2.6.3. Since $\nabla \times \mathbf{A} = -3\mathbf{k}$,

$$\iint_S \nabla \times \mathbf{A} \cdot \mathbf{n} \, da = -3 \iint_S \mathbf{k} \cdot \mathbf{n} \, da.$$

The surface is over a hemisphere. The geometry is shown in Fig. 2.10. The surface integral can be evaluated over the projection of the hemisphere on the xy plane using the relation

$$da = \frac{1}{\cos \theta} dx \, dy = \frac{1}{\mathbf{k} \cdot \mathbf{n}} dx \, dy.$$

The integration is simply over the disk of radius 2:

$$\iint_S \nabla \times \mathbf{A} \cdot \mathbf{n} \, da = -3 \iint \mathbf{k} \cdot \mathbf{n} \frac{1}{\mathbf{k} \cdot \mathbf{n}} dx \, dy = -3 \iint dx \, dy = -3(4\pi) = -12\pi.$$

To evaluate the line integral around the circle, it is convenient to write the circle in the parametric form

$$\mathbf{r} = x\mathbf{i} + y\mathbf{j}, \quad x = 2\cos\theta, \quad y = 2\sin\theta, \quad 0 \le \theta \le 2\pi.$$

$$\frac{d\mathbf{r}}{d\theta} = -2\sin\theta\mathbf{i} + 2\cos\theta\mathbf{j}, \quad \mathbf{A} = 4y\mathbf{i} + x\mathbf{j} + 2z\mathbf{k} = 8\sin\theta\mathbf{i} + 2\cos\theta\mathbf{j} + 2z\mathbf{k}.$$

$$\oint_C \mathbf{A} \cdot d\mathbf{r} = \oint_C \mathbf{A} \cdot \frac{d\mathbf{r}}{d\theta} d\theta = \int_0^{2\pi} (-16\sin^2\theta + 4\cos^2\theta)d\theta = -12\pi.$$

Thus, Stokes' theorem is verified,

$$\oint_C \mathbf{A} \cdot d\mathbf{r} = \iint_S \nabla \times \mathbf{A} \cdot \mathbf{n} \, da.$$

Example 2.6.4. Use Stokes' theorem to evaluate the line integral $\oint_C \mathbf{A} \cdot d\mathbf{r}$ with $\mathbf{A} = 2yz\mathbf{i} + x\mathbf{j} + z^2\mathbf{k}$ along the circle described by $x^2 + y^2 = 1$.

Solution 2.6.4. The curl of $\mathbf{A}$ is

$$\nabla \times \mathbf{A} = \begin{vmatrix} \mathbf{i} & \mathbf{j} & \mathbf{k} \\ \dfrac{\partial}{\partial x} & \dfrac{\partial}{\partial y} & \dfrac{\partial}{\partial z} \\ 2yz & x & z^2 \end{vmatrix} = 2y\mathbf{j} + (1 - 2z)\mathbf{k},$$

and according to Stokes' theorem

$$\oint_C \mathbf{A} \cdot d\mathbf{r} = \iint_S \mathbf{\nabla} \times \mathbf{A} \cdot \mathbf{n} \, da = \iint_S [2y\mathbf{j} + (1 - 2z)\mathbf{k}] \cdot \mathbf{n} \, da.$$

Since S can be any surface as long as it is bounded by the circle, the simplest way to do this problem is to use the flat surface inside the circle. In that case $z = 0$ and $\mathbf{n} = \mathbf{k}$. Hence,

$$\oint \mathbf{A} \cdot d\mathbf{r} = \iint_S da = \pi.$$

Connectivity of Space. Stokes' theorem is valid in a *simply connected region.* A region is simply connected if any closed loop in the region can be shrunk to a point without encountering any points not in the region. In a simply connected region, any two curves between two points can be distorted into each other within the region. The space inside a torus (doughnut) is multiply connected since a closed curve surrounds the hole cannot be shrunk to a point within the region. The space between two infinitely long concentric cylinders is also not simply connected. However, the region between two concentric spheres is simply connected.

If $\mathbf{\nabla} \times \mathbf{F} = 0$ in a simply connected region, we can use Stokes' theorem

$$\oint_C \mathbf{F} \cdot d\mathbf{r} = \iint_S \mathbf{\nabla} \times \mathbf{F} \cdot \mathbf{n} \, da = 0$$

to conclude that the line integral $\int_A^B \mathbf{F} \cdot d\mathbf{r}$ is independent of the path.

If $\mathbf{\nabla} \times \mathbf{F} = 0$ in a multiply connected region, then $\int_A^B \mathbf{F} \cdot d\mathbf{r}$ is not unique. In such a case, we often "cut" the region so as to make it simply connected. Then $\int_A^B \mathbf{F} \cdot d\mathbf{r}$ is independent of the path inside the simply connected region, but $\int_A^B \mathbf{F} \cdot d\mathbf{r}$ across the cut line may give a finite jump.

For example, consider the loop integral $\oint \mathbf{F} \cdot d\mathbf{r}$ with

$$\mathbf{F} = -\frac{y}{x^2 + y^2}\mathbf{i} + \frac{x}{x^2 + y^2}\mathbf{j}$$

around a unit circle centered at the origin. Since

$$\mathbf{\nabla} \times \mathbf{F} = \begin{vmatrix} \mathbf{i} & \mathbf{j} & \mathbf{k} \\ \dfrac{\partial}{\partial x} & \dfrac{\partial}{\partial y} & \dfrac{\partial}{\partial z} \\ -\dfrac{y}{x^2+y^2} & \dfrac{x}{x^2+y^2} & 0 \end{vmatrix} = \mathbf{0},$$

one might conclude that

$$\oint \mathbf{F} \cdot d\mathbf{r} = \iint_S \mathbf{\nabla} \times \mathbf{F} \cdot \mathbf{n} \, da = 0.$$

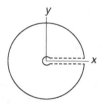

Fig. 2.17. If the function is singular on the z-axis, the region is multiply connected. A "cut" can be made to change it into a simply connected region. However, a lined integral across the cut line may give a sudden jump

This is incorrect, as one can readily see that

$$\oint \mathbf{F} \cdot d\mathbf{r} = \oint \left(-\frac{y}{x^2 + y^2} dx + \frac{x}{x^2 + y^2} dy \right).$$

With $x = \cos\theta$, $y = \sin\theta$, (so $dx = -\sin\theta \, d\theta$, $dy = \cos\theta \, d\theta$, and $x^2 + y^2 = 1$), this integral is seen to be

$$\oint \mathbf{F} \cdot d\mathbf{r} = \oint \left(\sin^2\theta + \cos^2\theta \right) d\theta = \oint d\theta = 2\pi,$$

which is certainly not zero. The source of the problem is that at $x = 0$ and $y = 0$ the function blows up. Thus we can only say that the curl of the function is zero except along the z-axis. If we try to exclude the z-axis from the region of integration, the region becomes multiply connected. In a multiply connected region, Stokes' theorem does not apply.

To make it simply connected, we can cut the region, such as along the $y = 0$ plane shown in Fig. 2.17 (or along any other direction). Within the simply connected region $\int_A^B \mathbf{F} \cdot d\mathbf{r} = \theta_B - \theta_A$. It will be equal to zero if A and B are infinitesimally close. However, if the integral is across the cut line, as long as A and B are on the different side of the cut, no matter how close are A and B, there is a sudden jump of 2π.

2.7 Further Vector Differential Operations

There are several combinations of vector operations involving the del ∇ operator which appear frequently in applications. They all follow the general rules of ordinary derivatives. The distributive rules are straightforward. With the definition the del operator ∇, one can readily verify

$$\nabla(\varphi_1 + \varphi_2) = \nabla\varphi_1 + \nabla\varphi_2; \tag{2.90}$$

$$\nabla \cdot (\mathbf{A} + \mathbf{B}) = \nabla \cdot \mathbf{A} + \nabla \cdot \mathbf{B}; \tag{2.91}$$

$$\nabla \times (\mathbf{A} + \mathbf{B}) = \nabla \times \mathbf{A} + \nabla \times \mathbf{B}. \tag{2.92}$$

However, the product rules are not so simple because there are more than one way to form a vector product.

2.7.1 Product Rules

The following is a list of useful product rules:

$$\nabla\left(\varphi\psi\right) = \varphi\nabla\psi + \psi\nabla\varphi, \tag{2.93}$$

$$\nabla\cdot\left(\varphi\mathbf{A}\right) = \nabla\varphi\cdot\mathbf{A} + \varphi\nabla\cdot\mathbf{A}, \tag{2.94}$$

$$\nabla\times\left(\varphi\mathbf{A}\right) = \nabla\varphi\times\mathbf{A} + \varphi\nabla\times\mathbf{A}, \tag{2.95}$$

$$\nabla\cdot\left(\mathbf{A}\times\mathbf{B}\right) = \left(\nabla\times\mathbf{A}\right)\cdot\mathbf{B} - \left(\nabla\times\mathbf{B}\right)\cdot\mathbf{A}, \tag{2.96}$$

$$\nabla\times\left(\mathbf{A}\times\mathbf{B}\right) = \left(\nabla\cdot\mathbf{B}\right)\mathbf{A} - \left(\nabla\cdot\mathbf{A}\right)\mathbf{B} + \left(\mathbf{B}\cdot\nabla\right)\mathbf{A} - \left(\mathbf{A}\cdot\nabla\right)\mathbf{B}, \tag{2.97}$$

$$\nabla(\mathbf{A}\cdot\mathbf{B}) = \left(\mathbf{A}\cdot\nabla\right)\mathbf{B} + \left(\mathbf{B}\cdot\nabla\right)\mathbf{A} + \mathbf{A}\times\left(\nabla\times\mathbf{B}\right) + \mathbf{B}\times\left(\nabla\times\mathbf{A}\right). \tag{2.98}$$

They can be verified by expanding both sides in terms of their Cartesian components. For example,

$$\nabla\left(\varphi\psi\right) = \mathbf{i}\frac{\partial}{\partial x}\left(\varphi\psi\right) + \mathbf{j}\frac{\partial}{\partial y}\left(\varphi\psi\right) + \mathbf{k}\frac{\partial}{\partial z}\left(\varphi\psi\right)$$

$$= \mathbf{i}\varphi\frac{\partial}{\partial x}\psi + \mathbf{j}\varphi\frac{\partial}{\partial y}\psi + \mathbf{k}\varphi\frac{\partial}{\partial z}\psi$$

$$+ \mathbf{i}\psi\frac{\partial}{\partial x}\varphi + \mathbf{j}\psi\frac{\partial}{\partial y}\varphi + \mathbf{k}\psi\frac{\partial}{\partial z}\varphi$$

$$= \varphi\nabla\psi + \psi\nabla\varphi. \tag{2.99}$$

Similarly,

$$\nabla\cdot\left(\varphi\mathbf{A}\right) = \frac{\partial}{\partial x}\left(\varphi A_x\right) + \frac{\partial}{\partial y}\left(\varphi A_y\right) + \frac{\partial}{\partial z}\left(\varphi A_z\right)$$

$$= \left(\frac{\partial\varphi}{\partial x}A_x + \varphi\frac{\partial A_x}{\partial x}\right) + \left(\frac{\partial\varphi}{\partial y}A_y + \varphi\frac{\partial A_y}{\partial y}\right) + \left(\frac{\partial\varphi}{\partial z}A_z + \varphi\frac{\partial A_z}{\partial z}\right)$$

$$= \left(\frac{\partial\varphi}{\partial x}A_x + \frac{\partial\varphi}{\partial y}A_y + \frac{\partial\varphi}{\partial z}A_z\right) + \varphi\left(\frac{\partial A_x}{\partial x} + \frac{\partial A_y}{\partial y} + \frac{\partial A_z}{\partial z}\right)$$

$$= \nabla\varphi\cdot\mathbf{A} + \varphi\nabla\cdot\mathbf{A}. \tag{2.100}$$

Clearly it will be very tedious to explicitly prove the rest of the product rules in this way. More "elegant" proofs will be given in the chapter of tensor analysis. Here, we will use the following formal procedure to establish these relations. The procedure consists of (1) first using ∇ as a differential operator and (2) then treating ∇ as if it were a regular vector. This procedure is a mnemonic device to give correct results.

Since ∇ is a linear combination of differential operators, we require it to obey the product rule of differentiation. When ∇ operates on a product, the result is the sum of two derivatives obtained by holding one of the factors constant and allowing the other to be operated on by ∇. As a matter of notation, we attach to ∇ a subscript indicating the one factor upon which it is currently allowed to operate, and the other factor is kept constant. For instance,

$$\nabla \times (\varphi \mathbf{A}) = \nabla_\varphi \times (\varphi \mathbf{A}) + \nabla_A \times (\varphi \mathbf{A}).$$

Since $\nabla_A \times (\varphi \mathbf{A})$ means that φ is a constant, it is then clear

$$\nabla_A \times (\varphi \mathbf{A}) = \varphi \nabla_A \times \mathbf{A} = \varphi \nabla \times \mathbf{A},$$

where the subscript A is omitted from the right-hand side, since it is clear what ∇ operates on when it is followed by just one factor. Similarly, $\nabla_\varphi \times (\varphi \mathbf{A})$ means $\mathbf{A}$ is constant. In this case it is easy to show

$$\nabla_\varphi \times (\varphi \mathbf{A}) = \begin{vmatrix} \mathbf{i} & \mathbf{j} & \mathbf{k} \\ \dfrac{\partial}{\partial x} & \dfrac{\partial}{\partial y} & \dfrac{\partial}{\partial z} \\ \varphi A_x & \varphi A_y & \varphi A_z \end{vmatrix} = \begin{vmatrix} \mathbf{i} & \mathbf{j} & \mathbf{k} \\ \dfrac{\partial \varphi}{\partial x} & \dfrac{\partial \varphi}{\partial y} & \dfrac{\partial \varphi}{\partial z} \\ A_x & A_y & A_z \end{vmatrix} = \nabla \varphi \times \mathbf{A}.$$

Thus,

$$\nabla \times (\varphi \mathbf{A}) = \nabla_\varphi \times (\varphi \mathbf{A}) + \nabla_A \times (\varphi \mathbf{A}) = \nabla \varphi \times \mathbf{A} + \varphi \nabla \times \mathbf{A}. \qquad (2.101)$$

For the divergence of a cross product, we start with

$$\nabla \cdot (\mathbf{A} \times \mathbf{B}) = \nabla_A \cdot (\mathbf{A} \times \mathbf{B}) + \nabla_B \cdot (\mathbf{A} \times \mathbf{B})$$
$$= \nabla_A \cdot (\mathbf{A} \times \mathbf{B}) - \nabla_B \cdot (\mathbf{B} \times \mathbf{A}).$$

Recall the scalar triple product $\mathbf{a} \cdot (\mathbf{b} \times \mathbf{c})$, the dot ($\cdot$) and the cross ($\times$) can be interchanged $\mathbf{a} \cdot (\mathbf{b} \times \mathbf{c}) = (\mathbf{a} \times \mathbf{b}) \cdot \mathbf{c}$. Treating ∇_A as a vector, we have

$$\nabla_A \cdot (\mathbf{A} \times \mathbf{B}) = (\nabla_A \times \mathbf{A}) \cdot \mathbf{B} = (\nabla \times \mathbf{A}) \cdot \mathbf{B},$$

where the subscript A is dropped in the last step because the meaning is clear without it. Similarly,

$$\nabla_B \cdot (\mathbf{B} \times \mathbf{A}) = (\nabla \times \mathbf{B}) \cdot \mathbf{A}.$$

Therefore,

$$\nabla \cdot (\mathbf{A} \times \mathbf{B}) = (\nabla \times \mathbf{A}) \cdot \mathbf{B} - (\nabla \times \mathbf{B}) \cdot \mathbf{A}. \qquad (2.102)$$

For the curl of a cross product, we will use the analogy of the vector triple product $\mathbf{a} \times (\mathbf{b} \times \mathbf{c}) = (\mathbf{a} \cdot \mathbf{c}) \mathbf{b} - (\mathbf{a} \cdot \mathbf{b}) \mathbf{c}$.

$$\nabla \times (\mathbf{A} \times \mathbf{B}) = \nabla_A \times (\mathbf{A} \times \mathbf{B}) + \nabla_B \times (\mathbf{A} \times \mathbf{B}),$$
$$\nabla_A \times (\mathbf{A} \times \mathbf{B}) = (\nabla_A \cdot \mathbf{B}) \mathbf{A} - (\nabla_A \cdot \mathbf{A}) \mathbf{B} = (\mathbf{B} \cdot \nabla_A) \mathbf{A} - (\nabla_A \cdot \mathbf{A}) \mathbf{B}.$$

In the last step, we have used the relation $(\nabla_A \cdot \mathbf{B}) \mathbf{A} = (\mathbf{B} \cdot \nabla_A) \mathbf{A}$, since $\mathbf{B}$ is regarded as a constant. Similarly,

$$\nabla_B \times (\mathbf{A} \times \mathbf{B}) = (\nabla_B \cdot \mathbf{B}) \mathbf{A} - (\nabla_B \cdot \mathbf{A}) \mathbf{B} = (\nabla_B \cdot \mathbf{B}) \mathbf{A} - (\mathbf{A} \cdot \nabla_B) \mathbf{B}.$$

Therefore

$$\nabla \times (\mathbf{A} \times \mathbf{B}) = (\mathbf{B} \cdot \nabla)\,\mathbf{A} - (\nabla \cdot \mathbf{A})\mathbf{B} + (\nabla \cdot \mathbf{B})\,\mathbf{A} - (\mathbf{A} \cdot \nabla)\mathbf{B}, \quad (2.103)$$

where we have dropped the subscripts because the meaning is clear without them.

The product rule of the gradient of a dot product v is more cumbersome,

$$\nabla(\mathbf{A} \cdot \mathbf{B}) = \nabla_A(\mathbf{A} \cdot \mathbf{B}) + \nabla_B(\mathbf{A} \cdot \mathbf{B}).$$

To work out $\nabla_A(\mathbf{A} \cdot \mathbf{B})$, we use the property of the vector triple product $\mathbf{a} \times (\mathbf{b} \times \mathbf{c}) = \mathbf{b}(\mathbf{a} \cdot \mathbf{c}) - (\mathbf{a} \cdot \mathbf{b})\mathbf{c}$,

$$\mathbf{A} \times (\nabla_B \times \mathbf{B}) = \nabla_B(\mathbf{A} \cdot \mathbf{B}) - (\mathbf{A} \cdot \nabla_B)\mathbf{B}.$$

Hence,

$$\nabla_B(\mathbf{A} \cdot \mathbf{B}) = (\mathbf{A} \cdot \nabla_B)\mathbf{B} + \mathbf{A} \times (\nabla_B \times \mathbf{B}).$$

Similarly,

$$\nabla_A(\mathbf{A} \cdot \mathbf{B}) = \nabla_A(\mathbf{B} \cdot \mathbf{A}) = (\mathbf{B} \cdot \nabla_A)\mathbf{A} + \mathbf{B} \times (\nabla_A \times \mathbf{A}).$$

Dropping the subscripts when they are not necessary, we have

$$\nabla(\mathbf{A} \cdot \mathbf{B}) = (\mathbf{B} \cdot \nabla)\mathbf{A} + \mathbf{B} \times (\nabla \times \mathbf{A}) + (\mathbf{A} \cdot \nabla)\mathbf{B} + \mathbf{A} \times (\nabla \times \mathbf{B}). \quad (2.104)$$

2.7.2 Second Derivatives

Several second derivatives can be constructed by applying ∇ twice. The following four identities of second derivatives are of great interests:

$$\nabla \times \nabla\varphi = \mathbf{0}, \quad (2.105)$$

$$\nabla \cdot \nabla \times \mathbf{A} = 0, \quad (2.106)$$

$$\nabla \times (\nabla \times \mathbf{A}) = \nabla(\nabla \cdot \mathbf{A}) - \nabla^2\mathbf{A}, \quad (2.107)$$

$$\nabla \cdot (\nabla\varphi \times \nabla\psi) = 0. \quad (2.108)$$

The first identity states that the curl of the gradient of a scalar function is identically equal to zero. This can be shown by direct expansion.

$$\nabla \times \nabla\varphi = \begin{vmatrix} \mathbf{i} & \mathbf{j} & \mathbf{k} \\ \dfrac{\partial}{\partial x} & \dfrac{\partial}{\partial y} & \dfrac{\partial}{\partial z} \\ \dfrac{\partial\varphi}{\partial x} & \dfrac{\partial\varphi}{\partial y} & \dfrac{\partial\varphi}{\partial z} \end{vmatrix} = \mathbf{i}\left(\dfrac{\partial^2\varphi}{\partial y\partial z} - \dfrac{\partial^2\varphi}{\partial z\partial y}\right)$$

$$+\mathbf{j}\left(\dfrac{\partial^2\varphi}{\partial z\partial x} - \dfrac{\partial^2\varphi}{\partial x\partial z}\right) + \mathbf{k}\left(\dfrac{\partial^2\varphi}{\partial x\partial y} - \dfrac{\partial^2\varphi}{\partial y\partial x}\right) = 0, \quad (2.109)$$

provided the second cross partial derivatives of φ are continuous (which are generally satisfied by functions of interests). In such a case, the order of differentiation is immaterial.

The second identity states that the divergence of curl of a vector function is identically equal to zero. This can also be shown by direct calculation.

$$\boldsymbol{\nabla} \cdot \boldsymbol{\nabla} \times \mathbf{A} = \left(\mathbf{i} \frac{\partial}{\partial x} + \mathbf{j} \frac{\partial}{\partial y} + \mathbf{k} \frac{\partial}{\partial z} \right) \cdot \begin{vmatrix} \mathbf{i} & \mathbf{j} & \mathbf{k} \\ \frac{\partial}{\partial x} & \frac{\partial}{\partial y} & \frac{\partial}{\partial z} \\ A_x & A_y & A_z \end{vmatrix}$$

$$= \begin{vmatrix} \frac{\partial}{\partial x} & \frac{\partial}{\partial y} & \frac{\partial}{\partial z} \\ \frac{\partial}{\partial x} & \frac{\partial}{\partial y} & \frac{\partial}{\partial z} \\ A_x & A_y & A_z \end{vmatrix} = 0. \tag{2.110}$$

It is understood that the determinant is to be expanded along the first row. Again if the partial derivatives are continuous, this determinant with two identical rows is equal to zero.

The curl curl identity is equally important and is worthwhile to commit to memory. For the mnemonic purpose, we can use the analogy of the vector triple product $\mathbf{a} \times (\mathbf{b} \times \mathbf{c}) = \mathbf{b}(\mathbf{a} \cdot \mathbf{c}) - (\mathbf{a} \cdot \mathbf{b})\mathbf{c}$, with $\boldsymbol{\nabla}, \boldsymbol{\nabla}, \mathbf{A}$ as $\mathbf{a}, \mathbf{b}, \mathbf{c}$, respectively. Thus, the vector triple product suggests

$$\boldsymbol{\nabla} \times (\boldsymbol{\nabla} \times \mathbf{A}) = \boldsymbol{\nabla} (\boldsymbol{\nabla} \cdot \mathbf{A}) - (\boldsymbol{\nabla} \cdot \boldsymbol{\nabla}) \mathbf{A}. \tag{2.111}$$

The $\boldsymbol{\nabla} \cdot \boldsymbol{\nabla}$ is a scalar operator. Because it appears often in physics, it has given a special name – the *Laplacian*, or just ∇^2

$$\boldsymbol{\nabla} \cdot \boldsymbol{\nabla} = \left(\mathbf{i} \frac{\partial}{\partial x} + \mathbf{j} \frac{\partial}{\partial y} + \mathbf{k} \frac{\partial}{\partial z} \right) \cdot \left(\mathbf{i} \frac{\partial}{\partial x} + \mathbf{j} \frac{\partial}{\partial y} + \mathbf{k} \frac{\partial}{\partial z} \right)$$

$$= \frac{\partial^2}{\partial x^2} + \frac{\partial^2}{\partial y^2} + \frac{\partial^2}{\partial z^2} = \nabla^2. \tag{2.112}$$

Therefore, (2.111) can be written as

$$\boldsymbol{\nabla} \times (\boldsymbol{\nabla} \times \mathbf{A}) = \boldsymbol{\nabla} (\boldsymbol{\nabla} \cdot \mathbf{A}) - \nabla^2 \mathbf{A}. \tag{2.113}$$

Expanding both sides of this equation in rectangular coordinates, one can readily verify that this is indeed an identity.

Since ∇^2 is a scalar operator, when it operates on a vector, it means the same operation on each component of the vector

$$\nabla^2 \mathbf{A} = \mathbf{i} \nabla^2 A_x + \mathbf{j} \nabla^2 A_y + \mathbf{k} \nabla^2 A_z. \tag{2.114}$$

The identity (2.108) follows from $\boldsymbol{\nabla} \cdot (\mathbf{A} \times \mathbf{B}) = \boldsymbol{\nabla} \times \mathbf{A} \cdot \mathbf{B} - \boldsymbol{\nabla} \times \mathbf{B} \cdot \mathbf{A}$. Since $\boldsymbol{\nabla}\varphi$ and $\boldsymbol{\nabla}\psi$ are two different vectors,

$$\boldsymbol{\nabla} \cdot (\boldsymbol{\nabla}\varphi \times \boldsymbol{\nabla}\psi) = \boldsymbol{\nabla} \times \boldsymbol{\nabla}\varphi \cdot \boldsymbol{\nabla}\psi - \boldsymbol{\nabla} \times \boldsymbol{\nabla}\psi \cdot \boldsymbol{\nabla}\varphi.$$

Now $\boldsymbol{\nabla} \times \boldsymbol{\nabla}\varphi = \boldsymbol{\nabla} \times \boldsymbol{\nabla}\psi = \mathbf{0}$, therefore

$$\boldsymbol{\nabla} \cdot (\boldsymbol{\nabla}\varphi \times \boldsymbol{\nabla}\psi) = 0. \tag{2.115}$$

Example 2.7.1. Show that $\boldsymbol{\nabla} \times \mathbf{A} = \mathbf{B}$, if $\mathbf{A} = \frac{1}{2}\mathbf{B} \times \mathbf{r}$ and $\mathbf{B}$ is a constant vector, first by direct expansion, then by the formula of the curl of a cross product.

Solution 2.7.1. *Method I*

$$\boldsymbol{\nabla} \times \mathbf{A} = \frac{1}{2}\boldsymbol{\nabla} \times (\mathbf{B} \times \mathbf{r}) = \frac{1}{2}\boldsymbol{\nabla} \times \begin{vmatrix} \mathbf{i} & \mathbf{j} & \mathbf{k} \\ B_x & B_y & B_z \\ x & y & z \end{vmatrix}$$

$$= \frac{1}{2}\boldsymbol{\nabla} \times [\mathbf{i}(B_y z - B_z y) + \mathbf{j}(B_z x - B_x z) + \mathbf{k}(B_x y - B_y x)]$$

$$= \frac{1}{2}\begin{vmatrix} \mathbf{i} & \mathbf{j} & \mathbf{k} \\ \dfrac{\partial}{\partial x} & \dfrac{\partial}{\partial y} & \dfrac{\partial}{\partial z} \\ (B_y z - B_z y) & (B_z x - B_x z) & (B_x y - B_y x) \end{vmatrix}$$

$$= \frac{1}{2}[\mathbf{i}2B_x + \mathbf{j}2B_y + \mathbf{k}2B_z] = \mathbf{B}.$$

Method II

$$\frac{1}{2}\boldsymbol{\nabla} \times (\mathbf{B} \times \mathbf{r}) = \frac{1}{2}[(\boldsymbol{\nabla} \cdot \mathbf{r})\mathbf{B} - (\boldsymbol{\nabla} \cdot \mathbf{B})\mathbf{r} + (\mathbf{r} \cdot \boldsymbol{\nabla})\mathbf{B} - (\mathbf{B} \cdot \boldsymbol{\nabla})\mathbf{r}]$$

$$= \frac{1}{2}[(\boldsymbol{\nabla} \cdot \mathbf{r})\mathbf{B} - (\mathbf{B} \cdot \boldsymbol{\nabla})\mathbf{r}] \quad \text{(since } \mathbf{B} \text{ is a constant)},$$

$$(\boldsymbol{\nabla} \cdot \mathbf{r})\mathbf{B} = 3\mathbf{B} \quad \text{(see Example 2.5.1)},$$
$$(\mathbf{B} \cdot \boldsymbol{\nabla})\mathbf{r} = \mathbf{B} \quad \text{(see Example 2.4.2)},$$

$$\frac{1}{2}\boldsymbol{\nabla} \times (\mathbf{B} \times \mathbf{r}) = \frac{1}{2}[3\mathbf{B} - \mathbf{B}] = \mathbf{B}.$$

Example 2.7.2. Show that $\nabla \times (\nabla^2 \mathbf{A}) = \nabla^2 (\nabla \times \mathbf{A})$.

Solution 2.7.2. Since $\nabla \times (\nabla \times \mathbf{A}) = \nabla(\nabla \cdot \mathbf{A}) - \nabla^2 \mathbf{A}$,

$$\nabla^2 \mathbf{A} = \nabla(\nabla \cdot \mathbf{A}) - \nabla \times (\nabla \times \mathbf{A}),$$

$$\nabla \times (\nabla^2 \mathbf{A}) = \nabla \times [\nabla(\nabla \cdot \mathbf{A}) - \nabla \times (\nabla \times \mathbf{A})].$$

Since curl gradient is equal to zero, $\nabla \times \nabla(\nabla \cdot \mathbf{A}) = \mathbf{0}$. Using curl curl formula again, we have

$$\nabla \times (\nabla^2 \mathbf{A}) = -\nabla \times \nabla \times (\nabla \times \mathbf{A}) = - \left\{ \nabla(\nabla \cdot (\nabla \times \mathbf{A}) - \nabla^2(\nabla \times \mathbf{A}) \right\}.$$

Since divergence of a curl is equal to zero, $\nabla \cdot (\nabla \times \mathbf{A}) = 0$, therefore

$$\nabla \times (\nabla^2 \mathbf{A}) = \nabla^2(\nabla \times \mathbf{A}).$$

Example 2.7.3. If

$$\nabla \cdot \mathbf{E} = 0, \quad \nabla \times \mathbf{E} = -\frac{\partial}{\partial t}\mathbf{H},$$

$$\nabla \cdot \mathbf{H} = 0, \quad \nabla \times \mathbf{H} = \frac{\partial}{\partial t}\mathbf{E},$$

show that

$$\nabla^2 \mathbf{E} = \frac{\partial^2}{\partial t^2}\mathbf{E}; \quad \nabla^2 \mathbf{H} = \frac{\partial^2}{\partial t^2}\mathbf{H}.$$

Solution 2.7.3.

$$\nabla \times (\nabla \times \mathbf{E}) = \nabla \times \left(-\frac{\partial}{\partial t}\mathbf{H} \right) = -\frac{\partial}{\partial t}(\nabla \times \mathbf{H}) = -\frac{\partial}{\partial t}\left(\frac{\partial}{\partial t}\mathbf{E} \right) = -\frac{\partial^2}{\partial t^2}\mathbf{E},$$

$$\nabla \times (\nabla \times \mathbf{E}) = \nabla(\nabla \cdot \mathbf{E}) - \nabla^2 \mathbf{E} = -\nabla^2 \mathbf{E} \quad (\text{since } \nabla \cdot \mathbf{E} = 0).$$

Therefore

$$\nabla^2 \mathbf{E} = \frac{\partial^2}{\partial t^2}\mathbf{E}.$$

Similarly,

$$\nabla \times (\nabla \times \mathbf{H}) = \nabla \times \left(\frac{\partial}{\partial t}\mathbf{E} \right) = \frac{\partial}{\partial t}(\nabla \times \mathbf{E}) = \frac{\partial}{\partial t}\left(-\frac{\partial}{\partial t}\mathbf{H} \right) = -\frac{\partial^2}{\partial t^2}\mathbf{H},$$

$$\nabla \times (\nabla \times \mathbf{H}) = \nabla(\nabla \cdot \mathbf{H}) - \nabla^2 \mathbf{H} = -\nabla^2 \mathbf{H} \quad (\text{since } \nabla \cdot \mathbf{H} = 0).$$

It follows that

$$\nabla^2 \mathbf{H} = \frac{\partial^2}{\partial t^2}\mathbf{H}.$$

2.8 Further Integral Theorems

There are many other integral identities that are useful in physical applications. They can be derived in a variety of ways. Here we discuss some of the most useful ones and show that they all follow from the fundamental theorems of gradient, divergence, and curl.

2.8.1 Green's Theorem

The following integral identities are all named after George Green (1793–1841). To distinguish them, we adopt the following terminology.

Green's Lemma:

$$\oint_C [f(x,y)\,dx + g(x,y)\,dy] = \iint_S \left(\frac{\partial g}{\partial x} - \frac{\partial f}{\partial y} \right) dx\,dy, \qquad (2.116)$$

Green's Theorem:

$$\oiint_S \varphi \nabla \psi \cdot \mathbf{n}\, da = \iiint_V (\nabla \varphi \cdot \nabla \psi + \varphi \nabla^2 \psi)\, dV, \qquad (2.117)$$

Symmetrical form of Green's Theorem:

$$\oiint_S (\varphi \nabla \psi - \psi \nabla \varphi) \cdot \mathbf{n}\, da = \iiint_V (\varphi \nabla^2 \psi - \psi \nabla^2 \varphi)\, dV. \qquad (2.118)$$

To prove Green's Lemma, we start with Stokes' theorem

$$\oint_C \mathbf{A} \cdot d\mathbf{r} = \int_S (\nabla \times \mathbf{A}) \cdot \mathbf{n}\, da.$$

With the curve C lying entirely on the xy plane,

$$\mathbf{A} \cdot d\mathbf{r} = (\mathbf{i} A_x + \mathbf{j} A_y + \mathbf{k} A_z) \cdot (\mathbf{i}\, dx + \mathbf{j}\, dy) = A_x dx + A_y dy,$$

and $\mathbf{n}$ is equal to $\mathbf{k}$, the unit vector in the z direction,

$$(\nabla \times \mathbf{A}) \cdot \mathbf{n}\, da = \begin{vmatrix} \mathbf{i} & \mathbf{j} & \mathbf{k} \\ \dfrac{\partial}{\partial x} & \dfrac{\partial}{\partial y} & \dfrac{\partial}{\partial z} \\ A_x & A_y & A_z \end{vmatrix} \cdot \mathbf{k} dx\, dy = \left(\frac{\partial A_y}{\partial x} - \frac{\partial A_x}{\partial y} \right) dx\, dy.$$

Thus we have

$$\oint_C (A_x dx + A_y dy) = \iint_S \left(\frac{\partial A_y}{\partial x} - \frac{\partial A_x}{\partial y} \right) dx\, dy.$$

Since **A** in Stokes' theorem can be any vector function, Green's Lemma follows with $A_x = f(x, y)$, and $A_y = g(x, y)$.

To prove Green's theorem, we start with the divergence theorem

$$\iiint_V \boldsymbol{\nabla} \cdot (\varphi \boldsymbol{\nabla} \psi) \mathrm{d}V = \oiint_S \varphi \boldsymbol{\nabla} \psi \cdot \mathbf{n} \, \mathrm{d}a.$$

Using the identity

$$\boldsymbol{\nabla} \cdot (\varphi \boldsymbol{\nabla} \psi) = \boldsymbol{\nabla} \varphi \cdot \boldsymbol{\nabla} \psi + \varphi \nabla^2 \psi,$$

we have

$$\iiint_V (\boldsymbol{\nabla} \varphi \cdot \boldsymbol{\nabla} \psi + \varphi \nabla^2 \psi) \mathrm{d}V = \oiint_S \varphi \boldsymbol{\nabla} \psi \cdot \mathbf{n} \, \mathrm{d}a, \qquad (2.119)$$

which is Green's theorem (2.117).

Clearly (2.119) is equally valid when φ and ψ are interchanged

$$\iiint_V (\boldsymbol{\nabla} \psi \cdot \boldsymbol{\nabla} \varphi + \psi \nabla^2 \varphi) \mathrm{d}V = \oiint_S \psi \boldsymbol{\nabla} \varphi \cdot \mathbf{n} \, \mathrm{d}a. \qquad (2.120)$$

Taking the difference of the last two equations, we obtain the symmetric form of the Green's theorem

$$\iiint_V (\varphi \nabla^2 \psi - \psi \nabla^2 \varphi) \, \mathrm{d}V = \oiint_S (\varphi \boldsymbol{\nabla} \psi - \psi \boldsymbol{\nabla} \varphi) \cdot \mathbf{n} \, \mathrm{d}a.$$

2.8.2 Other Related Integrals

The divergence theorem can take some other alternative forms. Let φ be a scalar function and **C** be an arbitrary constant vector. Then,

$$\iiint_V \boldsymbol{\nabla} \cdot (\varphi \mathbf{C}) \mathrm{d}V = \oiint_S \varphi \mathbf{C} \cdot \mathbf{n} \, \mathrm{d}a,$$

$$\boldsymbol{\nabla} \cdot (\varphi \mathbf{C}) = \boldsymbol{\nabla} \varphi \cdot \mathbf{C} + \varphi \boldsymbol{\nabla} \cdot \mathbf{C} = \boldsymbol{\nabla} \varphi \cdot \mathbf{C},$$

since **C** is constant and $\boldsymbol{\nabla} \cdot \mathbf{C} = 0$.

$$\iiint_V \boldsymbol{\nabla} \cdot (\varphi \mathbf{C}) \mathrm{d}V = \iiint_V \boldsymbol{\nabla} \varphi \cdot \mathbf{C} \, \mathrm{d}V = \mathbf{C} \cdot \iiint_V \boldsymbol{\nabla} \varphi \, \mathrm{d}V.$$

$$\oiint_S \varphi \mathbf{C} \cdot \mathbf{n} \, \mathrm{d}a = \mathbf{C} \cdot \oiint_S \varphi \mathbf{n} \, \mathrm{d}a.$$

Therefore the divergence theorem can be written as

$$\mathbf{C} \cdot \left[\iiint_V \boldsymbol{\nabla} \varphi \, \mathrm{d}V - \oiint_S \varphi \mathbf{n} \, \mathrm{d}a \right] = 0.$$

Since $\mathbf{C}$ is arbitrary, the terms in the brackets must be zero. Thus we have another interesting relation between volume integral and surface integral

$$\iiint_V \boldsymbol{\nabla}\varphi\,dV = \oiint_S \varphi\mathbf{n}\,da. \tag{2.121}$$

Similarly, let $\mathbf{A}$ be a vector function and $\mathbf{C}$, an arbitrary constant vector. $\mathbf{A} \times \mathbf{C}$ is another vector function. The divergence theorem can be written as

$$\iiint_V \boldsymbol{\nabla}\cdot(\mathbf{A}\times\mathbf{C})dV = \oiint_S (\mathbf{A}\times\mathbf{C})\cdot\mathbf{n}\,da.$$

Since

$$\boldsymbol{\nabla}\cdot(\mathbf{A}\times\mathbf{C}) = (\boldsymbol{\nabla}\times\mathbf{A})\cdot\mathbf{C} - (\boldsymbol{\nabla}\times\mathbf{C})\cdot\mathbf{A} = (\boldsymbol{\nabla}\times\mathbf{A})\cdot\mathbf{C},$$

$$(\mathbf{A}\times\mathbf{C})\cdot\mathbf{n} = -(\mathbf{C}\times\mathbf{A})\cdot\mathbf{n} = -\mathbf{C}\cdot\mathbf{A}\times\mathbf{n},$$

therefore

$$\iiint_V \boldsymbol{\nabla}\cdot(\mathbf{A}\times\mathbf{C})dV = \mathbf{C}\cdot\iiint_V \boldsymbol{\nabla}\times\mathbf{A}\,dV,$$

$$\oiint_S (\mathbf{A}\times\mathbf{C})\cdot\mathbf{n}\,da = -\mathbf{C}\cdot\oiint_S \mathbf{A}\times\mathbf{n}\,da.$$

Thus we have another form of the divergence theorem

$$\iiint_V \boldsymbol{\nabla}\times\mathbf{A}\,dV = -\oiint_S \mathbf{A}\times\mathbf{n}\,da. \tag{2.122}$$

This exploitation of the arbitrary nature of a part of a problem is a very useful technique. In the following examples some alternative forms of Stokes' theorem will be derived using this technique.

Example 2.8.1. Show that $\displaystyle\oint_C \varphi\,d\mathbf{r} = -\iint_S \boldsymbol{\nabla}\varphi\times\mathbf{n}\,da.$

Solution 2.8.1. Let $\mathbf{C}$ be an arbitrary constant vector. By Stokes' theorem we have

$$\oint_C \varphi\mathbf{C}\cdot d\mathbf{r} = \iint_S \boldsymbol{\nabla}\times(\varphi\mathbf{C})\cdot\mathbf{n}da.$$

Since $\mathbf{C}$ is a constant and $\boldsymbol{\nabla}\times\mathbf{C} = \mathbf{0}$,

$$\boldsymbol{\nabla}\times\varphi\mathbf{C} = \boldsymbol{\nabla}\varphi\times\mathbf{C} + \varphi\boldsymbol{\nabla}\times\mathbf{C} = \boldsymbol{\nabla}\varphi\times\mathbf{C},$$

$$\iint_S \boldsymbol{\nabla}\times(\varphi\mathbf{C})\cdot\mathbf{n}\,da = \iint_S \boldsymbol{\nabla}\varphi\times\mathbf{C}\cdot\mathbf{n}\,da.$$

Furthermore,

$$\nabla\varphi \times \mathbf{C} \cdot \mathbf{n} = -\mathbf{C} \times \nabla\varphi \cdot \mathbf{n} = -\mathbf{C} \cdot (\nabla\varphi \times \mathbf{n}).$$

Therefore

$$\iint_S \nabla \times (\varphi\mathbf{C}) \cdot \mathbf{n} \; da = -\mathbf{C} \cdot \iint_S \nabla\varphi \times \mathbf{n} \; da.$$

With

$$\oint_C \varphi\mathbf{C} \cdot d\mathbf{r} = \mathbf{C} \cdot \oint_C \varphi \, d\mathbf{r},$$

we can write Stokes' theorem as

$$\mathbf{C} \cdot \oint_C \varphi \, d\mathbf{r} = -\, \mathbf{C} \cdot \int_S \nabla\varphi \times \mathbf{n} \; da.$$

Again since $\mathbf{C}$ is an arbitrary constant vector, it follows that

$$\oint_C \varphi \, d\mathbf{r} = -\iint_S \nabla\varphi \times \mathbf{n} \; da. \tag{2.123}$$

Example 2.8.2. Show that $\oint_C \mathbf{r} \times d\mathbf{r} = 2\int_S \mathbf{n}\; da$ where $\mathbf{r}$ is the position vector from an origin that can be chosen at any point in space.

Solution 2.8.2. To prove this, we use an arbitrary constant vector $\mathbf{C}$ and start with Stokes' theorem,

$$\oint_C (\mathbf{C} \times \mathbf{r}) \cdot d\mathbf{r} = \iint_S \nabla \times (\mathbf{C} \times \mathbf{r}) \cdot \mathbf{n} \; da.$$

Since

$$\oint_C (\mathbf{C} \times \mathbf{r}) \cdot d\mathbf{r} = \oint_C \mathbf{C} \cdot \mathbf{r} \times d\mathbf{r} = \mathbf{C} \cdot \oint_C \mathbf{r} \times d\mathbf{r},$$

and

$$\nabla \times (\mathbf{C} \times \mathbf{r}) = 2\mathbf{C} \quad \text{(see example 2.7.1)},$$

$$\iint_S \nabla \times (\mathbf{C} \times \mathbf{r}) \cdot \mathbf{n} \; da = \iint_S 2\mathbf{C} \cdot \mathbf{n} \; da = 2\mathbf{C} \cdot \iint_S \mathbf{n} \; da,$$

it follows

$$\mathbf{C} \cdot \oint_C \mathbf{r} \times d\mathbf{r} = \mathbf{C} \cdot 2 \iint_S \mathbf{n} \; da.$$

Since $\mathbf{C}$ is an arbitrary constant vector, the integral identity

$$\oint_C \mathbf{r} \times d\mathbf{r} = 2 \iint_S \mathbf{n} \; da \tag{2.124}$$

must hold. This integral identity is of some importance in electrodynamics. This integral also shows that the area A of a flat surface S enclosed by a curve C is given by

$$A = \iint_S da = \frac{1}{2} \left| \oint_C \mathbf{r} \times d\mathbf{r} \right|.$$
(2.125)

2.9 Classification of Vector Fields

2.9.1 Irrotational Field and Scalar Potential

If $\boldsymbol{\nabla} \times \mathbf{F} = \mathbf{0}$ in a simply connected region, we say $\mathbf{F}$ is an *irrotational vector field*. An irrotational field is also known as a *conservative vector field*. We have seen that if $\boldsymbol{\nabla} \times \mathbf{F} = \mathbf{0}$, the line integral $\int_A^B \mathbf{F} \cdot d\mathbf{r}$ is independent of path. This means, as shown in Sect. 2.4.3, that $\mathbf{F}$ can be expressed as the gradient of a scalar function φ, known as the scalar potential.

Because of Stokes' theorem

$$\oint_C \mathbf{F} \cdot d\mathbf{r} = \iint_S \boldsymbol{\nabla} \times \mathbf{F} \cdot \mathbf{n} \, da,$$

an irrotational field $\mathbf{F}$ is characterized by any of the following equivalent conditions:

(a) $\boldsymbol{\nabla} \times \mathbf{F} = \mathbf{0}$,

(b) $\oint \mathbf{F} \cdot d\mathbf{r} = 0$ for any closed loop,

(c) $\int_A^B \mathbf{F} \cdot d\mathbf{r}$ is independent of path,

(d) $\mathbf{F} = -\boldsymbol{\nabla}\varphi$.

The sign in (d) is arbitrary, since φ is yet to be specified. In hydrodynamics, often a plus sign $(+)$ is chosen for the velocity potential. Here we have followed the convention in choosing a minus sign $(-)$ for the convenience of establishing the principle of conservation of energy.

Conservative Force Field. To see why an irrotational field is also called a conservative vector field, consider $\mathbf{F}(x, y, z)$ as the force in Newton's equation of motion

$$\mathbf{F}(x, y, z) = m\mathbf{a} = m\frac{d\mathbf{v}}{dt}.$$
(2.126)

Since $\mathbf{F}$ is irrotational, so

$$\mathbf{F}(x, y, z) = -\boldsymbol{\nabla}\varphi(x, y, z).$$
(2.127)

Therefore

$$m\frac{d\mathbf{v}}{dt} = -\boldsymbol{\nabla}\varphi. \tag{2.128}$$

Take dot product of both sides with dr and integrate. The left-hand side becomes

$$\int m\frac{d\mathbf{v}}{dt} \cdot d\mathbf{r} = \int m\frac{d\mathbf{v}}{dt} \cdot \frac{d\mathbf{r}}{dt}dt = \int m\frac{d\mathbf{v}}{dt} \cdot \mathbf{v}dt$$

$$= \int \frac{d}{dt}\left(\frac{1}{2}m\mathbf{v} \cdot \mathbf{v}\right) dt = \int d\left(\frac{1}{2}mv^2\right)$$

$$= \frac{1}{2}mv^2 + \text{constant}. \tag{2.129}$$

The right-hand side becomes

$$\int (-\boldsymbol{\nabla}\varphi) \cdot d\mathbf{r} = -\int d\varphi = -\varphi + \text{constant}. \tag{2.130}$$

Equating the results of the two sides of (2.128) gives

$$\frac{1}{2}mv^2 + \varphi = \text{constant}. \tag{2.131}$$

The expression $\frac{1}{2}mv^2$ is defined as the kinetic energy and $\varphi(x, y, z)$ is the potential energy in classical mechanics. The sum of the two is the total energy. The last equation says that no matter where and when the total energy is evaluated, it must be equal to the same constant. This is the principle of conservation of energy.

Although we have used classical mechanics to introduce the idea of conservative field, the idea can be generalized. Any vector field $\mathbf{v}(x, y, z)$ which can be expressed as the gradient of a scalar field $\varphi(x, y, z)$ is called a conservative field and the scalar function φ is called the scalar potential. Since $\boldsymbol{\nabla}\varphi = \boldsymbol{\nabla}(\varphi + \text{constant})$, the scalar potential is defined up to an additive constant.

Example 2.9.1. Determine which of the following is an irrotational (or conservative) field: (a) $\mathbf{F}_1 = 6xy\mathbf{i} + (3x^2 - 3y^2)\mathbf{j}$, (b) $\mathbf{F}_2 = xy\mathbf{i} - y\mathbf{j}$

Solution 2.9.1.

$$\boldsymbol{\nabla} \times \mathbf{F}_1 = \begin{vmatrix} \mathbf{i} & \mathbf{j} & \mathbf{k} \\ \dfrac{\partial}{\partial x} & \dfrac{\partial}{\partial y} & \dfrac{\partial}{\partial z} \\ 6xy & 3x^2 - 3y^2 & 0 \end{vmatrix} = -\mathbf{i}\frac{\partial}{\partial z}\left(3x^2 - 3y^2\right) + \mathbf{j}\frac{\partial}{\partial z}(6xy)$$

$$+ \mathbf{k}\left(\frac{\partial}{\partial x}\left(3x^2 - 3y^2\right) - \frac{\partial}{\partial y}(6xy)\right) = \mathbf{k}(6x - 6x) = \mathbf{0}.$$

$$\nabla \times \mathbf{F}_2 = \begin{vmatrix} \mathbf{i} & \mathbf{j} & \mathbf{k} \\ \dfrac{\partial}{\partial x} & \dfrac{\partial}{\partial y} & \dfrac{\partial}{\partial z} \\ xy & y & 0 \end{vmatrix} = -\mathbf{i}\dfrac{\partial}{\partial z}y + \mathbf{j}\dfrac{\partial}{\partial z}(xy)$$

$$+ \mathbf{k}\left(\dfrac{\partial}{\partial x}y - \dfrac{\partial}{\partial y}(xy)\right) = -x\mathbf{k} \neq \mathbf{0}.$$

Therefore $\mathbf{F}_1$ is an irrotational field and $\mathbf{F}_2$ is not an irrotational field. We have shown explicitly, in the examples of Sect. 2.4.3, that the line integral $\int_A^B \mathbf{F}_1 \cdot d\mathbf{r}$ is independent of path and $\int_A^B \mathbf{F}_2 \cdot d\mathbf{r}$ is dependent on the path.

Example 2.9.2. Show that the force field $\mathbf{F} = -(2ax + by)\mathbf{i} - bx\mathbf{j} - c\mathbf{k}$ is conservative, and find φ such that $-\nabla\varphi = \mathbf{F}$.

Solution 2.9.2.

$$\nabla \times \mathbf{F} = \begin{vmatrix} \mathbf{i} & \mathbf{j} & \mathbf{k} \\ \dfrac{\partial}{\partial x} & \dfrac{\partial}{\partial y} & \dfrac{\partial}{\partial z} \\ -(2ax+by) & -bx & -c \end{vmatrix} = \mathbf{0}.$$

Therefore, $\mathbf{F}$ is conservative, there must exist a φ such that $-\nabla\varphi = \mathbf{F}$.

$$-\dfrac{\partial\varphi}{\partial x} = F_x = -(2ax + by) \implies \varphi = ax^2 + bxy + f(y,z).$$

$$-\dfrac{\partial\varphi}{\partial y} = F_y = -bx, \text{ but } -\dfrac{\partial\varphi}{\partial y} = -bx - \dfrac{\partial}{\partial y}f(y,z)$$

$$\dfrac{\partial}{\partial y}f(y,z) = 0 \implies f(y,z) = g(z)$$

$$-\dfrac{\partial\varphi}{\partial z} = F_z = -c, \text{ but } -\dfrac{\partial\varphi}{\partial z} = -\dfrac{\partial}{\partial z}g(z)$$

$$\dfrac{\partial}{\partial z}g(z) = c \implies g(z) = cz + k.$$

$$\varphi = ax^2 + bxy + cz + k.$$

Example 2.9.3. Suppose a particle of mass m is moving in the force field of the last example, and at $t = 0$ the particle passes through the origin with speed v_0. What will the speed of the particle be if and when it passes through the point $\mathbf{r} = \mathbf{i} + 2\mathbf{j} + \mathbf{k}$?

Solution 2.9.3. The conservation of energy requires

$$\dfrac{1}{2}mv^2 + \varphi(\mathbf{r}) = \dfrac{1}{2}mv_0^2 + \varphi(\mathbf{0}).$$

$$v^2 = v_0^2 + \frac{2}{m}\left[k - \left(ax^2 + bxy + cz + k\right)\right].$$

At $x = 1$, $y = 2$, $z = 1$:

$$v^2 = v_0^2 + \frac{2}{m}\left(a + 2b + c\right).$$

2.9.2 Solenoidal Field and Vector Potential

If the field $\mathbf{F}$ is divergence-less (that is $\nabla \cdot \mathbf{F} = 0$) everywhere in a simply connected region, the field is called *solenoidal*. For a solenoidal field, the surface integral of $\mathbf{F} \cdot \mathbf{n}$ da over any closed surface is zero, since by the divergence theorem

$$\oiint_S \mathbf{F} \cdot \mathbf{n}\, da = \iiint \nabla \cdot \mathbf{F}\, dV = 0.$$

Furthermore, $\mathbf{F}$ can be expressed as the curl of another vector function $\mathbf{A}$,

$$\mathbf{F} = \nabla \times \mathbf{A}.$$

The vector function $\mathbf{A}$ is known as the *vector potential* of the field $\mathbf{F}$.

The existence of vector potentials for solenoidal fields can be shown in the following way. For any given solenoidal field $\mathbf{F}$ (that is, $F_x(x, y, z)$, $F_y(x, y, z)$, and $F_z(x, y, z)$ are known), we shall first show that it is possible to find a vector function $\mathbf{A}$ with one zero component to satisfy $\mathbf{F} = \nabla \times \mathbf{A}$. Then a general formula for all possible vector potentials can be found.

Let us take $A_z = 0$, and try to find A_x and A_y in $\mathbf{A} = A_x\mathbf{i} + A_y\mathbf{j}$ so that $\nabla \times \mathbf{A} = \mathbf{F}$:

$$\nabla \times \mathbf{A} = \begin{vmatrix} \mathbf{i} & \mathbf{j} & \mathbf{k} \\ \frac{\partial}{\partial x} & \frac{\partial}{\partial y} & \frac{\partial}{\partial z} \\ A_x & A_y & 0 \end{vmatrix} = -\mathbf{i}\frac{\partial}{\partial z}A_y + \mathbf{j}\frac{\partial}{\partial z}A_x + \mathbf{k}\left(\frac{\partial}{\partial x}A_y - \frac{\partial}{\partial y}A_x\right)$$

$$= \mathbf{i}F_x + \mathbf{j}F_y + \mathbf{k}F_z.$$

For this to hold, we must have

$$\frac{\partial}{\partial z}A_y = -F_x, \qquad \frac{\partial}{\partial z}A_x = F_y, \qquad \frac{\partial}{\partial x}A_y - \frac{\partial}{\partial y}A_x = F_z. \tag{2.132}$$

From the first two equations we have

$$A_y = -\int F_x(x, y, z)\mathrm{d}z + f(x, y), \tag{2.133}$$

$$A_x = \int F_y(x, y, z)\mathrm{d}z + g(x, y). \tag{2.134}$$

With A_y and A_x so obtained, if we can show $\frac{\partial}{\partial x}A_y - \frac{\partial}{\partial y}A_x = F_z$, then we would have proved that $\nabla \times \mathbf{A} = \mathbf{F}$.

Using (2.133) and (2.134), we have

$$\frac{\partial}{\partial x}A_y - \frac{\partial}{\partial y}A_x = -\int\left(\frac{\partial}{\partial x}F_x + \frac{\partial}{\partial y}F_y\right)\,\mathrm{d}z + h(x,y).$$

Since $\mathbf{F}$ is solenoidal, $\nabla \cdot \mathbf{F} = 0$ which can be written as

$$\frac{\partial}{\partial x}F_x + \frac{\partial}{\partial y}F_y = -\frac{\partial}{\partial z}F_z.$$

Thus,

$$\frac{\partial}{\partial x}A_y - \frac{\partial}{\partial y}A_x = \int\frac{\partial}{\partial z}F_z\,\mathrm{d}z + h(x,y).$$

With proper choice of $h(x,y)$, we can certainly make

$$\int\frac{\partial}{\partial z}F_z\,\mathrm{d}z + h(x,y) = F_z.$$

This proof clearly indicates that $\mathbf{A}$ is not unique. If $\mathbf{A}'$ is another vector potential, then both $\nabla \times \mathbf{A}$ and $\nabla \times \mathbf{A}'$ are equal to $\mathbf{F}$. Therefore $\nabla \times (\mathbf{A}' - \mathbf{A}) = 0$. Since $(\mathbf{A}' - \mathbf{A})$ is irrotational, it follows that $\mathbf{A}' - \mathbf{A} = \nabla\psi$. Thus we conclude that with one $\mathbf{A}$ obtained from the above procedure, all other vector potentials are of the form $\mathbf{A} + \nabla\psi$ where ψ is any scalar function.

It is also possible for us to require the vector potential to be solenoidal. If we find a vector potential $\mathbf{A}$ which is not solenoidal (that is, $\nabla \times \mathbf{A} = \mathbf{F}$ and $\nabla \cdot \mathbf{A} \neq 0$), we can construct another vector potential $\mathbf{A}'$ which is solenoidal ($\nabla \cdot \mathbf{A}' = 0$). Let

$$\mathbf{A}' = \mathbf{A} + \nabla\psi,$$
$$\nabla \times \mathbf{A}' = \nabla \times \mathbf{A} + \nabla \times \nabla\psi = \nabla \times \mathbf{A},$$
$$\nabla \cdot \mathbf{A}' = \nabla \cdot \mathbf{A} + \nabla^2\psi.$$

If we choose ψ such that $\nabla^2\psi + \nabla \cdot \mathbf{A} = 0$, then we will have $\nabla \cdot \mathbf{A}' = 0$. The following example will make this clear.

Example 2.9.4. Show that $\mathbf{F} = x^2\mathbf{i} + 3xz^2\mathbf{j} - 2xz\mathbf{k}$ is solenoidal, and find a vector potential $\mathbf{A}$ such that $\nabla \times \mathbf{A} = \mathbf{F}$ and $\nabla \cdot \mathbf{A} = 0$.

Solution 2.9.4. Since

$$\nabla \cdot \mathbf{F} = \frac{\partial}{\partial x}x^2 + \frac{\partial}{\partial y}\left(3xz^2\right) + \frac{\partial}{\partial z}\left(-2xz\right) = 0,$$

this shows that $\mathbf{F}$ is solenoidal. Let $\mathbf{A}_1 = A_x\mathbf{i} + A_y\mathbf{j}$ and $\nabla \times \mathbf{A}_1 = \mathbf{F}$. By (2.132)

$$\frac{\partial}{\partial z} A_y = -F_x = -x^2, \quad \Longrightarrow \quad A_y = -x^2 z + f(x, y),$$

$$\frac{\partial}{\partial z} A_x = F_y = 3xz^2, \quad \Longrightarrow \quad A_x = xz^3 + g(x, y),$$

$$\frac{\partial}{\partial x} A_y - \frac{\partial}{\partial y} A_x = F_z = -2xz, \quad \Longrightarrow \quad -2xy + \frac{\partial f}{\partial x} + \frac{\partial g}{\partial y} = -2xy.$$

Since f and g are arbitrary, the simplest choice is to make $f = g = 0$. Thus, $\mathbf{A}_1 = xz^3\mathbf{i} + -x^2 z\mathbf{j}$, but $\boldsymbol{\nabla} \cdot \mathbf{A}_1 = z^3 \neq 0$. Let

$$\mathbf{A} = \mathbf{A}_1 + \boldsymbol{\nabla}\psi, \quad \boldsymbol{\nabla} \cdot \mathbf{A} = \boldsymbol{\nabla} \cdot \mathbf{A}_1 + \nabla^2\psi = z^3 + \nabla^2\psi.$$

If $\boldsymbol{\nabla} \cdot \mathbf{A} = 0$, then $\nabla^2\psi = -z^3$. A simple solution of this equation is

$$\psi = -\frac{1}{20} z^5.$$

Since

$$\boldsymbol{\nabla}\psi = \boldsymbol{\nabla}\left(-\frac{1}{20}z^5\right) = -\frac{1}{4}z^4\mathbf{k},$$

$$\mathbf{A} = \mathbf{A}_1 + \boldsymbol{\nabla}\psi = xz^3\mathbf{i} - x^2 z\mathbf{j} - \frac{1}{4}z^4\mathbf{k}.$$

It can be readily verified that

$$\boldsymbol{\nabla} \cdot \mathbf{A} = \frac{\partial}{\partial x}\left(xz^3\right) + \frac{\partial}{\partial y}\left(-x^2 z\right) + \frac{\partial}{\partial z}(-\frac{1}{4}z^4) = 0,$$

$$\boldsymbol{\nabla} \times \mathbf{A} = \begin{vmatrix} \mathbf{i} & \mathbf{j} & \mathbf{k} \\ \dfrac{\partial}{\partial x} & \dfrac{\partial}{\partial y} & \dfrac{\partial}{\partial z} \\ xz^3 & -x^2 z & -\frac{1}{4}z^4 \end{vmatrix} = x^2\mathbf{i} + 3xz^2\mathbf{j} - 2xz\mathbf{k} = \mathbf{F}.$$

This vector potential is still not unique. For example, we can assume $\mathbf{A}_2 = A_y\mathbf{j} + A_z\mathbf{k}$ and $\boldsymbol{\nabla} \times \mathbf{A}_2 = \mathbf{F}$. Following the same procedure, we obtain

$$\mathbf{A}_2 = -x^2 z\mathbf{j} - \frac{3}{2}x^2 z^2\mathbf{k}.$$

Now, $\boldsymbol{\nabla} \cdot \mathbf{A}_2 = -3x^2 z \neq 0$. We can find $\mathbf{A}'$ such that $\mathbf{A}' = \mathbf{A}_2 + \boldsymbol{\nabla}\psi$ and $\boldsymbol{\nabla} \cdot \mathbf{A}' = 0$. It follows that $\nabla^2\psi = -\boldsymbol{\nabla} \cdot \mathbf{A}_2 = 3x^2 z$. A simple solution is $\psi = \frac{1}{4}x^4 z$. Therefore $\boldsymbol{\nabla}\psi = x^3 z\mathbf{i} + \frac{1}{4}x^4\mathbf{k}$, and

$$\mathbf{A}' = \mathbf{A}_2 + \boldsymbol{\nabla}\psi = x^3 z\mathbf{i} - x^2 z\mathbf{j} + \left(\frac{1}{4}x^4 - \frac{3}{2}x^2 z^2\right)\mathbf{k}.$$

Again, it can be readily verified that $\boldsymbol{\nabla} \times \mathbf{A}' = \mathbf{F}$ and $\boldsymbol{\nabla} \cdot \mathbf{A}' = 0$.

Clearly, $\mathbf{A}$ and $\mathbf{A}'$ are not identical. They must differ by an additive gradient

$$\mathbf{A}' = \mathbf{A} + \nabla \chi. \tag{2.135}$$

Now $\nabla \cdot \mathbf{A}' = \nabla \cdot \mathbf{A} + \nabla^2 \chi$ and $\nabla \cdot \mathbf{A}' = \nabla \cdot \mathbf{A} = 0$. Therefore

$$\nabla^2 \chi = 0. \tag{2.136}$$

In this particular case,

$$\nabla \chi = \mathbf{A}' - \mathbf{A} = \left(x^3 z - x z^3 \right) \mathbf{i} + \left(\frac{1}{4} x^4 - \frac{3}{2} x^2 z^2 + \frac{1}{4} z^4 \right) \mathbf{k},$$

$$\chi = \frac{1}{4} x^4 z - \frac{1}{2} x^2 z^3 + \frac{1}{20} z^5,$$

$$\nabla^2 \chi = \left(\frac{\partial^2}{\partial x^2} + \frac{\partial^2}{\partial y^2} + \frac{\partial^2}{\partial z^2} \right) \left(\frac{1}{4} x^4 z - \frac{1}{2} x^2 z^3 + \frac{1}{20} z^5 \right) = 0.$$

Equation (2.135) is an example of what is known as a *gauge transformation*. The requirement (2.136) leads to the so-called *Coulomb gauge*. The vector potential is not as useful as the scalar potential in computation. It is in the conceptual development of time-dependent problems, especially in electrodynamics, that the vector potential is essential.

2.10 Theory of Vector Fields

2.10.1 Functions of Relative Coordinates

Very often we deal with functions that depend only on the difference of the coordinates. For example, the electric field at the point (x, y, z) due to the a point charge at (x', y', z') is a function solely of $(x - x'), (y - y'), (z - z')$. The point (x, y, z) is called the field point and the point (x', y', z') is called source point. The relative position vector $\mathbf{R}$ shown in Fig. 2.18 can be written as

$$\mathbf{R} = \mathbf{r} - \mathbf{r}' = (x - x') \mathbf{i} + (y - y') \mathbf{j} + (z - z') \mathbf{k}. \tag{2.137}$$

The distance between these two points is

$$R = |\mathbf{r} - \mathbf{r}'| = [(x - x')^2 + (y - y')^2 + (z - z')^2]^{1/2}. \tag{2.138}$$

Let $f(\mathbf{R})$ be a function of the relative position vector. This function could be a scalar or a component of a vector. Functions of this type have some important properties. Let us define $X = (x - x'), Y = (y - y'), Z = (z - z')$. Using the chain rule of differentiation, we find

$$\frac{\partial f}{\partial x} = \frac{\partial f}{\partial X} \frac{\partial X}{\partial x} = \frac{\partial f}{\partial X}; \qquad \frac{\partial f}{\partial x'} = \frac{\partial f}{\partial X} \frac{\partial X}{\partial x'} = -\frac{\partial f}{\partial X}.$$

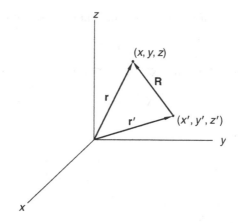

Fig. 2.18. Relative coordinates $\mathbf{R} = \mathbf{r} - \mathbf{r}'$

Similar expressions can be found for the y and z derivatives. It follows

$$\frac{\partial f}{\partial x} = -\frac{\partial f}{\partial x'}, \quad \frac{\partial f}{\partial y} = -\frac{\partial f}{\partial y'}, \quad \frac{\partial f}{\partial z} = -\frac{\partial f}{\partial z'}. \tag{2.139}$$

Corresponding to the gradient $\boldsymbol{\nabla}$ with respect to the field point

$$\boldsymbol{\nabla} f = \mathbf{i}\frac{\partial f}{\partial x} + \mathbf{j}\frac{\partial f}{\partial y} + \mathbf{k}\frac{\partial f}{\partial z},$$

we define the gradient $\boldsymbol{\nabla}'$ with respect to the source point

$$\boldsymbol{\nabla}' f = \mathbf{i}\frac{\partial f}{\partial x'} + \mathbf{j}\frac{\partial f}{\partial y'} + \mathbf{k}\frac{\partial f}{\partial z'}.$$

It follows from (2.139) that

$$\boldsymbol{\nabla} f = -\boldsymbol{\nabla}' f. \tag{2.140}$$

This shows that when we deal with functions of the relative coordinates the $\boldsymbol{\nabla}$ and $\boldsymbol{\nabla}'$ operator can be interchanged provided the sign is also changed. Similar calculations can be used to show

$$\boldsymbol{\nabla} \cdot \mathbf{A}(\mathbf{R}) = -\boldsymbol{\nabla}' \cdot \mathbf{A}(\mathbf{R}), \tag{2.141}$$

$$\boldsymbol{\nabla} \times \mathbf{A}(\mathbf{R}) = -\boldsymbol{\nabla}' \times \mathbf{A}(\mathbf{R}), \tag{2.142}$$

and

$$\nabla^2 f(\mathbf{R}) = \nabla'^2 f(\mathbf{R}). \tag{2.143}$$

Example 2.10.1. Show that (a) $\boldsymbol{\nabla} \cdot \mathbf{R} = 3$, (b) $\boldsymbol{\nabla} \times \mathbf{R} = \mathbf{0}$, (c) $\boldsymbol{\nabla} \times f(R)$ $\mathbf{R} = \mathbf{0}$, and (d) $\boldsymbol{\nabla} \cdot f(R)\mathbf{R} = \dfrac{d(R)}{dR}R + 3f(R)$.

Solution 2.10.1.

$$\boldsymbol{\nabla} \cdot \mathbf{R} = \frac{\partial}{\partial x}(x - x') + \frac{\partial}{\partial y}(y - y') + \frac{\partial}{\partial z}(z - z') = 3,$$

$$\boldsymbol{\nabla} \times \mathbf{R} = \begin{vmatrix} \mathbf{i} & \mathbf{j} & \mathbf{k} \\ \dfrac{\partial}{\partial x} & \dfrac{\partial}{\partial y} & \dfrac{\partial}{\partial z} \\ (x - x') & (y - y') & (z - z') \end{vmatrix} = \mathbf{0},$$

$$\boldsymbol{\nabla} \times f(R)\mathbf{R} = \boldsymbol{\nabla} f(R) \times \mathbf{R} + f(R)\boldsymbol{\nabla} \times \mathbf{R}$$
$$= \frac{df(R)}{dR}\widehat{\mathbf{R}} \times \mathbf{R} = \mathbf{0},$$

$$\boldsymbol{\nabla} \cdot f(R)\mathbf{R} = \boldsymbol{\nabla} f(R) \cdot \mathbf{R} + f(R)\boldsymbol{\nabla} \cdot \mathbf{R}$$
$$= \frac{df(R)}{dR}\widehat{\mathbf{R}} \cdot \mathbf{R} + 3f(R) = \frac{df(R)}{dR}R + 3f(R).$$

For functions that depend only on the distance between the two points, the gradient takes a simple form:

$$\boldsymbol{\nabla} f(R) = \mathbf{i}\frac{\partial f(R)}{\partial x} + \mathbf{j}\frac{\partial f(R)}{\partial y} + \mathbf{k}\frac{\partial f(R)}{\partial z}.$$

By the chain rule

$$\frac{\partial f(R)}{\partial x} = \frac{df(R)}{dR}\frac{\partial R}{\partial x},$$

$$\frac{\partial R}{\partial x} = \frac{\partial}{\partial x}\sqrt{(x - x')^2 + (y - y')^2 + (z - z')^2} = \frac{x - x'}{R}.$$

With similar expressions for y and z derivatives, we have

$$\boldsymbol{\nabla} f(R) = \frac{df(R)}{dR}\left(\mathbf{i}\frac{x - x'}{R} + \mathbf{j}\frac{y - y'}{R} + \mathbf{k}\frac{z - z'}{R}\right)$$
$$= \frac{df(R)}{dR}\frac{\mathbf{R}}{R} = \frac{df(R)}{dR}\widehat{\mathbf{R}}, \tag{2.144}$$

where $\widehat{\mathbf{R}}$ is the unit vector in the direction of $\mathbf{R}$. In particular

$$\nabla R = \widehat{\mathbf{R}}, \tag{2.145}$$

$$\nabla R^n = nR^{n-1}\widehat{\mathbf{R}}. \tag{2.146}$$

For $n = -1$

$$\nabla \frac{1}{R} = -\frac{1}{R^2}\widehat{\mathbf{R}}. \tag{2.147}$$

This last expression is an especially important case because $-\widehat{\mathbf{R}}/R^2$ is the "radial inverse-square-law" field. This vector field (with appropriate multiplicative constants) describes two of the most important fundamental forces in nature, namely the gravitational force field and the Coulomb force field of a static electric charge. The divergence of this field requires our special attention.

2.10.2 Divergence of $\widehat{\mathbf{R}}/\left|R\right|^2$ as a Delta Function

The divergence of $\widehat{\mathbf{R}}/R^2$ has some peculiar and important properties. Calculated directly

$$
\begin{aligned}
\nabla \cdot \frac{\widehat{\mathbf{R}}}{R^2} &= \nabla \cdot \frac{1}{R^3}\mathbf{R} = \left(\nabla \frac{1}{R^3}\right) \cdot \mathbf{R} + \frac{1}{R^3}\nabla \cdot \mathbf{R} \\
&= -3\frac{1}{R^4}\widehat{\mathbf{R}} \cdot \mathbf{R} + 3\frac{1}{R^3} = 0,
\end{aligned}
$$

we get zero. On the other hand, as we discussed earlier, the divergence is a measure of the strength of the source of the vector field. If it were zero everywhere, how could there be any gravitational and electric fields? Furthermore, if we apply the divergence theorem (2.72) to this function over a sphere of radius R around the point (x', y', z'), we will get a nonzero result,

$$
\iiint_V \nabla \cdot \frac{\widehat{\mathbf{R}}}{R^2}dV = \oiint_S \frac{\widehat{\mathbf{R}}}{R^2} \cdot \mathbf{n}\, da = \frac{1}{R^2}\oiint_S \widehat{\mathbf{R}} \cdot \widehat{\mathbf{R}}\, da
$$

$$
= \frac{1}{R^2}\oiint_S da = \frac{1}{R^2}4\pi R^2 = 4\pi. \tag{2.148}
$$

In the integral, we have used the facts that on the surface of a sphere, the unit normal $\mathbf{n}$ is equal to $\widehat{\mathbf{R}}$ and R is a constant. This integral would be zero if $\nabla \cdot (\widehat{\mathbf{R}}/R^2)$ were equal to zero everywhere.

The source of the problem is at the point $R = 0$ where $\widehat{\mathbf{R}}/R^2$ blows up and the derivative in the usual sense does not exist. Thus we can only say that the divergence is zero everywhere except at $R = 0$. To find out the divergence at $R = 0$, we note that the volume integral (2.148) of the divergence over a sphere is equal to 4π no matter how small R is. Evidently the entire contribution must be coming from the point $R = 0$. A useful way to describe this behavior is through the Dirac delta function $\delta^3(\mathbf{r} - \mathbf{r}')$.

A more detailed description of the delta function is given a later chapter. Here it suffices to know that the delta function $\delta^3(\mathbf{r} - \mathbf{r}')$ is a sharply peaked function at $\mathbf{r} = \mathbf{r}'$ with the properties

$$\delta^3(\mathbf{r} - \mathbf{r}') = \begin{cases} 0 & \mathbf{r} \neq \mathbf{r}' \\ \infty & \mathbf{r} = \mathbf{r}' \end{cases} \tag{2.149}$$

and

$$\iiint_{\text{all space}} \delta^3(\mathbf{r} - \mathbf{r}')d^3r = 1, \tag{2.150}$$

where $d^3\mathbf{r}$ is a commonly used symbol for the volume element around the field point $d^3\mathbf{r} = dV = dx\,dy\,dz$. It follows that the delta function is characterized by the shifting property

$$\iiint_{\text{all space}} f(\mathbf{r})\,\delta^3(\mathbf{r} - \mathbf{r}')d^3r = f(\mathbf{r}'), \tag{2.151}$$

because

$$\iiint_{\text{all space}} f(\mathbf{r})\,\delta^3(\mathbf{r} - \mathbf{r}')d^3r = \iiint_{\text{all space}} f(\mathbf{r}')\,\delta^3(\mathbf{r} - \mathbf{r}')d^3r,$$

since the value of $f(\mathbf{r})$ is immaterial for $\mathbf{r} \neq \mathbf{r}'$ as the integrand is going to be zero anyway. Furthermore,

$$\iiint_{\text{all space}} f(\mathbf{r}')\,\delta^3(\mathbf{r} - \mathbf{r}')d^3r = f(\mathbf{r}') \iiint_{\text{all space}} \delta^3(\mathbf{r} - \mathbf{r}')d^3r = f(\mathbf{r}'),$$

since the integration is over $d^3\mathbf{r}$. This property can also be written as

$$\iiint_{\text{all space}} f(\mathbf{r}')\,\delta^3(\mathbf{r} - \mathbf{r}')d^3r' = f(\mathbf{r}), \tag{2.152}$$

where $d^3\mathbf{r}' = dx'\,dy'\,dz'$.

With the delta function, the divergence of $\widehat{\mathbf{R}}/R^2$ can be precisely expressed as

$$\nabla \cdot \frac{\widehat{\mathbf{R}}}{R^2} = \nabla \cdot \frac{\mathbf{r} - \mathbf{r}'}{|\mathbf{r} - \mathbf{r}'|^3} = 4\pi\delta^3(\mathbf{r} - \mathbf{r}'). \tag{2.153}$$

With this understanding, we see that

$$\iiint_V \nabla \cdot \frac{\widehat{\mathbf{R}}}{R^2}dV = \iiint_V \nabla \cdot \frac{\mathbf{r} - \mathbf{r}'}{|\mathbf{r} - \mathbf{r}'|^3}d^3r = 4\pi \iiint_V \delta^3(\mathbf{r} - \mathbf{r}')d^3r$$

$$= \begin{cases} 4\pi & \text{if the volume includes } \mathbf{r}' \\ 0 & \text{if } \mathbf{r}' \text{ is outside the body.} \end{cases} \tag{2.154}$$

Since

$$\mathbf{\nabla}\frac{1}{R} = -\frac{\widehat{\mathbf{R}}}{R^2},$$

it follows that the Laplacian of $\left(\frac{1}{R}\right)$ is given by

$$\nabla^2\frac{1}{R} = \mathbf{\nabla}\cdot\mathbf{\nabla}\frac{1}{R} = -\mathbf{\nabla}\cdot\frac{\widehat{\mathbf{R}}}{R^2} = -4\pi\delta^3(\mathbf{r}-\mathbf{r}'). \qquad (2.155)$$

Example 2.10.2. Evaluate the integral

$$I = \iiint_V (r^3+1)\mathbf{\nabla}\cdot\frac{\widehat{\mathbf{r}}}{r^2}dV,$$

where V is a sphere of radius b centered at the origin.

Solution 2.10.2. *Method I.* Use the delta function. Since

$$\mathbf{\nabla}\cdot\frac{\widehat{\mathbf{r}}}{r^2} = 4\pi\delta^3(\mathbf{r}),$$

$$I = \iiint_V (r^3+1)4\pi\delta^3(\mathbf{r})dV = 4\pi\,(0+1) = 4\pi.$$

Method II. Use integration by parts. Since $f\mathbf{\nabla}\cdot\mathbf{A} = \mathbf{\nabla}\cdot(f\mathbf{A}) - \mathbf{\nabla}f\cdot\mathbf{A}$,

$$I = \iiint_V (r^3+1)\mathbf{\nabla}\cdot\frac{\widehat{\mathbf{r}}}{r^2}dV = \iiint_V \mathbf{\nabla}\cdot\left[(r^3+1)\frac{\widehat{\mathbf{r}}}{r^2}\right]dV - \iiint_V \mathbf{\nabla}(r^3+1)\cdot\frac{\widehat{\mathbf{r}}}{r^2}dV.$$

By the divergence theorem

$$\iiint_V \mathbf{\nabla}\cdot\left[(r^3+1)\frac{\widehat{\mathbf{r}}}{r^2}\right]dV = \oiint_S (r^3+1)\frac{\widehat{\mathbf{r}}}{r^2}\cdot\widehat{\mathbf{r}}\,da = \oiint_S (r+\frac{1}{r^2})da,$$

where S is the surface of the sphere of radius b. Since on this surface $r = b$ everywhere, therefore

$$\iiint_V \mathbf{\nabla}\cdot\left[(r^3+1)\frac{\widehat{\mathbf{r}}}{r^2}\right]dV = \left(b+\frac{1}{b^2}\right)\oiint_S da = \left(b+\frac{1}{b^2}\right)4\pi b^2 = 4\pi b^3 + 4\pi.$$

Since $\mathbf{\nabla}(r^3+1) = 3r^2\widehat{\mathbf{r}}$,

$$\iiint_V \mathbf{\nabla}(r^3+1)\cdot\frac{\widehat{\mathbf{r}}}{r^2}dV = \iiint_V 3r^2\widehat{\mathbf{r}}\cdot\frac{\widehat{\mathbf{r}}}{r^2}dV = 3\iiint_V dV = 3\frac{4}{3}\pi b^3 = 4\pi b^3.$$

Thus we have

$$I = 4\pi b^3 + 4\pi - 4\pi b^3 = 4\pi,$$

which is the same as the result of delta function method. This example illustrates the validity and power of the delta function method. If the volume is not a sphere, as long as it includes the origin, the delta function result is still valid, but the direct integration will be much more difficult to do.

2.10.3 Helmholtz's Theorem

The Helmholtz theorem deals with the question of what information we need to determine a vector field. Basically, the answer is that if the divergence and the curl of a vector field are known, with some boundary conditions the vector field can be found uniquely.

The Helmholtz theorem states that any vector field $\mathbf{F}$ may be decomposed into the sum of two vectors, one is the gradient of a scalar potential φ and the other the curl of a vector potential $\mathbf{A}$,

$$\mathbf{F} = -\nabla\varphi + \nabla \times \mathbf{A}. \tag{2.156}$$

Furthermore, if $\mathbf{F} \to \mathbf{0}$ on the surface at infinity faster than $1/R$ and $\nabla \cdot \mathbf{F}$ and $\nabla \times \mathbf{F}$ are known everywhere, then

$$\varphi(\mathbf{r}) = \frac{1}{4\pi} \iiint \frac{\nabla' \cdot \mathbf{F}(\mathbf{r}')}{|\mathbf{r} - \mathbf{r}'|} d^3 r', \tag{2.157}$$

$$\mathbf{A}(\mathbf{r}) = \frac{1}{4\pi} \iiint \frac{\nabla' \times \mathbf{F}(\mathbf{r}')}{|\mathbf{r} - \mathbf{r}'|} d^3 r'. \tag{2.158}$$

To prove this theorem, we first construct a vector function $\mathbf{G}$

$$\mathbf{G}(\mathbf{r}) = \iiint \frac{\mathbf{F}(\mathbf{r}')}{|\mathbf{r} - \mathbf{r}'|} d^3 r'. \tag{2.159}$$

Let us apply the Laplacian ∇^2 to both sides of this equation. Because ∇^2 operates only on $\mathbf{r}$ and only $|\mathbf{r} - \mathbf{r}'|^{-1}$ contains $\mathbf{r}$, we have

$$\nabla^2 \mathbf{G}(\mathbf{r}) = \iiint \left(\nabla^2 \frac{1}{|\mathbf{r} - \mathbf{r}'|}\right) \mathbf{F}(\mathbf{r}') d^3 r'. \tag{2.160}$$

Since by (2.155)

$$\nabla^2 \frac{1}{|\mathbf{r} - \mathbf{r}'|} = -4\pi\delta^3(\mathbf{r} - \mathbf{r}'),$$

it follows from the definition of the delta function that

$$\nabla^2 \mathbf{G}(\mathbf{r}) = \iiint \left(-4\pi\delta^3(\mathbf{r} - \mathbf{r}')\right) \mathbf{F}(\mathbf{r}') d^3 r' = -4\pi \mathbf{F}(\mathbf{r}). \tag{2.161}$$

Therefore

$$\mathbf{F}(\mathbf{r}) = -\frac{1}{4\pi}\nabla^2 \mathbf{G}(\mathbf{r}). \tag{2.162}$$

Using the vector identity $\nabla \times (\nabla \times \mathbf{G}) = \nabla(\nabla \cdot \mathbf{G}) - \nabla^2 \mathbf{G}$, we have

$$\nabla^2 \mathbf{G} = \nabla(\nabla \cdot \mathbf{G}) - \nabla \times (\nabla \times \mathbf{G}).$$

Thus with

$$\varphi = \frac{1}{4\pi}(\nabla \cdot \mathbf{G}), \qquad \mathbf{A} = \frac{1}{4\pi}(\nabla \times \mathbf{G}),$$

the first part of the theorem follows from (2.162)

$$\mathbf{F}(\mathbf{r}) = -\frac{1}{4\pi}\nabla^2 \mathbf{G}(\mathbf{r}) = -\nabla\varphi + \nabla \times \mathbf{A}.$$

To find the explicit expression for φ, we start with

$$\varphi(\mathbf{r}) = \frac{1}{4\pi}(\nabla \cdot \mathbf{G}) = \frac{1}{4\pi}\nabla \cdot \iiint \frac{\mathbf{F}(\mathbf{r}')}{|\mathbf{r} - \mathbf{r}'|}\mathrm{d}^3 r'.$$

Since ∇ operates only on $\mathbf{r}$, and only $|\mathbf{r} - \mathbf{r}'|$ contains $\mathbf{r}$,

$$\nabla \cdot \iiint \frac{\mathbf{F}(\mathbf{r}')}{|\mathbf{r} - \mathbf{r}'|}\mathrm{d}^3 r' = \iiint \nabla \cdot \frac{\mathbf{F}(\mathbf{r}')}{|\mathbf{r} - \mathbf{r}'|}\mathrm{d}^3 r' = \iiint \left(\nabla\frac{1}{|\mathbf{r} - \mathbf{r}'|}\right) \cdot \mathbf{F}(\mathbf{r}')\mathrm{d}^3 r'.$$

Now

$$\nabla\frac{1}{|\mathbf{r} - \mathbf{r}'|} = -\nabla'\frac{1}{|\mathbf{r} - \mathbf{r}'|}$$

and

$$\left(\nabla'\frac{1}{|\mathbf{r} - \mathbf{r}'|}\right) \cdot \mathbf{F}(\mathbf{r}') = \nabla' \cdot \frac{\mathbf{F}(\mathbf{r}')}{|\mathbf{r} - \mathbf{r}'|} - \frac{1}{|\mathbf{r} - \mathbf{r}'|}\nabla' \cdot \mathbf{F}(\mathbf{r}'),$$

so

$$\varphi(\mathbf{r}) = -\frac{1}{4\pi}\iiint \nabla' \cdot \frac{\mathbf{F}(\mathbf{r}')}{|\mathbf{r} - \mathbf{r}'|}\mathrm{d}^3 r' + \frac{1}{4\pi}\iiint \frac{1}{|\mathbf{r} - \mathbf{r}'|}\nabla' \cdot \mathbf{F}(\mathbf{r}')\mathrm{d}^3 r'. \quad (2.163)$$

The first integral on the right-hand side can be changed to a surface integral at infinity by the divergence theorem

$$\iiint_{\text{all space}} \nabla' \cdot \frac{\mathbf{F}(\mathbf{r}')}{|\mathbf{r} - \mathbf{r}'|}\mathrm{d}^3 r' = \iint_{S\to\infty} \frac{1}{|\mathbf{r} - \mathbf{r}'|}\mathbf{F}(\mathbf{r}') \cdot \mathbf{n}\,\mathrm{d}a'.$$

As $r' \to \infty$, $\mathbf{F}(\mathbf{r}')$ goes to zero faster than $1/r'$. Hence the surface integral is equal to zero. This follows from the fact that the surface is only proportional to r'^2, and $\mathbf{F}(\mathbf{r}')/|\mathbf{r} - \mathbf{r}'|$ goes to zero faster than $1/r'^2$. Thus only the second integral on the right-hand side of (2.163) remains

$$\varphi(\mathbf{r}) = \frac{1}{4\pi}\iiint \frac{1}{|\mathbf{r} - \mathbf{r}'|}\nabla' \cdot \mathbf{F}(\mathbf{r}')\mathrm{d}^3 r'.$$

Similarly, for the vector potential we start with

$$\mathbf{A}(\mathbf{r}) = \frac{1}{4\pi}(\nabla \times \mathbf{G}) = \frac{1}{4\pi}\iiint \nabla \times \frac{1}{|\mathbf{r} - \mathbf{r}'|}\mathbf{F}(\mathbf{r}')\mathrm{d}^3 r'$$

$$= \frac{1}{4\pi}\iiint \nabla\frac{1}{|\mathbf{r} - \mathbf{r}'|} \times \mathbf{F}(\mathbf{r}')\mathrm{d}^3 r'.$$

Using the identities

$$\nabla\frac{1}{|\mathbf{r}-\mathbf{r}'|}\times\mathbf{F}(\mathbf{r}')=-\nabla'\frac{1}{|\mathbf{r}-\mathbf{r}'|}\times\mathbf{F}(\mathbf{r}'),$$

$$\nabla'\frac{1}{|\mathbf{r}-\mathbf{r}'|}\times\mathbf{F}(\mathbf{r}')=\nabla'\times\frac{1}{|\mathbf{r}-\mathbf{r}'|}\mathbf{F}(\mathbf{r}')-\frac{1}{|\mathbf{r}-\mathbf{r}'|}\nabla'\times\mathbf{F}(r'),$$

we have

$$\mathbf{A}\left(\mathbf{r}\right)=-\frac{1}{4\pi}\iiint\nabla'\times\frac{1}{|\mathbf{r}-\mathbf{r}'|}\mathbf{F}(\mathbf{r}')d^3r'+\frac{1}{4\pi}\iiint\frac{1}{|\mathbf{r}-\mathbf{r}'|}\nabla'\times\mathbf{F}(\mathbf{r}')d^3r'.$$

$$(2.164)$$

By the integral theorem (2.122)

$$\iiint_V\nabla\times\mathbf{P}d^3r'=-\iint_S\mathbf{P}\times\mathbf{n}\,da,$$

the first integral on the right-hand side of (2.164) can be transformed into a surface integral

$$-\frac{1}{4\pi}\iiint_{\text{all space}}\nabla'\times\frac{1}{|\mathbf{r}-\mathbf{r}'|}\mathbf{F}(\mathbf{r}')d^3r'=\frac{1}{4\pi}\iint_{S\to\infty}\frac{1}{|\mathbf{r}-\mathbf{r}'|}\mathbf{F}(\mathbf{r}')\times\mathbf{n}\,da',$$

which is zero because $\mathbf{F}(\mathbf{r}')\to0$ on the surface at infinity faster than $1/r'$. Thus (2.164) becomes

$$\mathbf{A}\left(\mathbf{r}\right)=\frac{1}{4\pi}\iiint\frac{1}{|\mathbf{r}-\mathbf{r}'|}\nabla'\times\mathbf{F}(\mathbf{r}')d^3r'.\qquad(2.165)$$

This completes the proof. The divergence and curl of $\mathbf{F}$ are often called the sources of the field, since $\mathbf{F}$ can be found from the knowledge of them. The point $\mathbf{r}$ where we evaluate $\mathbf{F}$ is called the field point. The point $\mathbf{r}'$ where the sources are evaluated for the purpose of integration is called the source point. The volume element d^3r' is at the source point. The function φ and $\mathbf{A}$ are called scalar and vector potentials, respectively, because $\mathbf{F}$ is obtained from them by differentiation.

It should be noted that while the field $\mathbf{F}(\mathbf{r})$ so determined is unique, the potentials φ and $\mathbf{A}$ are not. Any constant can be added to φ, since $\nabla\left(\varphi+C\right)=\nabla\varphi$. The gradient of any scalar function can be added to $\mathbf{A}$, since $\nabla\times(\mathbf{A}+\nabla\psi)=\nabla\times\mathbf{A}$.

Example 2.10.3. If $\mathbf{A}\left(\mathbf{r}\right)=\frac{1}{4\pi}\iiint\frac{1}{|\mathbf{r}-\mathbf{r}'|}\nabla'\times\mathbf{F}(\mathbf{r}')d^3r'$ and $\mathbf{F}(\mathbf{r}')$ goes to zero on the surface at infinity faster than $1/r'$, show that $\nabla\cdot\mathbf{A}\left(\mathbf{r}\right)=0$.

Solution 2.10.3. Since ∇ operates only on $\mathbf{r}$,

$$\nabla \cdot \mathbf{A}\,(\mathbf{r}) = \frac{1}{4\pi} \iiint \nabla \cdot \frac{1}{|\mathbf{r} - \mathbf{r}'|} \nabla' \times \mathbf{F}(\mathbf{r}') d^3 r'$$

$$= \frac{1}{4\pi} \iiint \nabla \frac{1}{|\mathbf{r} - \mathbf{r}'|} \cdot \nabla' \times \mathbf{F}(\mathbf{r}') d^3 r'.$$

Now

$$\nabla \frac{1}{|\mathbf{r} - \mathbf{r}'|} \cdot \nabla' \times \mathbf{F}(\mathbf{r}') = -\nabla' \frac{1}{|\mathbf{r} - \mathbf{r}'|} \cdot \nabla' \times \mathbf{F}(\mathbf{r}')$$

and

$$\nabla' \cdot \left[\frac{1}{|\mathbf{r} - \mathbf{r}'|} \nabla' \times \mathbf{F}(\mathbf{r}') \right] = \nabla' \frac{1}{|\mathbf{r} - \mathbf{r}'|} \cdot \nabla' \times \mathbf{F}(\mathbf{r}') + \frac{1}{|\mathbf{r} - \mathbf{r}'|} \nabla' \cdot \nabla' \times \mathbf{F}(\mathbf{r}')$$

$$= \nabla' \frac{1}{|\mathbf{r} - \mathbf{r}'|} \cdot \nabla' \times \mathbf{F}(\mathbf{r}'),$$

because the divergence of a curl is equal to zero. Therefore we have

$$\nabla \cdot \mathbf{A}\,(\mathbf{r}) = -\frac{1}{4\pi} \iiint \left[\nabla' \cdot \frac{1}{|\mathbf{r} - \mathbf{r}'|} \nabla' \times \mathbf{F}(\mathbf{r}') \right] d^3 r'$$

$$= -\frac{1}{4\pi} \oiint_S \frac{1}{|\mathbf{r} - \mathbf{r}'|} \nabla' \times \mathbf{F}(\mathbf{r}') \cdot \mathbf{n}\, da.$$

As $S \to \infty$, $\nabla \cdot \mathbf{A}\,(\mathbf{r}) = 0$.

2.10.4 Poisson's and Laplace's Equations

The Helmholtz's theorem shows that the vector field is uniquely determined by its divergence and curl. To derive the expressions for the divergence and curl from experimental observations is therefore of great importance.

One of the most important vector fields is the radial inverse square law field, which is the mathematical statement of the gravitational law and the Coulomb's law, the two fundamental laws in nature. For example, together with the principle of superposition, the electric field $\mathbf{E}(\mathbf{r})$ produced by static charges can be written as

$$\mathbf{E}(\mathbf{r}) = \frac{1}{4\pi} \iiint \varrho(\mathbf{r}') \frac{\widehat{\mathbf{R}}}{R^2} d^3 r' = \frac{1}{4\pi} \iiint \varrho(\mathbf{r}') \frac{\mathbf{r} - \mathbf{r}'}{(\mathbf{r} - \mathbf{r}')^3} d^3 r', \qquad (2.166)$$

where $\varrho(\mathbf{r}')$ is the charge density (electric charge per unit volume) in the neighborhood of $\mathbf{r}'$. The constant $1/4\pi$ is a matter of units and need not concern us here. The divergence of $\mathbf{E}(\mathbf{r})$ is

$$\nabla \cdot \mathbf{E}(\mathbf{r}) = \frac{1}{4\pi} \iiint \varrho(\mathbf{r'}) \left(\nabla \cdot \frac{\widehat{\mathbf{R}}}{R^2} \right) d^3\mathbf{r'},$$

since ∇ operates only on $\mathbf{r}$. But,

$$\nabla \cdot \frac{\widehat{\mathbf{R}}}{R^2} = 4\pi\delta^3(\mathbf{r} - \mathbf{r'})$$

as shown in (2.153). Thus,

$$\nabla \cdot \mathbf{E}(\mathbf{r}) = \frac{1}{4\pi} \iiint \varrho(\mathbf{r'})4\pi\delta^3(\mathbf{r} - \mathbf{r'})d^3\mathbf{r'} = \varrho(\mathbf{r}). \qquad (2.167)$$

The fact that we can relate the divergence of $\mathbf{E}$ at $\mathbf{r}$ to the charge density at same point $\mathbf{r}$ is remarkable. Coulomb's law of (2.166) is the experimental result, which says that the electric field $\mathbf{E}$ at $\mathbf{r}$ is due to all other charges at different places $\mathbf{r'}$. Yet through vector analysis, we find $\nabla \cdot \mathbf{E}$ at $\mathbf{r}$ is equal to the charge density $\varrho(\mathbf{r})$ at the same place where $\mathbf{E}$ is to be evalued. This type of equation is called *field equation* which describes the property of the field at each point in space.

Since curl of $(\widehat{\mathbf{R}}/R^2)$ is equal to zero, $\mathbf{E}$ can be expressed as the gradient of scalar potential $\mathbf{E} = -\nabla\varphi$. Thus,

$$\nabla \cdot \mathbf{E} = -\nabla \cdot \nabla\varphi = \varrho.$$

Therefore,

$$\nabla^2\varphi = -\varrho. \qquad (2.168)$$

This result is known as *Poisson's equation* which specifies the relationship between the source density and the scalar potential for an irrotational field.

In that part of the space where there is no charge ($\varrho = 0$), the equation reduces to

$$\nabla^2\varphi = 0, \qquad (2.169)$$

which is known as *Laplace's equation.*

The equations of Poisson and Laplace are two of the most important equations in mathematical physics. They are encountered repeatedly in a variety of problems.

2.10.5 Uniqueness Theorem

In the following chapters, we shall describe various methods of solving Laplace's equation. It does not matter which method we use, as long as we can find a scalar function φ that satisfies the equation and the boundary conditions, the vector field derived from it is uniquely determined. This is known as *uniqueness theorem.*

Let the region of interests be surrounded by surface S, (if the boundary consists of many surfaces including the surface at infinity, then S represents all of them). There are two kinds of boundary conditions (1) the values of φ are specified on S, known as Dirichlet boundary condition and (2) the normal derivatives $\partial\varphi/\partial n$ over S are specified, known as Neumann boundary condition. The theorem says:

Two solutions φ_1 and φ_2 of the Laplace equation which satisfy the first kind of boundary conditions must be identical. Two solutions φ_1 and φ_2 of the Laplace equation which satisfy the second kind of boundary conditions can differ at most by an additive constant.

To prove this theorem, we define a new function $\Phi = \varphi_1 - \varphi_2$. Obviously, $\nabla^2\Phi = \nabla^2\varphi_1 - \nabla^2\varphi_2 = 0$. Furthermore, either Φ or $\partial\Phi/\partial n = \nabla\Phi \cdot \mathbf{n}$ vanishes on S. Applying the divergence theorem to $\Phi\nabla\Phi$, we have

$$\iiint \nabla \cdot (\Phi\nabla\Phi)\,\mathrm{d}V = \iint_S \Phi\nabla\Phi \cdot \mathbf{n}\,\mathrm{d}a = 0,$$

since the integral on the right-hand side vanishes. But

$$\nabla \cdot (\Phi\nabla\Phi) = \nabla\Phi \cdot \nabla\Phi + \Phi\nabla^2\Phi$$

and $\nabla^2\Phi = 0$ at all points, so the divergence theorem in this case becomes

$$\iiint \nabla\Phi \cdot \nabla\Phi\,\mathrm{d}V = 0.$$

Now $\nabla\Phi \cdot \nabla\Phi = (\nabla\Phi)^2$ must be positive or zero, and since the integral is zero, it follows that the only possibility is $\nabla\Phi = 0$ everywhere inside the volume. A function whose gradient is zero at all points cannot change, hence Φ has the same value that it has on the boundary S. For the first kind of boundary condition, $\Phi = 0$ on S, and Φ must equal to zero at every point in the region. Therefore $\varphi_1 = \varphi_2$. For the second kind of boundary conditions, $\nabla\Phi$ equal to zero at all points in the region and $\nabla\Phi \cdot \mathbf{n} = 0$ on S, the only possible solution is Φ equal to a constant. Thus φ_1 and φ_2 can differ at most by a constant. In either case, the vector field $\nabla\varphi$ is uniquely defined.

Exercises

1. Find $\dfrac{\mathrm{d}\mathbf{r}}{\mathrm{d}t}, \dfrac{\mathrm{d}^2\mathbf{r}}{\mathrm{d}t^2}, \left|\dfrac{\mathrm{d}\mathbf{r}}{\mathrm{d}t}\right|, \left|\dfrac{\mathrm{d}^2\mathbf{r}}{\mathrm{d}t^2}\right|$, if $\mathbf{r}(t) = \sin t\mathbf{i} + \cos t\mathbf{j} + t\mathbf{k}$.

   Ans. $\cos t\mathbf{i} - \sin t\mathbf{j} + \mathbf{k}$, $-\sin t\mathbf{i} - \cos t\mathbf{j}$, $\sqrt{2}$, 1.

2. Show that $\mathbf{A} \cdot \dfrac{\mathrm{d}\mathbf{A}}{\mathrm{d}t} = A\dfrac{\mathrm{d}A}{\mathrm{d}t}$.

3. A particle moves along the curve $\mathbf{r}\,(t) = 2t^2\mathbf{i} + \left(t^2 - 4t\right)\mathbf{j} + (3t - 5)\,\mathbf{k}$, where t is time. Find its velocity and acceleration at $t = 1$.
   Ans. $4\mathbf{i} - 2\mathbf{j} + 3\mathbf{k}$, $4\mathbf{i} + 2\mathbf{j}$.

4. A particle moves along the curve $\mathbf{r}\,(t) = \left(t^3 - 4t\right)\mathbf{i} + \left(t^2 + 4t\right)\mathbf{j} + \left(8t^2 - 3t^3\right)\mathbf{k}$, where t is time. Find the magnitudes of the tangential and normal components of its acceleration at $t = 2$.
   Ans. $16, 2\sqrt{73}$.

5. A velocity field is given by $\mathbf{v} = x^2\mathbf{i} - 2xy\mathbf{j} + 4t\mathbf{k}$. Determine the acceleration at the point $(2, 1, -4)$.
   Ans. $16\mathbf{i} + 8\mathbf{j} + 4\mathbf{k}$.

   Hint: $\mathbf{a} = \dfrac{\partial \mathbf{v}}{\partial x}\dfrac{dx}{dt} + \dfrac{\partial \mathbf{v}}{\partial y}\dfrac{dy}{dt} + \dfrac{\partial \mathbf{v}}{\partial z}\dfrac{dz}{dt} + \dfrac{\partial \mathbf{v}}{\partial t}$

6. A wheel of radius b rolls along the ground with a constant forward speed v_0. Find the acceleration of any point on the rim of the wheel.
   Ans. v_0^2/b toward the center of the wheel.

   Hint: Let the moving origin be at the center of the wheel with x' axis passing through the point in question, thus $\mathbf{r}' = b\mathbf{i}$, $\mathbf{v}' = 0$, $\mathbf{a}' = 0$. The angular velocity vector is $\boldsymbol{\omega} = (v_0/b)\mathbf{k}'$. Then use (2.44)

7. Find the arc length of $\mathbf{r}\,(t) = a\cos t\mathbf{i} + a\sin t\mathbf{j} + bt\mathbf{k}$ from $t = 0$ to $t = 2\pi$.
   Ans. $s = 2\pi\sqrt{a^2 + b^2}$.

   Hint: $ds = v\,dt = (\dot{\mathbf{r}} \cdot \dot{\mathbf{r}})^{1/2})dt$

8. Find the arc length of $\mathbf{r}\,(t) = (\cos t + t\sin t)\mathbf{i} + (\sin t - t\cos t)\mathbf{j}$ from $t = 0$ to $t = \pi$.
   Ans. $s = \pi^2/2$.

9. Given the space curve $\mathbf{r} = t\mathbf{i} + t^2\mathbf{j} + \frac{2}{3}t^3\mathbf{k}$, find (a) the curvature κ and (b) the torsion γ.
   Ans. $\dfrac{2}{(1+2t^2)^2}, \dfrac{2}{(1+2t^2)^2}$.

10. Show that $\ddot{\mathbf{r}} = \dot{v}\mathbf{t} + v^2\kappa\mathbf{n}$.

11. Show that the curvature κ of a space curve $\mathbf{r} = \mathbf{r}\,(t)$ is given by

$$\kappa = \left|\dot{\mathbf{r}} \times \ddot{\mathbf{r}}\right| / \left|\dot{\mathbf{r}}\right|^3,$$

where dots denote differentiation with respect to time t.
Hint: first show that $\dot{\mathbf{r}} \times \ddot{\mathbf{r}} = v\mathbf{t} \times (\dot{v}\mathbf{t} + v^2\kappa\mathbf{n})$

12. Show that the torsion γ of a space curve is given numerically by

$$\gamma = \left| \dot{\mathbf{r}} \cdot \ddot{\mathbf{r}} \times \dddot{\mathbf{r}} \right| / \left| \dot{\mathbf{r}} \times \ddot{\mathbf{r}} \right|^2 .$$

Hint: first show that $\dot{\mathbf{r}} \cdot \ddot{\mathbf{r}} \times \dddot{\mathbf{r}} = -v^6 \kappa^2 \gamma$, then use the result of the previous problem

13. Find the gradient of the scalar field $\phi = xyz$, and evaluate it at the point $(1, 2, 3)$, find the derivative of ϕ in the direction of $\mathbf{i} + \mathbf{j}$.
Ans. $yz\mathbf{i} + xz\mathbf{j} + xy\mathbf{k}$, $6\mathbf{i} + 3\mathbf{j} + 2\mathbf{k}$, $9/\sqrt{2}$.

14. Find the unit normal to each of the following surfaces at the point indicated: (a) $x^2 + y^2 - z = 0$ at $(1,1,2)$, (b) $x^2 + y^2 = 5$ at $(2, 1, 0)$, and (c) $y = x^2 + z^3$ at $(1, 2, 1)$.
Ans. $(2\mathbf{i} + 2\mathbf{j} - \mathbf{k})/3$, $(3\mathbf{i} + 4\mathbf{j})/5$, $(-\mathbf{i} - \mathbf{j} - 3\mathbf{k})/\sqrt{11}$.

15. The temperature T is given by $T = x^2 + xy + yz$. What is the unit vector that points in the direction of maximum change of temperature at $(2, 1, 4)$? What is the value of the derivative of the temperature in the x direction at that point?
Ans. $(5\mathbf{i} + 6\mathbf{j} + \mathbf{k})/\sqrt{62}$, 5.

16. Determine the equation of the plane tangent to the given surface at the point indicated: (a) $x^2 + y^2 + z^2 = 25$ $(3, 4, 0)$, and (b) $x^2 - 2xy = 0$ $(2, 2, 1)$.
Ans. $3x + 4y = 25$, $y = 2$.

17. Find the divergence of each of the following vector fields at the point $(2, 1, -1)$. (a) $\mathbf{F} = x^2\mathbf{i} + yz\mathbf{j} + y^2\mathbf{k}$, (b) $\mathbf{F} = x\mathbf{i} + y\mathbf{j} + y\mathbf{k}$, and (c) $\mathbf{F} = \mathbf{r}/r = (x\mathbf{i} + y\mathbf{j} + y\mathbf{k})/\sqrt{x^2 + y^2 + z^2}$.
Ans. 3, 3, $\sqrt{6}/3$.

18. Verify the divergence theorem by calculating both the volume integral and the surface integral for the vector field $\mathbf{F} = y\mathbf{i} + x\mathbf{j} + (z - x)\mathbf{k}$ and the volume of the unit cube $0 \leq x, y, z \leq 1$.

19. By using the divergence theorem, evaluate

$$(a) \quad \oiint_S (x\mathbf{i} + y\mathbf{j} + z\mathbf{k}) \cdot \mathbf{n} \, da,$$

where S is the surface of the sphere $x^2 + y^2 + z^2 = 9$;

$$(b) \quad \oiint_S (x\mathbf{i} + x\mathbf{j} + z^2\mathbf{k}) \cdot \mathbf{n} \, da,$$

where S is the surface of the cylinder $x^2 + y^2 = 4$, $0 \le z \le 8$;

$$(c) \quad \oiint_S \left(x \sin y\mathbf{i} + \cos^2 x\mathbf{j} - z \sin y\mathbf{k} \right) \cdot \mathbf{n} \, da,$$

where S is the surface of the sphere $x^2 + y^2 + (z-2)^2 = 1$.
Ans. 108π, 288π, 0.

20. Show that

$$\oiint_S \mathbf{r} \cdot \mathbf{n} \, da = 3V,$$

where V is the volume bounded by the closed surface S.

21. Recognizing that $\mathbf{i} \cdot \mathbf{n} \, da = dy \, dz$; $\mathbf{j} \cdot \mathbf{n} \, da = dx \, dz$; $\mathbf{k} \cdot \mathbf{n} \, da = dx \, dy$ (see Example 2.5.2), evaluating the following integral using the divergence theorem

$$\oiint_S (x \, dy \, dz + y \, dx \, dz + z \, dx \, dy),$$

where S is the surface of the cylindr $x^2 + y^2 = 9$, $0 \le z \le 3$.
Ans. 81π.

Hint: first show that $(x \, dy \, dz + y \, dx \, dz + z \, dx \, dy) = (x\mathbf{i} + y\mathbf{j} + z\mathbf{k}) \cdot \mathbf{n} \, da$ (see Example 2.5.2)

22. Evaluating the following integral using the divergence theorem

$$\oiint_S (x \, dy \, dz + 2y \, dx \, dz + y^2 \, dx \, dy),$$

where S is the surface of the sphere $x^2 + y^2 + z^2 = 4$.
Ans. 32π.

23. Use the divergence theorem to evaluate the surface integral

$$\iint_S [(x+y)\mathbf{i} + z^2\mathbf{j} + x^2\mathbf{k}] \cdot \mathbf{n} \, da,$$

where S is the surface of the hemisphere $x^2 + y^2 + z^2 = 1$ with $z > 0$ and $\mathbf{n}$ is the outward unit normal. Note that the surface is not closed.
Ans. $\frac{11}{12}\pi$.

Hint: the integral is equal to the closed surface integral over the hemisphere subtract the integral over the base.

24. Find the curl of each of the following vector fields at the point $(-2, 4, 1)$.
(a) $\mathbf{F} = x^2\mathbf{i} + y^2\mathbf{j} + z^2\mathbf{k}$ and (b) $\mathbf{F} = xy\mathbf{i} + y^2\mathbf{j} + xz\mathbf{k}$.
Ans. 0, $-\mathbf{j} + 2\mathbf{k}$.

25. Verify Stokes' theorem by evaluating both the line and surface integral for the vector field $\mathbf{A} = (2x - y)\,\mathbf{i} - y^2\mathbf{j} + y^2 z\mathbf{k}$ and the surface S given by the disc $z = 0$, $x^2 + y^2 \leq 1$.

26. Ampere's law states that the total flux of electric current flowing through a loop is proportional to the line integral of the magnetic field around the loop, that is $\oint_C \mathbf{B} \cdot \mathbf{dr} = \mu_0 \iint_S \mathbf{J} \cdot n \, da$ where B is the magnetic field, $\mathbf{J}$ is the current density and μ_0 is a proportional constant. If this is true for any loop C, show that $\boldsymbol{\nabla} \times \mathbf{B} = \mu_0 \mathbf{J}$.

27. Show that $\oint_C \mathbf{r} \cdot \mathbf{dr} = 0$ for any closed curve C.

28. Calculate the circulation of the vector $\mathbf{F} = y^2\mathbf{i} + xy\mathbf{j} + z^2\mathbf{k}$ $\left(\oint \mathbf{F} \cdot \mathbf{dr} \right)$ around a triangle with vertices at the origin, $(2, 2, 0)$, and $(0, 2, 0)$ by (a) direct integration, and (b) using Stokes' theorem.
    Ans. 8/3.

29. Calculate the circulation of $\mathbf{F} = y\mathbf{i} - x\mathbf{j} + z\mathbf{k}$ around a unit circle in the xy plane with center at the origin by (a) direct integration and (b) using Stokes' theorem.
    Ans. -2π.

30. Evaluate the circulation of the following vector fields around the curves specified. Use either direct integration or Stokes' theorem. (a) $\mathbf{F} = 2z\mathbf{i} + y\mathbf{j} + x\mathbf{k}$ around a triangle with vertices at the origin, $(1, 0, 0)$ and $(0, 0, 4)$. (b) $\mathbf{F} = x^2\mathbf{i} + y^2\mathbf{j} + z^2\mathbf{k}$ around a unit circle in the xy plane with center at the origin.
    Ans. $2, 0$.

31. Check the product rule

$$\boldsymbol{\nabla} \cdot (\mathbf{A} \times \mathbf{B}) = (\boldsymbol{\nabla} \times \mathbf{A}) \cdot \mathbf{B} - (\boldsymbol{\nabla} \times \mathbf{B}) \cdot \mathbf{A}$$

by calculating each term separately for the functions $\mathbf{A} = y^2\mathbf{i} + 2xy\mathbf{j} + z^2\mathbf{k}$, $\mathbf{B} = \sin y\mathbf{i} + \sin x\mathbf{j} + z^3\mathbf{k}$.

32. Check the relation

$$\boldsymbol{\nabla} \times (\boldsymbol{\nabla} \times \mathbf{A}) = \boldsymbol{\nabla} (\boldsymbol{\nabla} \cdot \mathbf{A}) - \nabla^2 \mathbf{A}$$

by calculating each term separately for the function $\mathbf{A} = y^2\mathbf{i} + 2xy\mathbf{j} + z^2\mathbf{k}$.

33. Show that $\boldsymbol{\nabla} \times (\varphi \boldsymbol{\nabla} \varphi) = \mathbf{0}$.

34. Show that

$$\iiint_V (\boldsymbol{\nabla} \times \mathbf{A}) \cdot \mathbf{B} \, dV = \iiint_V (\boldsymbol{\nabla} \times \mathbf{B}) \cdot \mathbf{A} \, dV + \oiint_S (\mathbf{A} \times \mathbf{B}) \cdot \mathbf{n} \, da,$$

where S is the surface bounding the volume V.

35. Show that for any closed surface S

$$\oiint_S (\boldsymbol{\nabla} \times \mathbf{B}) \cdot \mathbf{n} \, da = 0.$$

36. For what values, if any, of the constants a and b is the following vector field irrotational?
$\mathbf{F} = (y \cos x + axz)\mathbf{i} + (b \sin x + z)\mathbf{j} + (x^2 + y)\mathbf{k}$.
Ans. $a = 2$, $b = 1$.

37. (a) Show that $\mathbf{F} = (2xy + 3)\mathbf{i} + (x^2 - 4z)\mathbf{j} - 4y\mathbf{k}$ is a conservative field.
(b) Find a scalar potential φ such that $\boldsymbol{\nabla}\varphi = -\mathbf{F}$. (c) Evaluate the integral $\int_{3,-1,2}^{2,1,-1} \mathbf{F} \cdot d\mathbf{r}$.
Ans. $\boldsymbol{\nabla} \times \mathbf{F} = 0$, $\varphi = -x^2 y - 3x + 4yz$, 6.

38. (a) Show that $\mathbf{F} = y^2 z\mathbf{i} - (z^2 \sin y - 2xyz)\mathbf{j} + (2z \cos y + y^2 x)\mathbf{k}$ is irrotational.
(b) find a function φ such that $\boldsymbol{\nabla}\varphi = \mathbf{F}$.
(c) Evaluate the integral $\int_\Gamma \mathbf{F} \cdot d\mathbf{r}$ where Γ is along the curve $x = \sin(\pi t/2)$, $y = t^2 - t$, $z = t^4$, $0 \le t \le 1$.
Ans. $\boldsymbol{\nabla} \times \mathbf{F} = 0$, $\varphi = z^2 \cos y + xy^2 z$, 1.

39. If $\mathbf{A}$ is irrotational, show that $\mathbf{A} \times \mathbf{r}$ is solenoidal.

40. Vector $\mathbf{B}$ is formed by the product of two gradients

$$\mathbf{B} = (\boldsymbol{\nabla} u) \times (\boldsymbol{\nabla} v),$$

where u and v are scalar functions. (a) Show that $\mathbf{B}$ is solenoidal.
(b) Show that

$$\mathbf{A} = \frac{1}{2}(u\boldsymbol{\nabla} v - v\boldsymbol{\nabla} u)$$

is a vector potential for $\mathbf{B}$ in that $\mathbf{B} = \boldsymbol{\nabla} \times \mathbf{A}$.

41. Show that if $\nabla^2 \varphi = 0$ in the volume V, then

$$\oiint_S \boldsymbol{\nabla}\varphi \cdot \mathbf{n} \, da = 0,$$

where S is the surface bounding the volume.

42. Two fields f and g are related by Poisson's equation, $\nabla^2 f = g$. Show that

$$\iiint_V g \, dV = \oiint_S \nabla f \cdot \mathbf{n} \, da,$$

where S is the bounding surface of V.

43. Use Stokes' theorem to show that

$$\oint_C f \nabla g \cdot d\mathbf{r} = - \oint_C g \nabla f \cdot d\mathbf{r}$$

for any closed curve C and differentiable fields f and g.
Hint: first show $\oint_C f \nabla g \cdot d\mathbf{r} = \iint_S \nabla f \times \nabla g \cdot \mathbf{n} \, da$

3

Curved Coordinates

Up to now we have used only Cartesian (rectangular) coordinates with their constant unit vectors. Frequently, because of the geometry of the problems, other coordinate systems are much more convenient. There are many coordinate systems, each of them can be regarded as a particular case of the general curvilinear coordinate system. It would be most efficient if we first develop a theory of curvilinear coordinates and then introduce each coordinate system as a special example. However, for pedagogic reasons, we will do that after we first directly transform the vector expressions of the rectangular coordinates into the corresponding ones in the two most commonly used systems, namely cylindrical and spherical coordinates. This procedure has the advantage of emphasizing that the physical meaning of gradient, divergence, curl, and Laplacian operations remain the same in different coordinate systems. Their appearances are different only because they are expressed with different notations. Furthermore, expressions in cylindrical and spherical coordinates will serve as familiar examples to clarify the terms in the general curvilinear system. As a further example, the elliptical coordinate system is discussed in some detail because of its importance in dealing with two center problems. Within the framework of curvilinear coordinates, we introduce the Jacobian determinant for multiple integrals in Sect. 3.5.

3.1 Cylindrical Coordinates

The cylindrical coordinate system is formally known as the circular cylindrical or cylindrical polar coordinate system. In this system, the position of a point is specified by (ρ, φ, z) as shown in Fig. 3.1a: ρ is the perpendicular distance from the z-axis, φ is the angle between the x-axis and the projection of ρ on the xy-plane, and z is the same as in the rectangular coordinates. The three unit vectors, $\mathbf{e}_\rho$, $\mathbf{e}_\varphi$, $\mathbf{e}_z$, point in the direction of increase of the corresponding coordinates. The relation to Cartesian coordinates can be easily seen from Fig. 3.1b where we have moved $\mathbf{e}_\rho, \mathbf{e}_\varphi$ to the origin:

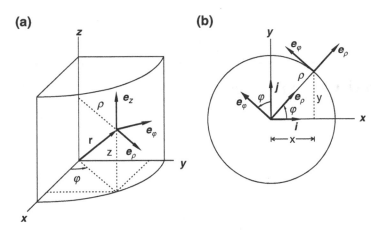

Fig. 3.1. Cylindrical coordinates. (a) A point is specified by (ρ, φ, z), the unit vectors $\mathbf{e}_\rho, \mathbf{e}_\varphi, \mathbf{e}_z$ are pointing in the direction of increase of the corresponding coordinates, (b) $\mathbf{e}_\rho, \mathbf{e}_\varphi$ are moved to the origin to find the relationships with $\mathbf{i}, \mathbf{j}$ of the rectangular coordinate system

$$x = \rho\cos\varphi, \quad y = \rho\sin\varphi, \quad \rho = (x^2 + y^2)^{1/2}, \quad \varphi = \tan^{-1}\frac{y}{x}. \tag{3.1}$$

$$\mathbf{e}_\rho = \cos\varphi\,\mathbf{i} + \sin\varphi\,\mathbf{j} = \frac{x}{(x^2 + y^2)^{1/2}}\mathbf{i} + \frac{y}{(x^2 + y^2)^{1/2}}\mathbf{j}, \tag{3.2}$$

$$\mathbf{e}_\varphi = -\sin\varphi\,\mathbf{i} + \cos\varphi\,\mathbf{j} = \frac{-y}{(x^2 + y^2)^{1/2}}\mathbf{i} + \frac{x}{(x^2 + y^2)^{1/2}}\mathbf{j}. \tag{3.3}$$

$$\mathbf{i} = \cos\varphi\,\mathbf{e}_\rho - \sin\varphi\,\mathbf{e}_\varphi, \tag{3.4}$$

$$\mathbf{j} = \sin\varphi\,\mathbf{e}_\rho + \cos\varphi\,\mathbf{e}_\varphi. \tag{3.5}$$

It follows

$$\frac{\partial x}{\partial \rho} = \cos\varphi, \quad \frac{\partial x}{\partial \varphi} = -\rho\sin\varphi, \quad \frac{\partial y}{\partial \rho} = \sin\varphi, \quad \frac{\partial y}{\partial \varphi} = \rho\cos\varphi, \tag{3.6}$$

$$\frac{\partial \rho}{\partial x} = \frac{\partial}{\partial x}(x^2 + y^2)^{1/2} = \frac{x}{(x^2 + y^2)^{1/2}} = \frac{\rho\cos\varphi}{\rho} = \cos\varphi, \tag{3.7}$$

$$\frac{\partial \rho}{\partial y} = \frac{\partial}{\partial y}(x^2 + y^2)^{1/2} = \frac{y}{(x^2 + y^2)^{1/2}} = \frac{\rho\sin\varphi}{\rho} = \sin\varphi, \tag{3.8}$$

$$\frac{\partial \varphi}{\partial x} = \frac{\partial}{\partial x}\tan^{-1}\left(\frac{y}{x}\right) = -\frac{y}{x^2 + y^2} = -\frac{\sin\varphi}{\rho}, \tag{3.9}$$

$$\frac{\partial \varphi}{\partial y} = \frac{\partial}{\partial y}\tan^{-1}\left(\frac{y}{x}\right) = \frac{x}{x^2 + y^2} = \frac{\cos\varphi}{\rho}. \tag{3.10}$$

The relationships between $\mathbf{e}_\rho, \mathbf{e}_\varphi, \mathbf{e}_z$ can be easily worked out, for example,

$$\mathbf{e}_\rho \cdot \mathbf{e}_\rho = (\cos\varphi\mathbf{i} + \sin\varphi\mathbf{j}) \cdot (\cos\varphi\mathbf{i} + \sin\varphi\mathbf{j}) = \cos^2\varphi + \sin^2\varphi = 1,$$
$$\mathbf{e}_\rho \cdot \mathbf{e}_\varphi = (\cos\varphi\mathbf{i} + \sin\varphi\mathbf{j}) \cdot (-\sin\varphi\mathbf{i} + \cos\varphi\mathbf{j}) = 0,$$
$$\mathbf{e}_\rho \times \mathbf{e}_\varphi = (\cos\varphi\mathbf{i} + \sin\varphi\mathbf{j}) \times (-\sin\varphi\mathbf{i} + \cos\varphi\mathbf{j}) = \cos^2\varphi\mathbf{k} + \sin^2\varphi\mathbf{k} = \mathbf{k} = \mathbf{e}_z.$$

Taken together, they form an orthonormal basis set

$$\mathbf{e}_\rho \cdot \mathbf{e}_\rho = \mathbf{e}_\varphi \cdot \mathbf{e}_\varphi = \mathbf{e}_z \cdot \mathbf{e}_z = 1,$$
$$\mathbf{e}_\rho \cdot \mathbf{e}_\varphi = \mathbf{e}_\varphi \cdot \mathbf{e}_z = \mathbf{e}_z \cdot \mathbf{e}_\rho = 0, \tag{3.11}$$
$$\mathbf{e}_\rho \times \mathbf{e}_\varphi = \mathbf{e}_z, \quad \mathbf{e}_\varphi \times \mathbf{e}_z = \mathbf{e}_\rho, \quad \mathbf{e}_z \times \mathbf{e}_\rho = \mathbf{e}_\varphi.$$

The position vector $\mathbf{r}$, from the origin to any point in space, is clearly seen in Fig. 3.1 to be

$$\mathbf{r} = \rho\mathbf{e}_\rho + z\mathbf{e}_z. \tag{3.12}$$

This expression can also be obtained from directly transforming $\mathbf{r} = x\mathbf{i} + y\mathbf{j} + z\mathbf{k}$ into the cylindrical coordinates.

Any vector can be expressed in terms of them. If the vector is a function of the position, then

$$\mathbf{A}(\rho, \varphi, z) = A_\rho(\rho, \varphi, z)\,\mathbf{e}_\rho + A_\varphi(\rho, \varphi, z)\,\mathbf{e}_\varphi + A_z(\rho, \varphi, z)\,\mathbf{e}_z. \tag{3.13}$$

In general, each component is a function of ρ, φ, z. Unlike the constant unit vector $\mathbf{i}, \mathbf{j}, \mathbf{k}$ in the rectangular coordinate system, only $\mathbf{e}_z = \mathbf{k}$ is fixed in space, the directions of $\mathbf{e}_\rho, \mathbf{e}_\varphi$ change as the point is moved around. Note that both $\mathbf{e}_\rho$ and $\mathbf{e}_\varphi$ depend on φ. In particular,

$$\frac{\partial}{\partial\varphi}\mathbf{e}_\rho = \frac{\partial}{\partial\varphi}(\cos\varphi\mathbf{i} + \sin\varphi\mathbf{j}) = -\sin\varphi\mathbf{i} + \cos\varphi\mathbf{j} = \mathbf{e}_\varphi,$$
$$\frac{\partial}{\partial\varphi}\mathbf{e}_\varphi = \frac{\partial}{\partial\varphi}(-\sin\varphi\mathbf{i} + \cos\varphi\mathbf{j}) = -(\cos\varphi\mathbf{i} + \sin\varphi\mathbf{j}) = -\mathbf{e}_\rho, \tag{3.14}$$
$$\frac{\partial}{\partial\rho}\mathbf{e}_\rho = \frac{\partial}{\partial\rho}\mathbf{e}_\varphi = 0.$$

Example 3.1.1. Show that the acceleration of a particle expressed in cylindrical coordinates is given by

$$\mathbf{a} = \left(\ddot{\rho} - \rho\dot{\varphi}^2\right)\mathbf{e}_\rho + \left(\rho\ddot{\varphi} + 2\dot{\rho}\dot{\varphi}\right)\mathbf{e}_\varphi + \ddot{z}\mathbf{e}_z,$$

where dots denote differentiation with respect to time t.

Solution 3.1.1. Since the position vector is given by $\mathbf{r} = \rho\mathbf{e}_\rho + z\mathbf{e}_z$, the velocity is $\mathbf{v} = \dot{\mathbf{r}}$,

$$\dot{\mathbf{r}} = \dot{\rho}\mathbf{e}_\rho + \rho\dot{\mathbf{e}}_\rho + \dot{z}\mathbf{e}_z,$$

where $\mathbf{e}_z$ is a constant unit vector and $\mathbf{e}_\rho$ depends on φ. Since by (3.14)

$$\dot{\mathbf{e}}_\rho = \frac{\mathrm{d}\mathbf{e}_\rho}{\mathrm{d}t} = \frac{\mathrm{d}\varphi}{\mathrm{d}t}\frac{\mathrm{d}\mathbf{e}_\rho}{\mathrm{d}\varphi} = \dot{\varphi}\mathbf{e}_\varphi,$$

$$\dot{\mathbf{r}} = \dot{\rho}\mathbf{e}_\rho + \rho\dot{\varphi}\mathbf{e}_\varphi + \dot{z}\mathbf{e}_z.$$

The acceleration is the rate of change of velocity, therefore $\mathbf{a} = \dot{\mathbf{v}} = \ddot{\mathbf{r}}$,

$$\ddot{\mathbf{r}} = \ddot{\rho}\mathbf{e}_\rho + \dot{\rho}\dot{\mathbf{e}}_\rho + \dot{\rho}\dot{\varphi}\mathbf{e}_\varphi + \rho\ddot{\varphi}\mathbf{e}_\varphi + \rho\dot{\varphi}\dot{\mathbf{e}}_\varphi + \ddot{z}\mathbf{e}_z.$$

Again by (3.14),

$$\dot{\mathbf{e}}_\varphi = \frac{\mathrm{d}\mathbf{e}_\varphi}{\mathrm{d}t} = \frac{\mathrm{d}\varphi}{\mathrm{d}t}\frac{\mathrm{d}\mathbf{e}_\varphi}{\mathrm{d}\varphi} = \dot{\varphi}(-\mathbf{e}_\rho),$$

$$\ddot{\mathbf{r}} = \ddot{\rho}\mathbf{e}_\rho + \dot{\rho}\dot{\varphi}\mathbf{e}_\varphi + \dot{\rho}\dot{\varphi}\mathbf{e}_\varphi + \rho\ddot{\varphi}\mathbf{e}_\varphi - \rho\dot{\varphi}^2\mathbf{e}_\rho + \ddot{z}\mathbf{e}_z.$$

Therefore

$$\mathbf{a} = \ddot{\mathbf{r}} = \left(\ddot{\rho} - \rho\dot{\varphi}^2\right)\mathbf{e}_\rho + (2\dot{\rho}\dot{\varphi} + \rho\ddot{\varphi})\mathbf{e}_\varphi + \ddot{z}\mathbf{e}_z.$$

3.1.1 Differential Operations

Gradient. Starting from the definition of gradient in the Cartesian coordinates, we can use the coordinate transformation to express it in terms of (ρ, φ, z). Using (3.4) and (3.5),

$$\boldsymbol{\nabla}\Phi = \mathbf{i}\frac{\partial\Phi}{\partial x} + \mathbf{j}\frac{\partial\Phi}{\partial y} + \mathbf{k}\frac{\partial\Phi}{\partial z}$$

$$= (\cos\varphi\,\mathbf{e}_\rho - \sin\varphi\,\mathbf{e}_\varphi)\frac{\partial\Phi}{\partial x} + (\sin\varphi\,\mathbf{e}_\rho + \cos\varphi\,\mathbf{e}_\varphi)\frac{\partial\Phi}{\partial y} + \mathbf{e}_z\frac{\partial\Phi}{\partial z}$$

$$= \left(\cos\varphi\frac{\partial\Phi}{\partial x} + \sin\varphi\frac{\partial\Phi}{\partial y}\right)\mathbf{e}_\rho + \left(-\sin\varphi\frac{\partial\Phi}{\partial x} + \cos\varphi\frac{\partial\Phi}{\partial y}\right)\mathbf{e}_\varphi + \frac{\partial\Phi}{\partial z}\mathbf{e}_z. \quad (3.15)$$

By chain rule and (3.6)

$$\frac{\partial\Phi}{\partial\rho} = \frac{\partial x}{\partial\rho}\frac{\partial\Phi}{\partial x} + \frac{\partial y}{\partial\rho}\frac{\partial\Phi}{\partial y} = \cos\varphi\frac{\partial\Phi}{\partial x} + \sin\varphi\frac{\partial\Phi}{\partial y}, \quad (3.16)$$

$$\frac{\partial\Phi}{\partial\varphi} = \frac{\partial x}{\partial\varphi}\frac{\partial\Phi}{\partial x} + \frac{\partial y}{\partial\varphi}\frac{\partial\Phi}{\partial y} = -\rho\sin\varphi\frac{\partial\Phi}{\partial x} + \rho\cos\varphi\frac{\partial\Phi}{\partial y}. \quad (3.17)$$

With these expressions, (3.15) becomes

$$\boldsymbol{\nabla}\Phi = \frac{\partial\Phi}{\partial\rho}\mathbf{e}_\rho + \frac{1}{\rho}\frac{\partial\Phi}{\partial\varphi}\mathbf{e}_\varphi + \frac{\partial\Phi}{\partial z}\mathbf{e}_z. \quad (3.18)$$

Thus, the gradient operator in the cylindrical coordinates can be written as

$$\nabla = \mathbf{e}_\rho \frac{\partial}{\partial \rho} + \mathbf{e}_\varphi \frac{1}{\rho} \frac{\partial}{\partial \varphi} + \mathbf{e}_z \frac{\partial}{\partial z}. \tag{3.19}$$

An immediate consequence is

$$\nabla \rho = \mathbf{e}_\rho, \quad \nabla \varphi = \frac{1}{\rho} \mathbf{e}_\varphi, \quad \nabla z = \mathbf{e}_z. \tag{3.20}$$

This is not a surprising result. After all, ∇u is a vector perpendicular to the surface $u = $ constant.

Divergence. The divergence of a vector

$$\nabla \cdot \mathbf{V} = \nabla \cdot (V_\rho \mathbf{e}_\rho + V_\varphi \mathbf{e}_\varphi + V_z \mathbf{e}_z)$$

can be expanded first by the distributive law of dot product. Now,

$$\begin{aligned} \nabla \cdot V_\rho \mathbf{e}_\rho &= \nabla \cdot V_\rho (\mathbf{e}_\varphi \times \mathbf{e}_z) = \nabla \cdot V_\rho (\rho \nabla \varphi \times \nabla z) \\ &= \nabla(\rho V_\rho) \cdot (\nabla \varphi \times \nabla z) + \rho V_\rho \nabla \cdot (\nabla \varphi \times \nabla z). \end{aligned}$$

But $\nabla \cdot (\nabla \varphi \times \nabla z) = \nabla \times \nabla \varphi \cdot \nabla z - \nabla \times \nabla z \cdot \nabla \varphi = 0$, so

$$\begin{aligned} \nabla \cdot V_\rho \mathbf{e}_\rho &= \nabla(\rho V_\rho) \cdot (\nabla \varphi \times \nabla z) = \nabla(\rho V_\rho) \cdot \left(\frac{1}{\rho} \mathbf{e}_\varphi \times \mathbf{e}_z \right) = \frac{1}{\rho} \nabla(\rho V_\rho) \cdot \mathbf{e}_\rho \\ &= \frac{1}{\rho} \left(\mathbf{e}_\rho \frac{\partial \rho V_\rho}{\partial \rho} + \mathbf{e}_\varphi \frac{1}{\rho} \frac{\partial \rho V_\rho}{\partial \varphi} + \mathbf{e}_z \frac{\partial \rho V_\rho}{\partial z} \right) \cdot \mathbf{e}_\rho = \frac{1}{\rho} \frac{\partial}{\partial \rho}(\rho V_\rho). \quad (3.21) \end{aligned}$$

$$\begin{aligned} \nabla \cdot V_\varphi \mathbf{e}_\varphi &= \nabla \cdot V_\varphi (\mathbf{e}_z \times \mathbf{e}_\rho) = \nabla \cdot V_\varphi (\nabla z \times \nabla \rho) \\ &= \nabla V_\varphi \cdot (\nabla z \times \nabla \rho) + V_\varphi \nabla \cdot (\nabla z \times \nabla \rho) \\ &= \nabla V_\varphi \cdot (\nabla z \times \nabla \rho) = \nabla V_\varphi \cdot (\mathbf{e}_z \times \mathbf{e}_\rho) = \nabla V_\varphi \cdot \mathbf{e}_\varphi \\ &= \left(\mathbf{e}_\rho \frac{\partial V_\varphi}{\partial \rho} + \mathbf{e}_\varphi \frac{1}{\rho} \frac{\partial V_\varphi}{\partial \varphi} + \mathbf{e}_z \frac{\partial V_\varphi}{\partial z} \right) \cdot \mathbf{e}_\varphi = \frac{1}{\rho} \frac{\partial V_\varphi}{\partial \varphi}. \quad (3.22) \end{aligned}$$

Therefore,

$$\nabla \cdot \mathbf{V} = \frac{1}{\rho} \frac{\partial}{\partial \rho}(\rho V_\rho) + \frac{1}{\rho} \frac{\partial V_\varphi}{\partial \varphi} + \frac{\partial V_z}{\partial z}. \tag{3.23}$$

Laplacian. By definition the Laplacian of Φ is given by

$$\nabla^2 \Phi = \nabla \cdot \nabla \Phi = \nabla \cdot \left(\frac{\partial \Phi}{\partial \rho} \mathbf{e}_\rho + \frac{1}{\rho} \frac{\partial \Phi}{\partial \varphi} \mathbf{e}_\varphi + \frac{\partial \Phi}{\partial z} \mathbf{e}_z \right). \tag{3.24}$$

Using the expression of the divergence, we have

$$\nabla \cdot \nabla \Phi = \frac{1}{\rho} \frac{\partial}{\partial \rho} \left(\rho \frac{\partial \Phi}{\partial \rho} \right) + \frac{1}{\rho} \frac{\partial}{\partial \varphi} \left(\frac{1}{\rho} \frac{\partial \Phi}{\partial \varphi} \right) + \frac{\partial}{\partial z} \left(\frac{\partial \Phi}{\partial z} \right).$$

Therefore

$$\nabla^2 \Phi = \frac{\partial^2 \Phi}{\partial \rho^2} + \frac{1}{\rho} \frac{\partial \Phi}{\partial \rho} + \frac{1}{\rho^2} \frac{\partial^2 \Phi}{\partial \varphi^2} + \frac{\partial^2 \Phi}{\partial z^2}. \tag{3.25}$$

Since the Laplacian is a scalar operator, it is instructive to convert it directly from its definition in the rectangular coordinates

$$\nabla^2 \Phi = \frac{\partial^2 \Phi}{\partial x^2} + \frac{\partial^2 \Phi}{\partial y^2} + \frac{\partial^2 \Phi}{\partial z^2}.$$

Now with chain rule and (3.7) and (3.9), we have

$$\frac{\partial \Phi}{\partial x} = \frac{\partial \rho}{\partial x} \frac{\partial \Phi}{\partial \rho} + \frac{\partial \varphi}{\partial x} \frac{\partial \Phi}{\partial \varphi} = \cos \varphi \frac{\partial \Phi}{\partial \rho} - \frac{\sin \varphi}{\rho} \frac{\partial \Phi}{\partial \varphi},$$

$$\begin{aligned}
\frac{\partial^2 \Phi}{\partial x^2} &= \frac{\partial}{\partial x} \left[\frac{\partial \Phi}{\partial x} \right] = \frac{\partial \rho}{\partial x} \frac{\partial}{\partial \rho} \left[\frac{\partial \Phi}{\partial x} \right] + \frac{\partial \varphi}{\partial x} \frac{\partial}{\partial \varphi} \left[\frac{\partial \Phi}{\partial x} \right] \\
&= \cos \varphi \frac{\partial}{\partial \rho} \left[\cos \varphi \frac{\partial \Phi}{\partial \rho} - \frac{\sin \varphi}{\rho} \frac{\partial \Phi}{\partial \varphi} \right] - \frac{\sin \varphi}{\rho} \frac{\partial}{\partial \varphi} \left[\cos \varphi \frac{\partial \Phi}{\partial \rho} - \frac{\sin \varphi}{\rho} \frac{\partial \Phi}{\partial \varphi} \right] \\
&= \cos^2 \varphi \frac{\partial^2 \Phi}{\partial \rho^2} + \frac{\cos \varphi \sin \varphi}{\rho^2} \frac{\partial \Phi}{\partial \varphi} - \frac{\cos \varphi \sin \varphi}{\rho} \frac{\partial^2 \Phi}{\partial \rho \partial \varphi} \\
&\quad + \frac{\sin^2 \varphi}{\rho} \frac{\partial \Phi}{\partial \rho} - \frac{\sin \varphi \cos \varphi}{\rho} \frac{\partial^2 \Phi}{\partial \varphi \partial \rho} + \frac{\sin \varphi \cos \varphi}{\rho^2} \frac{\partial \Phi}{\partial \varphi} + \frac{\sin^2 \varphi}{\rho^2} \frac{\partial^2 \Phi}{\partial \varphi^2}.
\end{aligned}$$

Similarly,

$$\frac{\partial \Phi}{\partial y} = \frac{\partial \rho}{\partial y} \frac{\partial \Phi}{\partial \rho} + \frac{\partial \varphi}{\partial y} \frac{\partial \Phi}{\partial \varphi} = \sin \varphi \frac{\partial \Phi}{\partial \rho} + \frac{\cos \varphi}{\rho} \frac{\partial \Phi}{\partial \varphi},$$

$$\begin{aligned}
\frac{\partial^2 \Phi}{\partial y^2} &= \frac{\partial}{\partial y} \left[\frac{\partial \Phi}{\partial y} \right] = \frac{\partial \rho}{\partial y} \frac{\partial}{\partial \rho} \left[\frac{\partial \Phi}{\partial y} \right] + \frac{\partial \varphi}{\partial y} \frac{\partial}{\partial \varphi} \left[\frac{\partial \Phi}{\partial y} \right] \\
&= \sin \varphi \frac{\partial}{\partial \rho} \left[\sin \varphi \frac{\partial \Phi}{\partial \rho} + \frac{\cos \varphi}{\rho} \frac{\partial \Phi}{\partial \varphi} \right] + \frac{\cos \varphi}{\rho} \frac{\partial}{\partial \varphi} \left[\sin \varphi \frac{\partial \Phi}{\partial \rho} + \frac{\cos \varphi}{\rho} \frac{\partial \Phi}{\partial \varphi} \right] \\
&= \sin^2 \varphi \frac{\partial^2 \Phi}{\partial \rho^2} - \frac{\sin \varphi \cos \varphi}{\rho^2} \frac{\partial \Phi}{\partial \varphi} + \frac{\sin \varphi \cos \varphi}{\rho} \frac{\partial^2 \Phi}{\partial \rho \partial \varphi} \\
&\quad + \frac{\cos^2 \varphi}{\rho} \frac{\partial \Phi}{\partial \rho} + \frac{\cos \varphi \sin \varphi}{\rho} \frac{\partial^2 \Phi}{\partial \varphi \partial \rho} - \frac{\cos \varphi \sin \varphi}{\rho^2} \frac{\partial \Phi}{\partial \varphi} + \frac{\cos^2 \varphi}{\rho^2} \frac{\partial^2 \Phi}{\partial \varphi^2}.
\end{aligned}$$

Thus,

$$\begin{aligned}
\frac{\partial^2 \Phi}{\partial x^2} + \frac{\partial^2 \Phi}{\partial y^2} &= (\cos^2 \varphi + \sin^2 \varphi) \frac{\partial^2 \Phi}{\partial \rho^2} + \frac{\sin^2 \varphi + \cos^2 \varphi}{\rho} \frac{\partial \Phi}{\partial \rho} \\
&\quad + \frac{\sin^2 \varphi + \cos^2 \varphi}{\rho^2} \frac{\partial^2 \Phi}{\partial \rho^2} = \frac{\partial^2 \Phi}{\partial \rho^2} + \frac{1}{\rho} \frac{\partial \Phi}{\partial \rho} + \frac{1}{\rho^2} \frac{\partial^2 \Phi}{\partial \varphi^2}.
\end{aligned}$$

Clearly the Laplacian obtained this way is identical to (3.25).

Curl. The curl of a vector can be written as

$$\mathbf{\nabla} \times \mathbf{V} = \mathbf{\nabla} \times (V_\rho \mathbf{e}_\rho + V_\varphi \mathbf{e}_\varphi + V_z \mathbf{e}_z). \tag{3.26}$$

Now

$$\mathbf{\nabla} \times V_\rho \mathbf{e}_\rho = \mathbf{\nabla} \times V_\rho \mathbf{\nabla}\rho = \mathbf{\nabla} V_\rho \times \mathbf{\nabla}\rho + V_\rho \mathbf{\nabla} \times \mathbf{\nabla}\rho.$$

Since $\mathbf{\nabla} \times \mathbf{\nabla}\rho = \mathbf{0}$,

$$
\begin{aligned}
\mathbf{\nabla} \times V_\rho \mathbf{e}_\rho &= \mathbf{\nabla} V_\rho \times \mathbf{\nabla}\rho = \mathbf{\nabla} V_\rho \times \mathbf{e}_\rho \\
&= \left(\mathbf{e}_\rho \frac{\partial V_\rho}{\partial \rho} + \mathbf{e}_\varphi \frac{1}{\rho} \frac{\partial V_\rho}{\partial \varphi} + \mathbf{e}_z \frac{\partial V_\rho}{\partial z} \right) \times \mathbf{e}_\rho \\
&= -\frac{1}{\rho} \frac{\partial V_\rho}{\partial \varphi} \mathbf{e}_z + \frac{\partial V_\rho}{\partial z} \mathbf{e}_\varphi,
\end{aligned} \tag{3.27}
$$

$$
\begin{aligned}
\mathbf{\nabla} \times V_\varphi \mathbf{e}_\varphi &= \mathbf{\nabla} \times V_\varphi (\rho \mathbf{\nabla}\varphi) = \mathbf{\nabla}(\rho V_\varphi) \times \mathbf{\nabla}\varphi + \rho V_\varphi \mathbf{\nabla} \times \mathbf{\nabla}\varphi \\
&= \mathbf{\nabla}(\rho V_\varphi) \times \mathbf{\nabla}\varphi = \mathbf{\nabla}(\rho V_\varphi) \times \frac{1}{\rho} \mathbf{e}_\varphi \\
&= \frac{1}{\rho} \left(\mathbf{e}_\rho \frac{\partial \rho V_\varphi}{\partial \rho} + \mathbf{e}_\varphi \frac{1}{\rho} \frac{\partial \rho V_\varphi}{\partial \varphi} + \mathbf{e}_z \frac{\partial \rho V_\varphi}{\partial z} \right) \times \mathbf{e}_\varphi \\
&= \frac{1}{\rho} \frac{\partial \rho V_\varphi}{\partial \rho} \mathbf{e}_z - \frac{1}{\rho} \frac{\partial \rho V_\varphi}{\partial z} \mathbf{e}_\rho,
\end{aligned} \tag{3.28}
$$

$$
\begin{aligned}
\mathbf{\nabla} \times V_z \mathbf{e}_z &= \mathbf{\nabla} V_z \times \mathbf{e}_z = \left(\mathbf{e}_\rho \frac{\partial V_z}{\partial \rho} + \mathbf{e}_\varphi \frac{1}{\rho} \frac{\partial V_z}{\partial \varphi} + \mathbf{e}_z \frac{\partial V_z}{\partial z} \right) \times \mathbf{e}_z \\
&= -\frac{\partial V_z}{\partial \rho} \mathbf{e}_\varphi + \frac{1}{\rho} \frac{\partial V_z}{\partial \varphi} \mathbf{e}_\rho.
\end{aligned}
$$

Thus,

$$
\begin{aligned}
\mathbf{\nabla} \times \mathbf{V} = {} &\left(\frac{1}{\rho} \frac{\partial V_z}{\partial \varphi} - \frac{1}{\rho} \frac{\partial \rho V_\varphi}{\partial z} \right) \mathbf{e}_\rho + \left(\frac{\partial V_\rho}{\partial z} - \frac{\partial V_z}{\partial \rho} \right) \mathbf{e}_\varphi \\
&+ \left(\frac{1}{\rho} \frac{\partial}{\partial \rho}(\rho V_\varphi) - \frac{1}{\rho} \frac{\partial V_\rho}{\partial \varphi} \right) \mathbf{e}_z.
\end{aligned} \tag{3.29}
$$

Example 3.1.2. (a) Show that the vector field

$$\mathbf{F} = \left(A - \frac{B}{\rho^2} \right) \cos \varphi \, \mathbf{e}_\rho - \left(A + \frac{B}{\rho^2} \right) \sin \varphi \, \mathbf{e}_\varphi$$

is irrotational $(\mathbf{\nabla} \times \mathbf{F} = \mathbf{0})$. (b) Find a scalar potential Φ such that $\mathbf{\nabla}\Phi = \mathbf{F}$. (c) Show that Φ satisfies the Laplace's equation $(\nabla^2 \Phi = 0)$.

Solution 3.1.2. (*a*) All derivatives with respect to z are equal to zero, since there is no z dependence. Furthermore, $V_z = 0$. Therefore

$$
\begin{aligned}
\nabla \times \mathbf{F} &= \left(\frac{1}{\rho} \frac{\partial}{\partial \rho} (\rho F_\varphi) - \frac{1}{\rho} \frac{\partial F_\rho}{\partial \varphi} \right) \mathbf{e}_z \\
&= \frac{1}{\rho} \frac{\partial}{\partial \rho} \left[\rho \left(-A - \frac{B}{\rho^2} \right) \sin \varphi \right] \mathbf{e}_z - \frac{1}{\rho} \frac{\partial}{\partial \varphi} \left[\left(A - \frac{B}{\rho^2} \right) \cos \varphi \right] \mathbf{e}_z \\
&= \frac{1}{\rho} \left[\left(-A + \frac{B}{\rho^2} \right) \sin \varphi + \left(A - \frac{B}{\rho^2} \right) \sin \varphi \right] \mathbf{e}_z = 0.
\end{aligned}
$$

(*b*)

$$
\begin{aligned}
\nabla \Phi &= \mathbf{e}_\rho \frac{\partial \Phi}{\partial \rho} + \mathbf{e}_\varphi \frac{1}{\rho} \frac{\partial \Phi}{\partial \varphi} + \mathbf{e}_z \frac{\partial \Phi}{\partial z} \\
&= \left(A - \frac{B}{\rho^2} \right) \cos \varphi \, \mathbf{e}_\rho - \left(A + \frac{B}{\rho^2} \right) \sin \varphi \, \mathbf{e}_\varphi,
\end{aligned}
$$

$$
\frac{\partial \Phi}{\partial \rho} = \left(A - \frac{B}{\rho^2} \right) \cos \varphi; \quad \frac{1}{\rho} \frac{\partial \Phi}{\partial \varphi} = - \left(A + \frac{B}{\rho^2} \right) \sin \varphi.
$$

It is clear, up to an additive constant,

$$
\Phi = \left(A\rho + \frac{B}{\rho} \right) \cos \varphi.
$$

(*c*)

$$
\begin{aligned}
\nabla^2 \Phi &= \frac{1}{\rho} \frac{\partial}{\partial \rho} \left(\rho \frac{\partial \Phi}{\partial \rho} \right) + \frac{1}{\rho^2} \frac{\partial^2 \Phi}{\partial \varphi^2} + \frac{\partial^2 \Phi}{\partial z^2} \\
&= \frac{1}{\rho} \frac{\partial}{\partial \rho} \left[\rho \frac{\partial}{\partial \rho} \left(A\rho + \frac{B}{\rho} \right) \cos \varphi \right] + \frac{1}{\rho^2} \frac{\partial^2}{\partial \varphi^2} \left(A\rho + \frac{B}{\rho} \right) \cos \varphi \\
&= \frac{1}{\rho} \left(A + \frac{B}{\rho^2} \right) \cos \varphi - \frac{1}{\rho} \left(A + \frac{B}{\rho^2} \right) \cos \varphi = 0.
\end{aligned}
$$

3.1.2 Infinitesimal Elements

When a point at (x, y, z) is moved to $(x + dx, y + dy, z + dz)$, the infinitesimal displacement vector is $d\mathbf{r} = \mathbf{i} dx + \mathbf{j} dy + \mathbf{k} dz$. Similarly, when the point at (ρ, φ, z) in the cylindrical coordinates is moved to $(\rho + d\rho, \varphi + d\varphi, z + dz)$, the infinitesimal displacement vector is

$$
d\mathbf{r} = \mathbf{e}_\rho d\rho + \mathbf{e}_\varphi \rho \, d\varphi + \mathbf{e}_z \, dz. \tag{3.30}
$$

Notice the distance in $\mathbf{e}_\varphi$ direction is $\rho \, d\varphi$ as shown in Fig. 3.2. The infinitesimal length element is

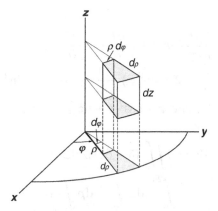

Fig. 3.2. Differential elements in cylindrical coordinates. Note that the differential length in the direction of increasing φ is $\rho \, d\varphi$. The differential volume element is $\rho \, d\varphi \, d\rho \, dz$.

$$ds = (d\mathbf{r} \cdot d\mathbf{r})^{1/2} = \left[(d\rho)^2 + (\rho \, d\varphi)^2 + (dz)^2 \right]^{1/2}. \tag{3.31}$$

The gradient is defined as a vector of derivatives with respect to the distances in three perpendicular directions. Thus the gradient in cylindrical coordinates should be

$$\boldsymbol{\nabla} = \mathbf{e}_\rho \frac{\partial}{\partial \rho} + \mathbf{e}_\varphi \frac{1}{\rho} \frac{\partial}{\partial \varphi} + \mathbf{e}_z \frac{\partial}{\partial z},$$

which is, of course, identical to (3.19) obtained from direct transformation.

The infinitesimal volume element dV is the product of the perpendicular infinitesimal displacements

$$dV = (d\rho)(\rho \, d\varphi)(dz) = \rho \, d\rho \, d\varphi \, dz. \tag{3.32}$$

The possible range of ρ is 0 to ∞, φ goes from 0 to 2π, and z from $-\infty$ to ∞. The infinitesimal surface element depends on the orientation of the surface. For example, on the side surface of a cylinder parallel to z-axis and of constant radius ρ, the surface element directed outward is $\mathbf{n} \, da = \rho \, d\varphi \, dz \mathbf{e}_\rho$. The surface element on the xy-plane directed upward is $\mathbf{n} \, da = \rho \, d\varphi \, d\rho \mathbf{e}_z$.

Example 3.1.3. Verify the divergence theorem

$$\iiint_V \boldsymbol{\nabla} \cdot \mathbf{F} \, dV = \oiint_S \mathbf{F} \cdot \mathbf{n} \, da$$

with a vector field

$$\mathbf{F} = \rho \left(2 + \sin^2 \varphi \right) \mathbf{e}_\rho + \rho \sin \varphi \cos \varphi \, \mathbf{e}_\varphi + 3z^2 \, \mathbf{e}_z$$

over a cylinder with base of radius 2 and height 5.

Solution 3.1.3.

$$\mathbf{\nabla} \cdot \mathbf{F} = \frac{1}{\rho} \frac{\partial}{\partial \rho} (\rho F_\rho) + \frac{1}{\rho} \frac{\partial F_\varphi}{\partial \varphi} + \frac{\partial F_z}{\partial z}$$
$$= \frac{1}{\rho} 2\rho \left(2 + \sin^2 \varphi\right) + \frac{1}{\rho} \rho \left(\cos^2 \varphi - \sin^2 \varphi\right) + 6z$$
$$= 4 + \sin^2 \varphi + \cos^2 \varphi + 6z = 5 + 6z.$$

$$\iiint_V \mathbf{\nabla} \cdot \mathbf{F} \, dV = \iiint_V (5 + 6z) \, \rho \, d\varphi \, d\rho \, dz$$
$$= \int_0^{2\pi} d\varphi \int_0^2 \rho \, d\rho \int_0^5 (5 + 6z) \, dz = 400\pi.$$

$$\oiint_S \mathbf{F} \cdot \mathbf{n} \, da = \iint_{S_1} \mathbf{F} \cdot \mathbf{n} \, da + \iint_{S_2} \mathbf{F} \cdot \mathbf{n} \, da + \iint_{S_3} \mathbf{F} \cdot \mathbf{n} \, da,$$

where S_1 is the side surface of the cylinder, S_2 and S_3 are, respectively, the bottom and top surfaces of the cylinder.

$$\iint_{S_1} \mathbf{F} \cdot \mathbf{n} \, da = \iint_{S_1} \mathbf{F} \cdot \mathbf{e}_\rho \, da = \int_0^{2\pi} \int_0^5 \left[\rho \left(2 + \sin^2 \varphi\right) \rho \right]_{\rho=2} d\varphi \, dz = 100\pi.$$

$$\iint_{S_2} \mathbf{F} \cdot \mathbf{n} \, da = \iint_{S_2} \mathbf{F} \cdot (-\mathbf{e}_z) \, da = \iint_{S_2} \left[-3z^2 \right]_{z=0} da = 0,$$

$$\iint_{S_3} \mathbf{F} \cdot \mathbf{n} \, da = \iint_{S_3} \mathbf{F} \cdot (\mathbf{e}_z) \, da = \int_0^{2\pi} \int_0^2 \left[3z^2 \right]_{z=5} \rho \, d\varphi \, d\rho = 300\pi.$$

Therefore,

$$\oiint_S \mathbf{F} \cdot \mathbf{n} \, da = 100\pi + 300\pi = 400\pi.$$

Clearly,

$$\iiint_V \mathbf{\nabla} \cdot \mathbf{F} \, dV = \oiint_S \mathbf{F} \cdot \mathbf{n} \, da.$$

3.2 Spherical Coordinates

The spherical polar coordinate system is commonly known just as the spherical coordinates. The location of a point is specified by (r, θ, φ) as shown in Fig. 3.3, where r is the distance from the origin, θ is the angle made by the position vector $\mathbf{r}$ with the positive z-axis which is often called polar angle, and φ is the angle made with the positive x-axis by the projection of $\mathbf{r}$ on the xy-plane, this angle is known as azimuthal angle. The relations between the rectangular and spherical coordinates are seen from Fig. 3.3b and c.

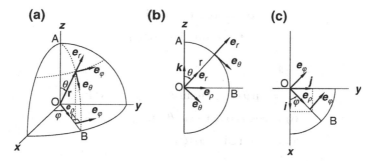

Fig. 3.3. Spherical coordinates. **(a)** A point is specified by (r, θ, φ) where r is the distance from the origin, θ is the angle made by **r** with the positive z-axis, and φ is the angle made with the positive x-axis by the projection of **r** on the xy-plane. The three unit vectors $\mathbf{e}_r, \mathbf{e}_\theta, \mathbf{e}_\varphi$ are in the direction of increasing r, θ, φ, respectively. The auxiliary unit vector $\mathbf{e}_\varrho$ is in the direction of the projection of **r** on the xy-plane. **(b)** The unit vectors $\mathbf{e}_r$ and $\mathbf{e}_\theta$ are moved to the origin in the AOB plane to find their relationships with **k** and $\mathbf{e}_\varrho$. **(c)** In the xy-plane, $\mathbf{e}_\varphi$ is moved to the origin to find the relationships between $\mathbf{e}_\varphi, \mathbf{e}_\varrho$ and **i, j**

$$x = r \sin \theta \cos \varphi, \qquad y = r \sin \theta \sin \varphi, \qquad z = r \cos \theta \qquad (3.33)$$

and

$$r = \left(x^2 + y^2 + z^2\right)^{1/2}, \quad \tan \theta = \frac{\left(x^2 + y^2\right)^{1/2}}{z}, \quad \tan \varphi = \frac{y}{x}, \qquad (3.34)$$

$$\sin \theta = \frac{\left(x^2 + y^2\right)^{1/2}}{\left(x^2 + y^2 + z^2\right)^{1/2}}, \qquad \cos \theta = \frac{z}{\left(x^2 + y^2 + z^2\right)^{1/2}}, \qquad (3.35)$$

$$\sin \varphi = \frac{y}{\left(x^2 + y^2\right)^{1/2}}, \qquad \cos \varphi = \frac{x}{\left(x^2 + y^2\right)^{1/2}}. \qquad (3.36)$$

Figure 3.3 also shows a set of mutually perpendicular unit vectors $\mathbf{e}_r, \mathbf{e}_\theta, \mathbf{e}_\varphi$ in the sense of increasing r, θ, φ, respectively. In this system, the position vector **r** is simply

$$\mathbf{r} = r\hat{\mathbf{r}} = r\mathbf{e}_r. \qquad (3.37)$$

The relations between the unit vectors in the spherical coordinates and those in the Cartesian coordinates can be seen from Fig. 3.3b. In the plane AOB, we have drawn $\mathbf{e}_r$ and $\mathbf{e}_\theta$ from the origin. It can be seen

$$\mathbf{e}_r = \sin \theta \mathbf{e}_\varrho + \cos \theta \mathbf{k},$$
$$\mathbf{e}_\theta = \cos \theta \mathbf{e}_\varrho - \sin \theta \mathbf{k},$$

where $\mathbf{e}_\varrho$ is a unit vector along OB. In Fig. 3.3c, $\mathbf{e}_\varrho$ and $\mathbf{e}_\theta$ are drawn from the origin, clearly

$$\mathbf{e}_\varrho = \cos\varphi\mathbf{i} + \sin\varphi\mathbf{j},$$
$$\mathbf{e}_\varphi = -\sin\varphi\mathbf{i} + \cos\varphi\mathbf{j}.$$

Thus,

$$\mathbf{e}_r = \sin\theta\cos\varphi\mathbf{i} + \sin\theta\sin\varphi\mathbf{j} + \cos\theta\mathbf{k},$$
$$\mathbf{e}_\theta = \cos\theta\cos\varphi\mathbf{i} + \cos\theta\sin\varphi\mathbf{j} - \sin\theta\mathbf{k}, \qquad (3.38)$$
$$\mathbf{e}_\varphi = -\sin\varphi\mathbf{i} + \cos\varphi\mathbf{j}.$$

The inverse relations can either be read from the same figures, or be solved for $\mathbf{i}$, $\mathbf{j}$, $\mathbf{k}$ from the above equations,

$$\mathbf{i} = \sin\theta\cos\varphi\mathbf{e}_r + \cos\theta\cos\varphi\mathbf{e}_\theta - \sin\varphi\mathbf{e}_\varphi,$$
$$\mathbf{j} = \sin\theta\sin\varphi\mathbf{e}_r + \cos\theta\sin\varphi\mathbf{e}_\theta + \cos\varphi\mathbf{e}_\varphi, \qquad (3.39)$$
$$\mathbf{k} = \cos\theta\mathbf{e}_r - \sin\theta\mathbf{e}_\theta.$$

It can easily be verified that $(\mathbf{e}_r, \mathbf{e}_\theta, \mathbf{e}_\varphi)$ form an orthonormal basis set and satisfy the following relations

$$\mathbf{e}_r \cdot \mathbf{e}_r = \mathbf{e}_\theta \cdot \mathbf{e}_\theta = \mathbf{e}_\varphi \cdot \mathbf{e}_\varphi = 1,$$
$$\mathbf{e}_r \cdot \mathbf{e}_\theta = \mathbf{e}_\theta \cdot \mathbf{e}_\varphi = \mathbf{e}_\varphi \cdot \mathbf{e}_r = 0, \qquad (3.40)$$
$$\mathbf{e}_r \times \mathbf{e}_\theta = \mathbf{e}_\varphi, \quad \mathbf{e}_\theta \times \mathbf{e}_\varphi = \mathbf{e}_r, \quad \mathbf{e}_\varphi \times \mathbf{e}_r = \mathbf{e}_\theta.$$

Any vector can be expressed in terms of them

$$\mathbf{A} = A_r\mathbf{e}_r + A_\theta\mathbf{e}_\theta + A_\varphi\mathbf{e}_\varphi, \qquad (3.41)$$

where A_r, A_θ, A_φ are the radial, polar, and azimuthal components of $\mathbf{A}$. The derivatives of the unit vectors are easily obtained from (3.38):

$$\frac{\partial\mathbf{e}_r}{\partial r} = \frac{\partial\mathbf{e}_\theta}{\partial r} = \frac{\partial\mathbf{e}_\varphi}{\partial r} = \frac{\partial\mathbf{e}_\varphi}{\partial\theta} = 0, \qquad (3.42)$$

$$\frac{\partial\mathbf{e}_r}{\partial\theta} = \cos\theta\cos\varphi\mathbf{i} + \cos\theta\sin\varphi\mathbf{j} - \sin\theta\mathbf{k} = \mathbf{e}_\theta, \qquad (3.43)$$

$$\frac{\partial\mathbf{e}_\theta}{\partial\theta} = -\sin\theta\cos\varphi\mathbf{i} - \sin\theta\sin\varphi\mathbf{j} - \cos\theta\mathbf{k} = -\mathbf{e}_r, \qquad (3.44)$$

$$\frac{\partial\mathbf{e}_r}{\partial\varphi} = -\sin\theta\sin\varphi\mathbf{i} + \sin\theta\cos\varphi\mathbf{j} = \sin\theta\mathbf{e}_\varphi, \qquad (3.45)$$

$$\frac{\partial\mathbf{e}_\theta}{\partial\varphi} = -\cos\theta\sin\varphi\mathbf{i} + \cos\theta\cos\varphi\mathbf{j} = \cos\theta\mathbf{e}_\varphi, \qquad (3.46)$$

$$\frac{\partial\mathbf{e}_\varphi}{\partial\varphi} = -\cos\varphi\mathbf{i} - \sin\varphi\mathbf{j} = -(\sin\theta\mathbf{e}_r + \cos\theta\mathbf{e}_\theta). \qquad (3.47)$$

3.2.1 Differential Operations

Gradient. We are ready to express the gradient operator ∇ in the spherical coordinates. Using (3.39), we have

$$\nabla = \mathbf{i}\frac{\partial}{\partial x} + \mathbf{j}\frac{\partial}{\partial y} + \mathbf{k}\frac{\partial}{\partial z} = (\sin\theta\cos\varphi\mathbf{e}_r + \cos\theta\cos\varphi\mathbf{e}_\theta - \sin\varphi\mathbf{e}_\varphi)\frac{\partial}{\partial x}$$

$$+ (\sin\theta\sin\varphi\mathbf{e}_r + \cos\theta\sin\varphi\mathbf{e}_\theta + \cos\varphi\mathbf{e}_\varphi)\frac{\partial}{\partial y} + (\cos\theta\mathbf{e}_r - \sin\theta\mathbf{e}_\theta)\frac{\partial}{\partial z}$$

$$= \mathbf{e}_r\left[\sin\theta\cos\varphi\frac{\partial}{\partial x} + \sin\theta\sin\varphi\frac{\partial}{\partial y} + \cos\theta\frac{\partial}{\partial z}\right]$$

$$+ \mathbf{e}_\theta\left[\cos\theta\cos\varphi\frac{\partial}{\partial x} + \cos\theta\sin\varphi\frac{\partial}{\partial y} - \sin\theta\frac{\partial}{\partial z}\right]$$

$$+ \mathbf{e}_\varphi\left[-\sin\varphi\frac{\partial}{\partial x} + \cos\varphi\frac{\partial}{\partial y}\right]. \tag{3.48}$$

The quantities in the brackets can be recognized if we use (3.33) and the chain rule of derivatives

$$\frac{\partial}{\partial r} = \frac{\partial x}{\partial r}\frac{\partial}{\partial x} + \frac{\partial y}{\partial r}\frac{\partial}{\partial y} + \frac{\partial z}{\partial r}\frac{\partial}{\partial z}$$

$$= \sin\theta\cos\varphi\frac{\partial}{\partial x} + \sin\theta\sin\varphi\frac{\partial}{\partial y} + \cos\theta\frac{\partial}{\partial z}, \tag{3.49}$$

$$\frac{\partial}{\partial\theta} = \frac{\partial x}{\partial\theta}\frac{\partial}{\partial x} + \frac{\partial y}{\partial\theta}\frac{\partial}{\partial y} + \frac{\partial z}{\partial\theta}\frac{\partial}{\partial z}$$

$$= r\cos\theta\cos\varphi\frac{\partial}{\partial x} + r\cos\theta\sin\varphi\frac{\partial}{\partial y} - r\sin\theta\frac{\partial}{\partial z}, \tag{3.50}$$

$$\frac{\partial}{\partial\varphi} = \frac{\partial x}{\partial\varphi}\frac{\partial}{\partial x} + \frac{\partial y}{\partial\varphi}\frac{\partial}{\partial y} + \frac{\partial z}{\partial\varphi}\frac{\partial}{\partial z}$$

$$= -r\sin\theta\sin\varphi\frac{\partial}{\partial x} + r\sin\theta\cos\varphi\frac{\partial}{\partial y}. \tag{3.51}$$

Thus (3.48) can be written as

$$\nabla = \mathbf{e}_r\frac{\partial}{\partial r} + \mathbf{e}_\theta\frac{1}{r}\frac{\partial}{\partial\theta} + \mathbf{e}_\varphi\frac{1}{r\sin\theta}\frac{\partial}{\partial\varphi}. \tag{3.52}$$

It follows that

$$\nabla r = \mathbf{e}_r, \quad \nabla\theta = \frac{1}{r}\mathbf{e}_\theta, \quad \nabla\varphi = \frac{1}{r\sin\theta}\mathbf{e}_\varphi. \tag{3.53}$$

Divergence. The divergence of a vector in the spherical coordinates is

$$\nabla \cdot \mathbf{V} = \nabla \cdot (V_r \mathbf{e}_r + V_\theta \mathbf{e}_\theta + V_\varphi \mathbf{e}_\varphi)$$
$$= \nabla V_r \cdot \mathbf{e}_r + V_r \nabla \cdot \mathbf{e}_r + \nabla V_\theta \cdot \mathbf{e}_\theta + V_\theta \nabla \cdot \mathbf{e}_\theta$$
$$+ \nabla V_\varphi \cdot \mathbf{e}_\varphi + V_\varphi \nabla \cdot \mathbf{e}_\varphi. \tag{3.54}$$

Although the divergence in the spherical coordinates can be worked out just as we did for in the cylindrical coordinates, it is instructive to find the expression by using the derivatives of (3.42)–(3.47),

$$\nabla \cdot \mathbf{e}_r = \left(\mathbf{e}_r \frac{\partial}{\partial r} + \mathbf{e}_\theta \frac{1}{r} \frac{\partial}{\partial \theta} + \mathbf{e}_\varphi \frac{1}{r \sin \theta} \frac{\partial}{\partial \varphi} \right) \cdot \mathbf{e}_r$$
$$= \mathbf{e}_r \cdot \frac{\partial \mathbf{e}_r}{\partial r} + \frac{1}{r} \mathbf{e}_\theta \cdot \frac{\partial \mathbf{e}_r}{\partial \theta} + \frac{1}{r \sin \theta} \mathbf{e}_\varphi \cdot \frac{\partial \mathbf{e}_r}{\partial \varphi}$$
$$= \frac{1}{r} \mathbf{e}_\theta \cdot \mathbf{e}_\theta + \frac{1}{r \sin \theta} \mathbf{e}_\varphi \cdot \sin \theta \mathbf{e}_\varphi = \frac{2}{r}, \tag{3.55}$$

$$\nabla \cdot \mathbf{e}_\theta = \left(\mathbf{e}_r \frac{\partial}{\partial r} + \mathbf{e}_\theta \frac{1}{r} \frac{\partial}{\partial \theta} + \mathbf{e}_\varphi \frac{1}{r \sin \theta} \frac{\partial}{\partial \varphi} \right) \cdot \mathbf{e}_\theta$$
$$= \mathbf{e}_r \cdot \frac{\partial \mathbf{e}_\theta}{\partial r} + \frac{1}{r} \mathbf{e}_\theta \cdot \frac{\partial \mathbf{e}_\theta}{\partial \theta} + \frac{1}{r \sin \theta} \mathbf{e}_\varphi \cdot \frac{\partial \mathbf{e}_\theta}{\partial \varphi}$$
$$= \frac{1}{r} \mathbf{e}_\theta \cdot (-\mathbf{e}_r) + \frac{1}{r \sin \theta} \mathbf{e}_\varphi \cdot \cos \theta \mathbf{e}_\varphi = \frac{1}{r} \frac{\cos \theta}{\sin \theta}, \tag{3.56}$$

$$\nabla \cdot \mathbf{e}_\varphi = \left(\mathbf{e}_r \frac{\partial}{\partial r} + \mathbf{e}_\theta \frac{1}{r} \frac{\partial}{\partial \theta} + \mathbf{e}_\varphi \frac{1}{r \sin \theta} \frac{\partial}{\partial \varphi} \right) \cdot \mathbf{e}_\varphi$$
$$= \mathbf{e}_r \cdot \frac{\partial \mathbf{e}_\varphi}{\partial r} + \frac{1}{r} \mathbf{e}_\theta \cdot \frac{\partial \mathbf{e}_\varphi}{\partial \theta} + \frac{1}{r \sin \theta} \mathbf{e}_\varphi \cdot \frac{\partial \mathbf{e}_\varphi}{\partial \varphi}$$
$$= \frac{1}{r \sin \theta} \mathbf{e}_\varphi \cdot (-\sin \theta \mathbf{e}_r + \cos \theta \mathbf{e}_\theta) = 0. \tag{3.57}$$

Furthermore,

$$\nabla V_r \cdot \mathbf{e}_r = \left(\mathbf{e}_r \frac{\partial V_r}{\partial r} + \mathbf{e}_\theta \frac{1}{r} \frac{\partial V_r}{\partial \theta} + \mathbf{e}_\varphi \frac{1}{r \sin \theta} \frac{\partial V_r}{\partial \varphi} \right) \cdot \mathbf{e}_r = \frac{\partial V_r}{\partial r}. \tag{3.58}$$

Similarly,

$$\nabla V_\theta \cdot \mathbf{e}_\theta = \frac{1}{r} \frac{\partial V_\theta}{\partial \theta}, \qquad \nabla V_\varphi \cdot \mathbf{e}_\varphi = \frac{1}{r \sin \theta} \frac{\partial V_\varphi}{\partial \varphi}. \tag{3.59}$$

Thus,

$$\nabla \cdot \mathbf{V} = \frac{\partial V_r}{\partial r} + \frac{2}{r} V_r + \frac{1}{r} \frac{\partial V_\theta}{\partial \theta} + \frac{1}{r} \frac{\cos \theta}{\sin \theta} V_\theta + \frac{1}{r \sin \theta} \frac{\partial V_\varphi}{\partial \varphi}$$
$$= \frac{1}{r^2} \frac{\partial}{\partial r} \left(r^2 V_r \right) + \frac{1}{r \sin \theta} \frac{\partial}{\partial \theta} (\sin \theta \, V_\theta) + \frac{1}{r \sin \theta} \frac{\partial}{\partial \varphi} V_\varphi. \tag{3.60}$$

Laplacian. The Laplacian in spherical coordinates can be written as

$$\nabla^2 \Phi = \nabla \cdot \nabla \Phi = \nabla \cdot \left(\mathbf{e}_r \frac{\partial \Phi}{\partial r} + \mathbf{e}_\theta \frac{1}{r} \frac{\partial \Phi}{\partial \theta} + \mathbf{e}_\varphi \frac{1}{r \sin \theta} \frac{\partial \Phi}{\partial \varphi} \right).$$

Regarding $\nabla \Phi$ as a vector and using the expression of divergence, we have

$$\nabla \cdot \nabla \Phi = \frac{1}{r^2} \frac{\partial}{\partial r} \left(r^2 \frac{\partial \Phi}{\partial r} \right) + \frac{1}{r \sin \theta} \frac{\partial}{\partial \theta} \left(\sin \theta \frac{1}{r} \frac{\partial \Phi}{\partial \theta} \right) + \frac{1}{r \sin \theta} \frac{\partial}{\partial \varphi} \left(\frac{1}{r \sin \theta} \frac{\partial \Phi}{\partial \varphi} \right).$$
$$(3.61)$$

Therefore the Laplacian operator can be written as

$$\nabla^2 = \frac{1}{r^2} \frac{\partial}{\partial r} \left(r^2 \frac{\partial}{\partial r} \right) + \frac{1}{r^2 \sin \theta} \frac{\partial}{\partial \theta} \left(\sin \theta \frac{\partial}{\partial \theta} \right) + \frac{1}{r^2 \sin^2 \theta} \frac{\partial^2}{\partial \varphi^2}. \qquad (3.62)$$

Curl. The curl of a vector in the spherical coordinates can be written as

$$\begin{aligned}
\nabla \times \mathbf{V} &= \nabla \times (V_r \mathbf{e}_r + V_\theta \mathbf{e}_\theta + V_\varphi \mathbf{e}_\varphi) \\
&= \nabla V_r \times \mathbf{e}_r + V_r \nabla \times \mathbf{e}_r + \nabla V_\theta \times \mathbf{e}_\theta + V_\theta \nabla \times \mathbf{e}_\theta \\
&\quad + \nabla V_\varphi \times \mathbf{e}_\varphi + V_\varphi \nabla \times \mathbf{e}_\varphi.
\end{aligned} \qquad (3.63)$$

Again we will derive the expression of curl in the spherical coordinates with the derivatives of (3.42)–(3.47).

$$\begin{aligned}
\nabla \times \mathbf{e}_r &= \left(\mathbf{e}_r \frac{\partial}{\partial r} + \mathbf{e}_\theta \frac{1}{r} \frac{\partial}{\partial \theta} + \mathbf{e}_\varphi \frac{1}{r \sin \theta} \frac{\partial}{\partial \varphi} \right) \times \mathbf{e}_r \\
&= \mathbf{e}_r \times \frac{\partial \mathbf{e}_r}{\partial r} + \frac{1}{r} \mathbf{e}_\theta \times \frac{\partial \mathbf{e}_r}{\partial \theta} + \frac{1}{r \sin \theta} \mathbf{e}_\varphi \times \frac{\partial \mathbf{e}_r}{\partial \varphi} \\
&= \frac{1}{r} \mathbf{e}_\theta \times \mathbf{e}_\theta + \frac{1}{r \sin \theta} \mathbf{e}_\varphi \times \sin \theta \mathbf{e}_\varphi = 0, \qquad (3.64)
\end{aligned}$$

$$\begin{aligned}
\nabla \times \mathbf{e}_\theta &= \left(\mathbf{e}_r \frac{\partial}{\partial r} + \mathbf{e}_\theta \frac{1}{r} \frac{\partial}{\partial \theta} + \mathbf{e}_\varphi \frac{1}{r \sin \theta} \frac{\partial}{\partial \varphi} \right) \times \mathbf{e}_\theta \\
&= \mathbf{e}_r \times \frac{\partial \mathbf{e}_\theta}{\partial r} + \frac{1}{r} \mathbf{e}_\theta \times \frac{\partial \mathbf{e}_\theta}{\partial \theta} + \frac{1}{r \sin \theta} \mathbf{e}_\varphi \times \frac{\partial \mathbf{e}_\theta}{\partial \varphi} \\
&= \frac{1}{r} \mathbf{e}_\theta \times (-\mathbf{e}_r) + \frac{1}{r \sin \theta} \mathbf{e}_\varphi \times \cos \theta \mathbf{e}_\varphi = \frac{1}{r} \mathbf{e}_\varphi, \qquad (3.65)
\end{aligned}$$

$$\begin{aligned}
\nabla \times \mathbf{e}_\varphi &= \left(\mathbf{e}_r \frac{\partial}{\partial r} + \mathbf{e}_\theta \frac{1}{r} \frac{\partial}{\partial \theta} + \mathbf{e}_\varphi \frac{1}{r \sin \theta} \frac{\partial}{\partial \varphi} \right) \times \mathbf{e}_\varphi \\
&= \mathbf{e}_r \times \frac{\partial \mathbf{e}_\varphi}{\partial r} + \frac{1}{r} \mathbf{e}_\theta \times \frac{\partial \mathbf{e}_\varphi}{\partial \theta} + \frac{1}{r \sin \theta} \mathbf{e}_\varphi \times \frac{\partial \mathbf{e}_\varphi}{\partial \varphi} \\
&= \frac{1}{r \sin \theta} \mathbf{e}_\varphi \times (-\sin \theta \mathbf{e}_r + \cos \theta \mathbf{e}_\theta) = -\frac{1}{r} \mathbf{e}_\theta + \frac{1}{r} \frac{\cos \theta}{\sin \theta} \mathbf{e}_r. \quad (3.66)
\end{aligned}$$

Furthermore

$$\nabla V_r \times \mathbf{e}_r = \left(\mathbf{e}_r \frac{\partial V_r}{\partial r} + \mathbf{e}_\theta \frac{1}{r} \frac{\partial V_r}{\partial \theta} + \mathbf{e}_\varphi \frac{1}{r \sin \theta} \frac{\partial V_r}{\partial \varphi} \right) \times \mathbf{e}_r$$

$$= -\frac{1}{r} \frac{\partial V_r}{\partial \theta} \mathbf{e}_\varphi + \frac{1}{r \sin \theta} \frac{\partial V_r}{\partial \varphi} \mathbf{e}_\theta, \tag{3.67}$$

$$\nabla V_\theta \times \mathbf{e}_\theta = \left(\mathbf{e}_r \frac{\partial V_\theta}{\partial r} + \mathbf{e}_\theta \frac{1}{r} \frac{\partial V_\theta}{\partial \theta} + \mathbf{e}_\varphi \frac{1}{r \sin \theta} \frac{\partial V_\theta}{\partial \varphi} \right) \times \mathbf{e}_\theta$$

$$= \frac{\partial V_\theta}{\partial r} \mathbf{e}_\varphi - \frac{1}{r \sin \theta} \frac{\partial V_\theta}{\partial \varphi} \mathbf{e}_r, \tag{3.68}$$

$$\nabla V_\varphi \times \mathbf{e}_\varphi = \left(\mathbf{e}_r \frac{\partial V_\varphi}{\partial r} + \mathbf{e}_\theta \frac{1}{r} \frac{\partial V_\varphi}{\partial \theta} + \mathbf{e}_\varphi \frac{1}{r \sin \theta} \frac{\partial V_\varphi}{\partial \varphi} \right) \times \mathbf{e}_\varphi$$

$$= -\frac{\partial V_\varphi}{\partial r} \mathbf{e}_\theta + \frac{1}{r} \frac{\partial V_\varphi}{\partial \theta} \mathbf{e}_r. \tag{3.69}$$

Combining these six terms, we have

$$\nabla \times \mathbf{V} = \left(\frac{1}{r} \frac{\partial V_\varphi}{\partial \theta} - \frac{1}{r \sin \theta} \frac{\partial V_\theta}{\partial \varphi} + \frac{1}{r} \frac{\cos \theta}{\sin \theta} V_\varphi \right) \mathbf{e}_r$$

$$+ \left(\frac{1}{r \sin \theta} \frac{\partial V_r}{\partial \varphi} - \frac{\partial V_\varphi}{\partial r} - \frac{1}{r} V_\varphi \right) \mathbf{e}_\theta + \left(\frac{\partial V_\theta}{\partial r} - \frac{1}{r} \frac{\partial V_r}{\partial \theta} + \frac{1}{r} V_\theta \right) \mathbf{e}_\varphi. \tag{3.70}$$

3.2.2 Infinitesimal Elements

In spherical coordinates, the infinitesimal displacement vector between a point at (r, θ, φ) and at $(r + dr, \theta + d\theta, \varphi + d\varphi)$ is

$$d\mathbf{r} = \mathbf{e}_r \, dr + \mathbf{e}_\theta r \, d\theta + \mathbf{e}_\varphi r \sin \theta \, d\varphi. \tag{3.71}$$

Note from Fig. 3.4 that only in the $\mathbf{e}_r$ direction, the increment dr is an element of length. Both $d\theta$ and $d\varphi$ are infinitesimal angles. They do not even have the units of length. The element of length in the $\mathbf{e}_\theta$ direction is $r \, d\theta$ and in the $\mathbf{e}_\varphi$ direction is $r \sin \theta \, d\varphi$. Thus, one would expect the gradient in the spherical coordinates to be

$$\nabla = \mathbf{e}_r \frac{\partial}{\partial r} + \mathbf{e}_\theta \frac{1}{r} \frac{\partial}{\partial r} + \mathbf{e}_\varphi \frac{1}{r \sin \theta} \frac{\partial}{\partial \varphi},$$

which is indeed the case as shown in (3.52).

The infinitesimal volume element is the product of the three perpendicular infinitesimal displacements

$$dV = (dr)(r \, d\theta)(r \sin \theta \, d\varphi) = r^2 \sin \theta \, dr \, d\theta \, d\varphi. \tag{3.72}$$

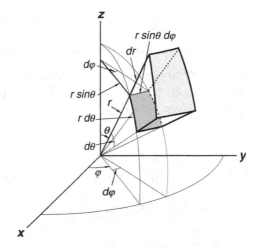

Fig. 3.4. Differential elements in the spherical coordinates. The differential length in the direction of increasing θ is $r\,d\theta$. The differential length in the direction of increasing φ is $r\sin\theta\,d\varphi$. The differential volume element is $r^2\sin\theta\,dr\,d\theta\,d\varphi$

The possible range of r is 0 to ∞, θ from 0 to π, and φ from 0 to 2π. Note that θ goes from 0 to only π, and not 2π. If it goes to 2π, then every point would be counted twice.

Example 3.2.1. Use spherical coordinates to find

$$\boldsymbol{\nabla} r, \quad \boldsymbol{\nabla}\cdot\mathbf{r}, \quad \boldsymbol{\nabla} r^n, \quad \boldsymbol{\nabla}\cdot r^n\mathbf{e}_r, \quad \nabla^2 r^n, \quad \boldsymbol{\nabla}\times f\left(r\right)\mathbf{e}_r.$$

(We have found them in previous chapter with Cartesian coordinates. Using spherical coordinates, the results can be easily obtained, almost by inspection.)

Solution 3.2.1. Since these functions depend only on r, we need to retain only terms involving r variable:

$$\boldsymbol{\nabla} r = \mathbf{e}_r \frac{\partial}{\partial r} r = \mathbf{e}_r = \widehat{\mathbf{r}},$$

$$\boldsymbol{\nabla}\cdot\mathbf{r} = \frac{1}{r^2}\frac{\partial}{\partial r}\left(r^2 r\right) = 3,$$

$$\boldsymbol{\nabla} r^n = \mathbf{e}_r \frac{\partial}{\partial r} r^n = \mathbf{e}_r n r^{n-1},$$

$$\boldsymbol{\nabla}\cdot r^n\mathbf{e}_r = \frac{1}{r^2}\frac{\partial}{\partial r}\left(r^2 r^n\right) = (n+2)r^{n-1},$$

$$\nabla^2 r^n = \frac{1}{r^2}\frac{\partial}{\partial r}\left(r^2\frac{\partial}{\partial r} r^n\right) = n\frac{1}{r^2}\frac{\partial}{\partial r} r^{n+1} = n\left(n+1\right)r^{n-2},$$

$$\nabla \times f\left(r\right) \mathbf{e}_r = \frac{1}{r \sin \theta} \frac{\partial f(r)}{\partial \varphi} \mathbf{e}_\theta + \frac{1}{r} \frac{\partial f(r)}{\partial \theta} \mathbf{e}_\varphi = 0.$$

Example 3.2.2. Express $\mathbf{r} \times \nabla$ in spherical coordinates. (In quantum mechanics, the angular momentum operator $\mathbf{L}$ is defined as $\mathbf{L} = \mathbf{r} \times \mathbf{p}$, where $\mathbf{p}$ is the linear momentum operator, given by $-i\hbar\nabla$.)

Solution 3.2.2.

$$\mathbf{r} \times \nabla = r\mathbf{e}_r \times \left[\mathbf{e}_r \frac{\partial}{\partial r} + \mathbf{e}_\theta \frac{1}{r} \frac{\partial}{\partial \theta} + \mathbf{e}_\varphi \frac{1}{r \sin \theta} \frac{\partial}{\partial \varphi} \right]$$

$$= r \left[\mathbf{e}_\varphi \frac{1}{r} \frac{\partial}{\partial \theta} - \mathbf{e}_\theta \frac{1}{r \sin \theta} \frac{\partial}{\partial \varphi} \right]$$

$$= \mathbf{e}_\varphi \frac{\partial}{\partial \theta} - \mathbf{e}_\theta \frac{1}{\sin \theta} \frac{\partial}{\partial \varphi}.$$

3.3 General Curvilinear Coordinate System

3.3.1 Coordinate Surfaces and Coordinate Curves

In this section, we will develop the general theory of a curvilinear coordinate system. Suppose there is a one to one relationship between the Cartesian coordinate system (x, y, z) and another curvilinear system (u_1, u_2, u_3). This means that (x, y, z) can be written as functions of u_i,

$$x = x\left(u_1, u_2, u_3\right), \quad y = y\left(u_1, u_2, u_3\right), \quad z = z\left(u_1, u_2, u_3\right), \tag{3.73}$$

and conversely,

$$u_1 = u_1\left(x, y, z\right), \quad u_2 = u_2\left(x, y, z\right), \quad u_3 = u_3\left(x, y, z\right). \tag{3.74}$$

The surfaces u_i = constant are referred to as *coordinate surfaces* and the intersections of these surfaces define the *coordinate curves*. For example, if the curvilinear system is the cylindrical coordinate system, then $u_1 = \rho$, $u_2 = \varphi$, $u_3 = z$ as shown in Fig. 3.5. Thus, u_1 = constant is the surface of the cylinder, u_2 = constant is the vertical plane, and u_3 =constant is the horizontal plane shown in the figure. The intersection of the vertical plane and the horizontal plane is the u_1 curve which is the line shown as the ρ curve. The intersection of the horizontal plane and the surface of the cylinder is the u_2 curve which is the circle shown as the φ curve. The intersection of the surface of the cylinder and the vertical plane is the u_1 curve which is the vertical line shown as the z curve.

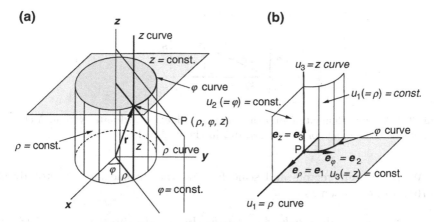

Fig. 3.5. Coordinate surfaces and coordinate curves of cylindrical coordinate system. **(a)** The side surface of the cylinder ($\rho = $ constant), the horizontal plane ($z = $ constant), and the plane containing z axis ($\varphi = $ constant) are the coordinate surfaces. The intersections of them are the coordinate curves. **(b)** Locally, the three unit vectors along the coordinate curves form an orthogonal basis set

Now the position vector $\mathbf{r}$ can be expressed as a function of u_i,

$$\mathbf{r} = \mathbf{r}\,(u_1, u_2, u_3), \tag{3.75}$$

$$d\mathbf{r} = \frac{\partial \mathbf{r}}{\partial u_1} du_1 + \frac{\partial \mathbf{r}}{\partial u_2} du_2 + \frac{\partial \mathbf{r}}{\partial u_3} du_3. \tag{3.76}$$

The partial derivative $\dfrac{\partial \mathbf{r}}{\partial u_1}$ means the rate of variation of $\mathbf{r}$ with u_1, while u_2 and u_3 are held fixed. So, the vector $\dfrac{\partial \mathbf{r}}{\partial u_1}$ lies in the u_2 and u_3 coordinate surfaces and is, therefore, along the u_1 coordinate curve. This enables a unit vector $\mathbf{e}_1$ to be defined in the direction of the u_1 curve,

$$\mathbf{e}_1 = \frac{\partial \mathbf{r}}{\partial u_1}\Big/ h_1 \tag{3.77}$$

where h_1 is the magnitude of $\dfrac{\partial \mathbf{r}}{\partial u_1}$

$$h_1 = \left| \frac{\partial \mathbf{r}}{\partial u_1} \right|, \tag{3.78}$$

known as the scale factor. The unit vectors $\mathbf{e}_2$ and $\mathbf{e}_3$ and the corresponding scale factors h_2 and h_3 are defined in a similar way. In the case of cylindrical coordinates, $\mathbf{e}_1$ is a unit vector along the ρ curve, which is previously defined as $\mathbf{e}_\rho$, $\mathbf{e}_2$ is a unit vector tangent to the φ curve, which is previously defined as $\mathbf{e}_\varphi$, and $\mathbf{e}_3$ is a unit vector along the z curve, which is previously defined as $\mathbf{e}_z = \mathbf{k}$.

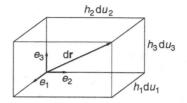

Fig. 3.6. Volume element of an orthogonal curvilinear coordinate system. A change in u_i leads to a change of distance $h_i \, du_i$ in the $\mathbf{e}_i$ direction

With the unit vectors and scale factors, the displacement vector $\mathbf{dr}$ of (3.76) can be written as

$$\mathbf{dr} = \mathbf{e}_1 h_1 \, du_1 + \mathbf{e}_2 h_2 \, du_2 + \mathbf{e}_3 h_3 \, du_3. \qquad (3.79)$$

If the unit vectors are orthogonal, that is

$$\mathbf{e}_i \cdot \mathbf{e}_j = \begin{cases} 1 & i = j \\ 0 & i \neq j \end{cases}, \qquad (3.80)$$

then the coordinate curves are perpendicular to each other where they intersect. Such coordinate systems are known as orthogonal curvilinear coordinates. It will also be assumed that the coordinate system is right handed, so that

$$\mathbf{e}_1 \times \mathbf{e}_2 = \mathbf{e}_3, \quad \mathbf{e}_2 \times \mathbf{e}_3 = \mathbf{e}_1, \quad \mathbf{e}_3 \times \mathbf{e}_1 = \mathbf{e}_2. \qquad (3.81)$$

Thus, locally $\mathbf{e}_1, \mathbf{e}_2, \mathbf{e}_3$ form a set of unit orthogonal basis vectors for the coordinate system (u_1, u_2, u_3), although they may change directions from point to point. In this coordinate system, a change in u_i of size du_i leads to a change of distance $h_i \, du_i$ in the $\mathbf{e}_i$ direction. Schematically this is shown in Fig. 3.6.

It follows from (3.79) and Fig. 3.6 that the arc length ds of a line element along $\mathbf{dr}$ is given by

$$ds = (\mathbf{dr} \cdot \mathbf{dr})^{1/2} = [(h_1 \, du_1)^2 + (h_2 \, du_2)^2 + (h_3 \, du_3)^2]^{1/2}. \qquad (3.82)$$

The directed surface element along $\mathbf{e}_1$ generated by the displacements du_2 and du_3 is

$$\mathbf{e}_1 \, da = \mathbf{e}_2 h_2 \, du_2 \times \mathbf{e}_3 h_3 \, du, \qquad (3.83)$$

and similarly for surface elements $\mathbf{e}_2 \, da$ and $\mathbf{e}_3 \, da$. Finally, the volume elements dV produced by the displacements du_1, du_2, du_3 are given by

$$dV = |\mathbf{e}_1 h_1 \, du_1 \cdot (\mathbf{e}_2 h_2 \, du_2 \times \mathbf{e}_3 h_3 \, du)| = h_1 h_2 h_3 \, du_1 \, du_2 \, du_3, \qquad (3.84)$$

since $\mathbf{e}_1 \cdot (\mathbf{e}_2 \times \mathbf{e}_3) = 1$.

3.3.2 Differential Operations in Curvilinear Coordinate Systems

Gradient. The gradient $\nabla\Phi$ of a scalar function is a vector perpendicular to the surface $\Phi = \text{constant}$, defined by the equation

$$d\Phi = \nabla\Phi \cdot d\mathbf{r} \tag{3.85}$$

To find the expression of $\nabla\Phi$ in a curvilinear coordinate system, let us assume

$$\nabla\Phi = f_1\mathbf{e}_1 + f_2\mathbf{e}_2 + f_3\mathbf{e}_3. \tag{3.86}$$

Since

$$d\mathbf{r} = h_1\,du_1\mathbf{e}_1 + h_2\,du_2\mathbf{e}_2 + h_3\,du_3\mathbf{e}_3,$$

it follows that

$$\nabla\Phi \cdot d\mathbf{r} = f_1 h_1\,du_1 + f_2 h_2\,du_2 + f_3 h_3\,du_3. \tag{3.87}$$

On the other hand

$$d\Phi = \frac{\partial\Phi}{\partial u_1}du_1 + \frac{\partial\Phi}{\partial u_2}du_2 + \frac{\partial\Phi}{\partial u_3}du_3. \tag{3.88}$$

Equating the last two equations, we have

$$f_1 h_1 = \frac{\partial\Phi}{\partial u_1}, \quad f_2 h_2 = \frac{\partial\Phi}{\partial u_2}, \quad f_3 h_3 = \frac{\partial\Phi}{\partial u_3}. \tag{3.89}$$

Thus (3.86) becomes

$$\nabla\Phi = \mathbf{e}_1\frac{1}{h_1}\frac{\partial\Phi}{\partial u_1} + \mathbf{e}_2\frac{1}{h_2}\frac{\partial\Phi}{\partial u_2} + \mathbf{e}_3\frac{1}{h_3}\frac{\partial\Phi}{\partial u_3}. \tag{3.90}$$

Therefore the del operator in curvilinear coordinates can be written as

$$\nabla = \mathbf{e}_1\frac{1}{h_1}\frac{\partial}{\partial u_1} + \mathbf{e}_2\frac{1}{h_2}\frac{\partial}{\partial u_2} + \mathbf{e}_3\frac{1}{h_3}\frac{\partial}{\partial u_3}. \tag{3.91}$$

In particular,

$$\nabla u_1 = \mathbf{e}_1\frac{1}{h_1}\frac{\partial u_1}{\partial u_1} + \mathbf{e}_2\frac{1}{h_2}\frac{\partial u_1}{\partial u_2} + \mathbf{e}_3\frac{1}{h_3}\frac{\partial u_1}{\partial u_3}.$$

Since u_1, u_2, u_3 are independent variables,

$$\frac{\partial u_1}{\partial u_1} = 1, \quad \frac{\partial u_1}{\partial u_2} = 0, \quad \frac{\partial u_1}{\partial u_3} = 0.$$

Therefore

$$\nabla u_1 = \mathbf{e}_1\frac{1}{h_1}. \tag{3.92}$$

Similarly,

$$\nabla u_2 = \mathbf{e}_2\frac{1}{h_2}, \quad \nabla u_3 = \mathbf{e}_3\frac{1}{h_3}. \tag{3.93}$$

Divergence. The expression of the divergence of a vector field $\mathbf{A} = A_1\mathbf{e}_1 + A_2\mathbf{e}_2 + A_3\mathbf{e}_3$ in a curvilinear coordinates can be found by direct calculation using the del operator.

$$\boldsymbol{\nabla} \cdot \mathbf{A} = \boldsymbol{\nabla} \cdot (A_1\mathbf{e}_1 + A_2\mathbf{e}_2 + A_3\mathbf{e}_3),$$

$$\begin{aligned}
\boldsymbol{\nabla} \cdot A_1\mathbf{e}_1 &= \boldsymbol{\nabla} \cdot A_1(\mathbf{e}_2 \times \mathbf{e}_3) = \boldsymbol{\nabla} \cdot A_1 h_2 h_3 (\boldsymbol{\nabla} u_2 \times \boldsymbol{\nabla} u_3) \\
&= (\boldsymbol{\nabla} A_1 h_2 h_3) \cdot (\boldsymbol{\nabla} u_2 \times \boldsymbol{\nabla} u_3) + A_1 h_2 h_3 \boldsymbol{\nabla} \cdot (\boldsymbol{\nabla} u_2 \times \boldsymbol{\nabla} u_3).
\end{aligned}$$

The term $\boldsymbol{\nabla} \cdot (\boldsymbol{\nabla} u_2 \times \boldsymbol{\nabla} u_3) = \boldsymbol{\nabla} \times \boldsymbol{\nabla} u_2 \cdot \boldsymbol{\nabla} u_3 - \boldsymbol{\nabla} \times \boldsymbol{\nabla} u_3 \cdot \boldsymbol{\nabla} u_2$ vanishes because $\boldsymbol{\nabla} \times \boldsymbol{\nabla} f = 0$. Thus,

$$\begin{aligned}
\boldsymbol{\nabla} \cdot A_1\mathbf{e}_1 &= (\boldsymbol{\nabla} A_1 h_2 h_3) \cdot (\boldsymbol{\nabla} u_2 \times \boldsymbol{\nabla} u_3) \\
&= (\boldsymbol{\nabla} A_1 h_2 h_3) \cdot \frac{\mathbf{e}_2 \times \mathbf{e}_3}{h_2 h_3} = (\boldsymbol{\nabla} A_1 h_2 h_3) \cdot \frac{\mathbf{e}_1}{h_2 h_3}.
\end{aligned}$$

Using the del operator of (3.91), we have

$$\boldsymbol{\nabla}(A_1 h_2 h_3) = \mathbf{e}_1 \frac{1}{h_1} \frac{\partial(A_1 h_2 h_3)}{\partial u_1} + \mathbf{e}_2 \frac{1}{h_2} \frac{\partial(A_1 h_2 h_3)}{\partial u_2} + \mathbf{e}_3 \frac{1}{h_3} \frac{\partial(A_1 h_2 h_3)}{\partial u_3},$$

$$(\boldsymbol{\nabla} A_1 h_2 h_3) \cdot \frac{\mathbf{e}_1}{h_2 h_3} = \frac{1}{h_1 h_2 h_3} \frac{\partial(A_1 h_2 h_3)}{\partial u_1}.$$

Therefore,

$$\boldsymbol{\nabla} \cdot A_1\mathbf{e}_1 = \frac{1}{h_1 h_2 h_3} \frac{\partial(A_1 h_2 h_3)}{\partial u_1}. \tag{3.94}$$

With similar expressions

$$\boldsymbol{\nabla} \cdot A_2\mathbf{e}_2 = \frac{1}{h_1 h_2 h_3} \frac{\partial(A_2 h_3 h_1)}{\partial u_2},$$

$$\boldsymbol{\nabla} \cdot A_3\mathbf{e}_3 = \frac{1}{h_1 h_2 h_3} \frac{\partial(A_3 h_1 h_2)}{\partial u_3},$$

we obtain

$$\boldsymbol{\nabla} \cdot \mathbf{A} = \frac{1}{h_1 h_2 h_3} \left[\frac{\partial(A_1 h_2 h_3)}{\partial u_1} + \frac{\partial(A_2 h_3 h_1)}{\partial u_2} + \frac{\partial(A_3 h_1 h_2)}{\partial u_3} \right]. \tag{3.95}$$

Laplacian. The Laplacian follows from its definition

$$\nabla^2 \Phi = \boldsymbol{\nabla} \cdot \boldsymbol{\nabla} \Phi.$$

Since the $\boldsymbol{\nabla} \Phi$ is given by

$$\boldsymbol{\nabla} \Phi = \mathbf{e}_1 \frac{1}{h_1} \frac{\partial \Phi}{\partial u_1} + \mathbf{e}_2 \frac{1}{h_2} \frac{\partial \Phi}{\partial u_2} + \mathbf{e}_3 \frac{1}{h_3} \frac{\partial \Phi}{\partial u_3},$$

the divergence of this vector is

$$\nabla \cdot \nabla \Phi = \frac{1}{h_1 h_2 h_3} \left[\frac{\partial}{\partial u_1} \left(\frac{1}{h_1} \frac{\partial \Phi}{\partial u_1} h_2 h_3 \right) \right.$$
$$\left. + \frac{\partial}{\partial u_2} \left(\frac{1}{h_2} \frac{\partial \Phi}{\partial u_2} h_3 h_1 \right) + \frac{\partial}{\partial u_3} \left(\frac{1}{h_3} \frac{\partial \Phi}{\partial u_3} h_1 h_2 \right) \right].$$

Hence

$$\nabla^2 \Phi = \frac{1}{h_1 h_2 h_3} \left[\frac{\partial}{\partial u_1} \left(\frac{h_2 h_3}{h_1} \frac{\partial \Phi}{\partial u_1} \right) + \frac{\partial}{\partial u_2} \left(\frac{h_3 h_1}{h_2} \frac{\partial \Phi}{\partial u_2} \right) + \frac{\partial}{\partial u_3} \left(\frac{h_1 h_2}{h_3} \frac{\partial \Phi}{\partial u_3} \right) \right].$$
$$(3.96)$$

Curl. The curl of a vector field in a curvilinear coordinates can also be calculated directly,

$$\nabla \times \mathbf{A} = \nabla \times (A_1 \mathbf{e}_1 + A_2 \mathbf{e}_2 + A_3 \mathbf{e}_3),$$

$$\nabla \times A_1 \mathbf{e}_1 = \nabla \times A_1 h_1 \nabla u_1 = \nabla (A_1 h_1) \times \nabla u_1 + A_1 h_1 \nabla \times \nabla u_1.$$

Since $\nabla \times \nabla u_1 = 0$,

$$\nabla \times A_1 \mathbf{e}_1 = \nabla (A_1 h_1) \times \nabla u_1 = \nabla (A_1 h_1) \times \frac{\mathbf{e}_1}{h_1}$$

Now

$$\nabla (A_1 h_1) = \mathbf{e}_1 \frac{1}{h_1} \frac{\partial (A_1 h_1)}{\partial u_1} + \mathbf{e}_2 \frac{1}{h_2} \frac{\partial (A_1 h_1)}{\partial u_2} + \mathbf{e}_3 \frac{1}{h_3} \frac{\partial (A_1 h_1)}{\partial u_3},$$

$$\nabla (A_1 h_1) \times \frac{\mathbf{e}_1}{h_1} = -\mathbf{e}_3 \frac{1}{h_2 h_1} \frac{\partial (A_1 h_1)}{\partial u_2} + \mathbf{e}_2 \frac{1}{h_3 h_1} \frac{\partial (A_1 h_1)}{\partial u_3}.$$

Therefore,

$$\nabla \times A_1 \mathbf{e}_1 = -\mathbf{e}_3 \frac{1}{h_2 h_1} \frac{\partial (A_1 h_1)}{\partial u_2} + \mathbf{e}_2 \frac{1}{h_3 h_1} \frac{\partial (A_1 h_1)}{\partial u_3}.$$

With similar expressions for $\nabla \times A_2 \mathbf{e}_2$ and $\nabla \times A_3 \mathbf{e}_3$, we have

$$\nabla \times \mathbf{A} = \mathbf{e}_1 \left[\frac{1}{h_2 h_3} \frac{\partial (A_3 h_3)}{\partial u_2} - \frac{1}{h_2 h_3} \frac{\partial (A_2 h_2)}{\partial u_3} \right]$$
$$+ \mathbf{e}_2 \left[\frac{1}{h_1 h_3} \frac{\partial (A_1 h_1)}{\partial u_3} - \frac{1}{h_1 h_3} \frac{\partial (A_3 h_3)}{\partial u_1} \right]$$
$$+ \mathbf{e}_3 \left[\frac{1}{h_2 h_1} \frac{\partial (A_2 h_2)}{\partial u_1} - \frac{1}{h_2 h_1} \frac{\partial (A_1 h_1)}{\partial u_2} \right].$$

This expression can be put in a more symmetrical form, which is easier to remember,

$$\nabla \times \mathbf{A} = \frac{h_1 \mathbf{e}_1}{h_1 h_2 h_3} \left[\frac{\partial (A_3 h_3)}{\partial u_2} - \frac{\partial (A_2 h_2)}{\partial u_3} \right] + \frac{h_2 \mathbf{e}_2}{h_1 h_2 h_3} \left[\frac{\partial (A_1 h_1)}{\partial u_3} - \frac{\partial (A_3 h_3)}{\partial u_1} \right]$$

$$+ \frac{h_3 \mathbf{e}_3}{h_1 h_2 h_3} \left[\frac{\partial (A_2 h_2)}{\partial u_1} - \frac{\partial (A_1 h_1)}{\partial u_2} \right]$$

$$= \frac{1}{h_1 h_2 h_3} \begin{vmatrix} h_1 \mathbf{e}_1 & h_2 \mathbf{e}_2 & h_3 \mathbf{e}_3 \\ \dfrac{\partial}{\partial u_1} & \dfrac{\partial}{\partial u_2} & \dfrac{\partial}{\partial u_3} \\ A_1 h_1 & A_2 h_2 & A_3 h_3 \end{vmatrix}. \tag{3.97}$$

Example 3.3.1. For the cylindrical coordinates, $x = \rho \cos \varphi$, $y = \rho \sin \varphi$, $z = z$. With $u_1 = \rho$, $u_2 = \varphi$, $u_3 = z$, (a) find the scale factors h_1, h_2, and h_3, (b) find the gradient, divergence, Laplacian, and curl in the cylindrical coordinates from the general formulas derived in this section.

Solution 3.3.1. (a) Since $\mathbf{r} = x\mathbf{i} + y\mathbf{j} + z\mathbf{k}$ and x, y, z are functions of u_1, u_2, u_3, so

$$\frac{\partial \mathbf{r}}{\partial u_i} = \frac{\partial x}{\partial u_i}\mathbf{i} + \frac{\partial y}{\partial u_i}\mathbf{j} + \frac{\partial z}{\partial u_i}\mathbf{k},$$

$$h_i = \left| \frac{\partial \mathbf{r}}{\partial u_i} \right| = \left| \frac{\partial \mathbf{r}}{\partial u_i} \cdot \frac{\partial \mathbf{r}}{\partial u_i} \right|^{1/2} = \left[\left(\frac{\partial x}{\partial u_i} \right)^2 + \left(\frac{\partial y}{\partial u_i} \right)^2 + \left(\frac{\partial z}{\partial u_i} \right)^2 \right]^{1/2}.$$

Now

$$\frac{\partial x}{\partial u_1} = \frac{\partial x}{\partial \rho} = \cos \varphi, \qquad \frac{\partial y}{\partial u_1} = \frac{\partial y}{\partial \rho} = \sin \varphi, \qquad \frac{\partial z}{\partial u_1} = \frac{\partial z}{\partial \rho} = 0,$$

$$\frac{\partial x}{\partial u_2} = \frac{\partial x}{\partial \varphi} = -\rho \sin \varphi, \qquad \frac{\partial y}{\partial u_2} = \frac{\partial y}{\partial \varphi} = \rho \cos \varphi, \qquad \frac{\partial z}{\partial u_2} = \frac{\partial z}{\partial \varphi} = 0,$$

$$\frac{\partial x}{\partial u_3} = \frac{\partial x}{\partial z} = 0, \qquad \frac{\partial y}{\partial u_3} = \frac{\partial y}{\partial z} = 0, \qquad \frac{\partial z}{\partial u_3} = \frac{\partial z}{\partial z} = 1.$$

$$h_1 = \left(\cos^2 \varphi + \sin^2 \varphi \right)^{1/2} = 1,$$

$$h_2 = \left(\rho^2 \cos^2 \varphi + \rho^2 \sin^2 \varphi \right)^{1/2} = \rho,$$

$$h_3 = (1)^{1/2} = 1.$$

(b)

$$\nabla \Phi = \mathbf{e}_1 \frac{1}{h_1} \frac{\partial \Phi}{\partial u_1} + \mathbf{e}_2 \frac{1}{h_2} \frac{\partial \Phi}{\partial u_2} + \mathbf{e}_3 \frac{1}{h_3} \frac{\partial \Phi}{\partial u_3}$$

$$= \mathbf{e}_\rho \frac{\partial \Phi}{\partial \rho} + \mathbf{e}_\varphi \frac{1}{\rho} \frac{\partial \Phi}{\partial \varphi} + \mathbf{e}_z \frac{\partial \Phi}{\partial z}.$$

$$\boldsymbol{\nabla} \cdot \mathbf{A} = \frac{1}{h_1 h_2 h_3} \left[\frac{\partial (A_1 h_2 h_3)}{\partial u_1} + \frac{\partial (A_2 h_3 h_1)}{\partial u_2} + \frac{\partial (A_3 h_1 h_2)}{\partial u_3} \right]$$

$$= \frac{1}{\rho} \left[\frac{\partial (A_\rho \rho)}{\partial \rho} + \frac{\partial (A_\varphi)}{\partial \varphi} + \frac{\partial (A_z \rho)}{\partial z} \right] = \frac{1}{\rho} \frac{\partial}{\partial \rho} (\rho A_\rho) + \frac{1}{\rho} \frac{\partial A_\varphi}{\partial \varphi} + \frac{\partial A_z}{\partial z}.$$

$$\nabla^2 \Phi = \frac{1}{h_1 h_2 h_3} \left[\frac{\partial}{\partial u_1} \left(\frac{h_2 h_3}{h_1} \frac{\partial \Phi}{\partial u_1} \right) + \frac{\partial}{\partial u_2} \left(\frac{h_3 h_1}{h_2} \frac{\partial \Phi}{\partial u_2} \right) + \frac{\partial}{\partial u_3} \left(\frac{h_1 h_2}{h_3} \frac{\partial \Phi}{\partial u_3} \right) \right]$$

$$= \frac{1}{\rho} \left[\frac{\partial}{\partial \rho} \left(\rho \frac{\partial \Phi}{\partial \rho} \right) + \frac{\partial}{\partial \varphi} \left(\frac{1}{\rho} \frac{\partial \Phi}{\partial \varphi} \right) + \frac{\partial}{\partial z} \left(\rho \frac{\partial \Phi}{\partial z} \right) \right]$$

$$= \frac{\partial^2 \Phi}{\partial \rho^2} + \frac{1}{\rho} \frac{\partial \Phi}{\partial \rho} + \frac{1}{\rho^2} \frac{\partial^2 \Phi}{\partial \varphi^2} + \frac{\partial^2 \Phi}{\partial z^2}.$$

$$\boldsymbol{\nabla} \times \mathbf{A} = \frac{1}{h_1 h_2 h_3} \begin{vmatrix} h_1 \mathbf{e}_1 & h_2 \mathbf{e}_2 & h_3 \mathbf{e}_3 \\ \frac{\partial}{\partial u_1} & \frac{\partial}{\partial u_2} & \frac{\partial}{\partial u_3} \\ A_1 h_1 & A_2 h_2 & A_3 h_3 \end{vmatrix} = \frac{1}{\rho} \begin{vmatrix} \mathbf{e}_\rho & \rho \mathbf{e}_\varphi & \mathbf{e}_z \\ \frac{\partial}{\partial \rho} & \frac{\partial}{\partial \varphi} & \frac{\partial}{\partial z} \\ A_\rho & \rho A_\rho & A_z \end{vmatrix}$$

$$= \left(\frac{1}{\rho} \frac{\partial A_z}{\partial \varphi} - \frac{1}{\rho} \frac{\partial}{\partial z} (\rho A_\varphi) \right) \mathbf{e}_\rho + \left(\frac{\partial A_\rho}{\partial z} - \frac{\partial A_z}{\partial \rho} \right) \mathbf{e}_\varphi$$

$$+ \left(\frac{1}{\rho} \frac{\partial}{\partial \rho} (\rho A_\varphi) - \frac{1}{\rho} \frac{\partial A_\rho}{\partial \varphi} \right) \mathbf{e}_z.$$

Example 3.3.2. For the spherical coordinates, $u_1 = r$, $u_2 = \theta$, $u_3 = \varphi$, and $x = r \sin\theta \cos\varphi$, $\quad y = r \sin\theta \sin\varphi$, $\quad z = r\cos\theta$. (a) Find the scale factors h_1, h_2, and h_3, (b) find the gradient, divergence, Laplacian, and curl in the spherical coordinates from the general formulas derived in this section.

Solution 3.3.2. (a)

$$h_i = \left| \frac{\partial \mathbf{r}}{\partial u_i} \right| = \left| \frac{\partial \mathbf{r}}{\partial u_i} \cdot \frac{\partial \mathbf{r}}{\partial u_i} \right|^{1/2} = \left[\left(\frac{\partial x}{\partial u_i} \right)^2 + \left(\frac{\partial y}{\partial u_i} \right)^2 + \left(\frac{\partial z}{\partial u_i} \right)^2 \right]^{1/2}.$$

Now

$$\frac{\partial x}{\partial u_1} = \frac{\partial x}{\partial r} = \sin\theta \cos\varphi, \qquad \frac{\partial y}{\partial u_1} = \frac{\partial y}{\partial r} = \sin\theta \sin\varphi, \qquad \frac{\partial z}{\partial u_1} = \frac{\partial z}{\partial r} = \cos\theta,$$

$$\frac{\partial x}{\partial u_2} = \frac{\partial x}{\partial \theta} = r\cos\theta \cos\varphi, \qquad \frac{\partial y}{\partial u_2} = \frac{\partial y}{\partial \theta} = r\cos\theta \sin\varphi, \qquad \frac{\partial z}{\partial u_2} = \frac{\partial z}{\partial \theta} = -r\sin\theta,$$

$$\frac{\partial x}{\partial u_3} = \frac{\partial x}{\partial \varphi} = -r\sin\theta \sin\varphi, \qquad \frac{\partial y}{\partial u_3} = \frac{\partial y}{\partial \varphi} = r\sin\theta \cos\varphi, \qquad \frac{\partial z}{\partial u_3} = \frac{\partial z}{\partial \varphi} = 0.$$

$$h_1 = (\sin^2\theta\cos^2\varphi + \sin^2\theta\sin^2\varphi + \cos^2\theta)^{1/2} = 1,$$

$$h_2 = \left(r^2\cos^2\theta\cos^2\varphi + r^2\cos^2\theta\sin^2\varphi + r^2\sin^2\theta\right)^{1/2} = r.$$

$$h_3 = \left(r^2\sin^2\theta\sin^2\varphi + r^2\sin^2\theta\cos^2\varphi\right)^{1/2} = r\sin\theta.$$

(b)

$$\nabla\Phi = \mathbf{e}_1\frac{1}{h_1}\frac{\partial\Phi}{\partial u_1} + \mathbf{e}_2\frac{1}{h_2}\frac{\partial\Phi}{\partial u_2} + \mathbf{e}_3\frac{1}{h_3}\frac{\partial\Phi}{\partial u_3}$$

$$= \mathbf{e}_r\frac{\partial\Phi}{\partial r} + \mathbf{e}_\theta\frac{1}{r}\frac{\partial\Phi}{\partial\theta} + \mathbf{e}_\varphi\frac{1}{r\sin\theta}\frac{\partial\Phi}{\partial\varphi}.$$

$$\nabla\cdot\mathbf{A} = \frac{1}{h_1h_2h_3}\left[\frac{\partial(A_1h_2h_3)}{\partial u_1} + \frac{\partial(A_2h_3h_1)}{\partial u_2} + \frac{\partial(A_3h_1h_2)}{\partial u_3}\right]$$

$$= \frac{1}{r^2\sin\theta}\left[\frac{\partial(A_r r^2\sin\theta)}{\partial r} + \frac{\partial(A_\theta r\sin\theta)}{\partial\theta} + \frac{\partial(A_\varphi r)}{\partial\varphi}\right]$$

$$= \frac{1}{r^2}\frac{\partial}{\partial r}\left(r^2 A_r\right) + \frac{1}{r\sin\theta}\frac{\partial}{\partial\theta}(\sin\theta A_\theta) + \frac{1}{r\sin\theta}\frac{\partial}{\partial\varphi}(A_\varphi).$$

$$\nabla^2\Phi = \frac{1}{h_1h_2h_3}\left[\frac{\partial}{\partial u_1}\left(\frac{h_2h_3}{h_1}\frac{\partial\Phi}{\partial u_1}\right) + \frac{\partial}{\partial u_2}\left(\frac{h_3h_1}{h_2}\frac{\partial\Phi}{\partial u_2}\right) + \frac{\partial}{\partial u_3}\left(\frac{h_1h_2}{h_3}\frac{\partial\Phi}{\partial u_3}\right)\right]$$

$$= \frac{1}{r^2\sin\theta}\left[\frac{\partial}{\partial r}\left(r^2\sin\theta\frac{\partial\Phi}{\partial r}\right) + \frac{\partial}{\partial\theta}\left(\sin\theta\frac{\partial\Phi}{\partial\theta}\right) + \frac{\partial}{\partial\varphi}\left(\frac{1}{\sin\theta}\frac{\partial\Phi}{\partial\varphi}\right)\right]$$

$$= \frac{1}{r^2}\frac{\partial}{\partial r}\left(r^2\frac{\partial}{\partial r}\right) + \frac{1}{r^2\sin\theta}\frac{\partial}{\partial\theta}\left(\sin\theta\frac{\partial}{\partial\theta}\right) + \frac{1}{r^2\sin^2\theta}\frac{\partial^2}{\partial\varphi^2}.$$

$$\nabla\times\mathbf{A} = \frac{1}{h_1h_2h_3}\begin{vmatrix} h_1\mathbf{e}_1 & h_2\mathbf{e}_2 & h_3\mathbf{e}_3 \\ \frac{\partial}{\partial u_1} & \frac{\partial}{\partial u_2} & \frac{\partial}{\partial u_3} \\ A_1h_1 & A_2h_2 & A_3h_3 \end{vmatrix} = \frac{1}{r^2\sin\theta}\begin{vmatrix} \mathbf{e}_r & r\mathbf{e}_\theta & r\sin\theta\mathbf{e}_\varphi \\ \frac{\partial}{\partial r} & \frac{\partial}{\partial\theta} & \frac{\partial}{\partial\varphi} \\ A_r & rA_\theta & r\sin\theta A_\varphi \end{vmatrix}$$

$$= \frac{1}{r^2\sin\theta}\left(\frac{\partial}{\partial\theta}(r\sin\theta A_\varphi) - \frac{\partial}{\partial\varphi}(rA_\theta)\right)\mathbf{e}_r$$

$$+ \frac{1}{r\sin\theta}\left(\frac{\partial}{\partial\varphi}(A_r) - \frac{\partial}{\partial r}(r\sin\theta A_\varphi)\right)\mathbf{e}_\theta + \frac{1}{r}\left(\frac{\partial}{\partial r}(rA_\theta) - \frac{\partial}{\partial\theta}(A_r)\right)\mathbf{e}_\varphi.$$

3.4 Elliptical Coordinates

There are many coordinate systems. In the classical text of Morse and Feshbach, "Methods of Theoretical Physics," no less than 13 coordinate systems are discussed. Each of them is particularly convenient for certain special problems. However, because of the development of high-speed computers, the need for most of them has diminished. In this section, we will introduce only the elliptical coordinate system as one more example of special coordinate systems. The elliptical coordinate system is important in dealing with the two center problems in diatomic molecules.

3.4.1 Coordinate Surfaces

Elliptical coordinates are families of confocal ellipses and hyperbolas in two dimensions. Rotating them around the major axis of the ellipses, surfaces of prolate spheroids and hyperboloids are generated. These surfaces together with the planes containing the major axis form a three dimensional coordinate system which is commonly known as the elliptical coordinates.

The coordinate surfaces are shown in Fig. 3.7a. Let r_1 and r_2 be the distances from the two focal points which are separated by a distance $2c$ on the z axis as shown in Fig. 3.7b. A point in space is determined by r_1 and r_2, the distances from the two focal points, and the angle φ around the z axis. The coordinates of the point are λ, μ and φ with

$$\lambda = \frac{r_2 + r_1}{2c}, \tag{3.98}$$

$$\mu = \frac{r_2 - r_1}{2c}. \tag{3.99}$$

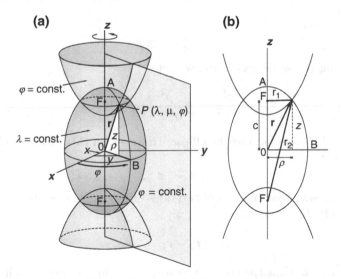

Fig. 3.7. Elliptical Coordinate System. (a) The coordinate surfaces generated by an ellipse, two hyperbolas and a plane containing the major axis of the ellipse. (b) The confocal ellipse given by $r_2 + r_1 =$ constant, and the confocal hyperbolas given by $r_2 - r_1 =$ constant

For λ = constant, (3.98) maps out a prolate spheroid in space, on any φ = constant plane, it is just an ellipse as shown in Fig. 3.7b. This can be seen as follows:

$$r_2 = \left[(z+c)^2 + \rho^2 \right]^{1/2}, \tag{3.100}$$

$$r_1 = \left[(z-c)^2 + \rho^2 \right]^{1/2}, \tag{3.101}$$

$$r_2 + r_1 = 2c\lambda. \tag{3.102}$$

Square both sides of $r_2 = 2c\lambda - r_1$, and collect terms, it becomes

$$z = c\lambda^2 - \lambda r_1.$$

Square both sides again, we find

$$(\lambda^2 - 1)z^2 + \lambda^2 \rho^2 = c^2 \lambda^2 \left(\lambda^2 - 1 \right).$$

This equation can be written in the standard form of an ellipse,

$$\frac{z^2}{c^2 \lambda^2} + \frac{\rho^2}{c^2 \left(\lambda^2 - 1 \right)} = 1, \tag{3.103}$$

which cuts the z axis at $\pm c\lambda$, the ρ axis at $\pm c \left(\lambda^2 - 1 \right)^{1/2}$. The range of λ is clearly $\infty \geq \lambda \geq 1$. When $\lambda = 1$, the ellipse reduces to the line between the two focal points.

Starting with

$$r_2 - r_1 = 2c\mu$$

and following the same procedure, we get

$$\frac{z^2}{c^2 \mu^2} + \frac{\rho^2}{c^2 \left(\mu^2 - 1 \right)} = 1,$$

which is of the same form as the ellipse. However, in this case it is clear from Fig. 3.7b that

$$r_1 + 2c \geq r_2,$$

which simply says that the sum of two sides of a triangle must be greater than the third side. It follows that

$$2c \geq r_2 - r_1 = 2c\mu.$$

Therefore $1 \geq \mu$. Thus the equation is seen to be in the form of a hyperbola:

$$\frac{z^2}{c^2 \mu^2} - \frac{\rho^2}{c^2 \left(1 - \mu^2 \right)} = 1, \tag{3.104}$$

which cuts the z axis at $\pm c\mu$. There are two sheets of hyperbola, one corresponds to positive value of μ and the other, negative μ. Therefore the range of μ is $1 \geq \mu \geq -1$. When $\mu = 0$, the hyperbola reduces to a straight line perpendicular to the z axis through the origin. When $\mu = 1$, it reduces to a line from $z = c$ along the z axis to ∞. When $\mu = -1$, it reduces to a line from $z = -c$ along the z axis to $-\infty$.

Surfaces of hyperboloids are generated by rotating this family of hyperbolas around the z axis. The range of the angle of rotation φ is of course, $0 \leq \varphi \leq 2\pi$.

3.4.2 Relations with Rectangular Coordinates

The transformation between (x, y, z) and (λ, u, φ) can be seen from (3.100) and (3.101):

$$r_2^2 = z^2 + 2zc + c^2 + \rho^2,$$
$$r_1^2 = z^2 - 2zc + c^2 + \rho^2.$$

It follows that

$$r_2^2 - r_1^2 = 4zc.$$

Since

$$r_2^2 - r_1^2 = (r_2 - r_1)(r_2 + r_1) = (2c\mu)(2c\lambda),$$

therefore $4zc = 4c^2\mu\lambda$ which gives

$$z = c\mu\lambda. \tag{3.105}$$

Putting this into (3.103), we have

$$\frac{c^2\mu^2\lambda^2}{c^2\lambda^2} + \frac{\rho^2}{c^2(\lambda^2 - 1)} = 1,$$

which gives

$$\rho^2 = c^2(\lambda^2 - 1)(1 - \mu^2). \tag{3.106}$$

Now, from Fig. 3.7a

$$x = \rho\cos\varphi, \quad y = \rho\sin\varphi. \tag{3.107}$$

Therefore

$$x = c\left[(\lambda^2 - 1)(1 - \mu^2)\right]^{1/2}\cos\varphi,$$
$$y = c\left[(\lambda^2 - 1)(1 - \mu^2)\right]^{1/2}\sin\varphi, \tag{3.108}$$
$$z = c\mu\lambda.$$

From the position vector

$$\mathbf{r} = x(\lambda, \mu, \varphi)\,\mathbf{i} + y(\lambda, \mu, \varphi)\,\mathbf{j} + z(\lambda, \mu, \varphi)\,\mathbf{k}, \tag{3.109}$$

we can find the unit vectors along the λ, μ, φ coordinate curves. The three unit vectors are defined as

$$\mathbf{e}_\lambda = \frac{\partial \mathbf{r}}{\partial \lambda}/h_\lambda, \quad \mathbf{e}_\mu = \frac{\partial \mathbf{r}}{\partial \mu}/h_\mu, \quad \mathbf{e}_\varphi = \frac{\partial \mathbf{r}}{\partial \varphi}/h_\varphi. \tag{3.110}$$

Since

$$\frac{\partial \mathbf{r}}{\partial \lambda} = \frac{\partial x}{\partial \lambda}\mathbf{i} + \frac{\partial y}{\partial \lambda}\mathbf{j} + \frac{\partial x}{\partial \lambda}\mathbf{k} = \frac{c\lambda\left(1 - \mu^2\right)}{\left[\left(\lambda^2 - 1\right)\left(1 - \mu^2\right)\right]^{1/2}} \cos\varphi\mathbf{i}$$

$$+ \frac{c\lambda\left(1 - \mu^2\right)}{\left[\left(\lambda^2 - 1\right)\left(1 - \mu^2\right)\right]^{1/2}} \sin\varphi\mathbf{j} + c\mu\mathbf{k}, \tag{3.111}$$

$$\frac{\partial \mathbf{r}}{\partial \mu} = \frac{\partial x}{\partial \mu}\mathbf{i} + \frac{\partial y}{\partial \mu}\mathbf{j} + \frac{\partial x}{\partial \mu}\mathbf{k} = \frac{-c\mu\left(\lambda^2 - 1\right)}{\left[\left(\lambda^2 - 1\right)\left(1 - \mu^2\right)\right]^{1/2}} \cos\varphi\mathbf{i}$$

$$+ \frac{-c\mu\left(\lambda^2 - 1\right)}{\left[\left(\lambda^2 - 1\right)\left(1 - \mu^2\right)\right]^{1/2}} \sin\varphi\mathbf{j} + c\lambda\mathbf{k}, \tag{3.112}$$

$$\frac{\partial \mathbf{r}}{\partial \varphi} = \frac{\partial x}{\partial \varphi}\mathbf{i} + \frac{\partial y}{\partial \varphi}\mathbf{j} + \frac{\partial x}{\partial \varphi}\mathbf{k} = -c\left[\left(\lambda^2 - 1\right)\left(1 - \mu^2\right)\right]^{1/2} \sin\varphi\mathbf{i}$$

$$+ c\left[\left(\lambda^2 - 1\right)\left(1 - \mu^2\right)\right]^{1/2} \cos\varphi\mathbf{j}, \tag{3.113}$$

the scale factors are seen to be

$$h_\lambda = \left|\frac{\partial \mathbf{r}}{\partial \lambda}\right| = \left[\left(\frac{\partial x}{\partial \lambda}\right)^2 + \left(\frac{\partial y}{\partial \lambda}\right)^2 + \left(\frac{\partial z}{\partial \lambda}\right)^2\right]^{1/2}$$

$$= \left[\frac{c^2\left(\lambda^2 - \mu^2\right)}{\lambda^2 - 1}\right]^{1/2}, \tag{3.114}$$

$$h_\mu = \left[\frac{c^2\left(\lambda^2 - \mu^2\right)}{1 - \mu^2}\right]^{1/2}, \quad h_\varphi = \left[c^2\left(\lambda^2 - 1\right)\left(1 - \mu^2\right)\right]^{1/2}. \tag{3.115}$$

It can be readily verified that

$$\mathbf{e}_\lambda \times \mathbf{e}_\varphi = \mathbf{e}_\mu, \quad \mathbf{e}_\varphi \times \mathbf{e}_\mu = \mathbf{e}_\lambda, \quad \mathbf{e}_\mu \times \mathbf{e}_\lambda = \mathbf{e}_\varphi. \tag{3.116}$$

Therefore $\mathbf{e}_\lambda, \mathbf{e}_\varphi, \mathbf{e}_\mu$ form an orthogonal basis set. Note that in the right-hand convention, the sequence is $(\mathbf{e}_\lambda, \mathbf{e}_\varphi, \mathbf{e}_\mu)$ and not $(\mathbf{e}_\lambda, \mathbf{e}_\mu, \mathbf{e}_\varphi)$.

The volume element in this system is

$$dV = h_\lambda h_\varphi h_\mu \, d\lambda \, d\varphi \, d\mu = c^3\left(\lambda^2 - \mu^2\right) d\lambda \, d\varphi \, d\mu. \tag{3.117}$$

Example 3.4.1. Use the elliptical coordinates to find the volume of the prolate spheroid generated by rotating the ellipse

$$\frac{z^2}{a^2} + \frac{\rho^2}{b^2} = 1$$

around its major axis z.

Solution 3.4.1. In terms of elliptical coordinates, the ellipse is given by

$$\frac{z^2}{c^2\lambda^2} + \frac{\rho^2}{c^2(\lambda^2 - 1)} = 1,$$

where $2c$ is the distance between the two focal points. To find the upper limit of λ, we note $a^2 = c^2\lambda^2$, or

$$\lambda = a/c$$

Furthermore,

$$b^2 = c^2(\lambda^2 - 1) = c^2[(a/c)^2 - 1] = a^2 - c^2.$$

The volume of the prolate spheroid is

$$V = \iiint dV = \int_0^{2\pi} \int_{-1}^{1} \int_1^{a/c} c^3 \left(\lambda^2 - u^2\right) d\lambda \, d\mu \, d\varphi$$

$$= 2\pi c^3 \left[\int_{-1}^{1} d\mu \int_1^{a/c} \lambda^2 \, d\lambda - \int_1^{a/c} d\lambda \int_{-1}^{1} \mu^2 \, d\mu \right],$$

$$\int_{-1}^{1} d\mu \int_1^{a/c} \lambda^2 \, d\lambda = \frac{2}{3}\left[\left(\frac{a}{c}\right)^3 - 1\right], \qquad \int_1^{a/c} d\lambda \int_{-1}^{1} \mu^2 \, d\mu = \frac{2}{3}\left[\frac{a}{c} - 1\right].$$

$$V = \frac{4\pi}{3} c^3 \left[\left(\frac{a}{c}\right)^3 - \frac{a}{c}\right] = \frac{4\pi}{3} a \left(a^2 - c^2\right) = \frac{4\pi}{3} ab^2.$$

Example 3.4.2. Evaluate the following integral over all space

$$I = \iiint e^{-r_1} e^{-r_2} \, dV,$$

where r_1 and r_2 are distances from two fixed points separated by a distance R. (This happens to be the overlap integral of the H_2^+ molecular ion.)

Solution 3.4.2.

$$I = \iiint e^{-r_1} e^{-r_2} \, dV = \iiint e^{-(r_1 + r_2)} \, dV.$$

Using elliptical coordinates

$$r_1 + r_2 = 2c\lambda = R\lambda,$$

$$
\begin{aligned}
I &= \left(\frac{R}{2}\right)^3 \int_0^{2\pi} \int_{-1}^{1} \int_{1}^{\infty} e^{-R\lambda}(\lambda^2 - \mu^2)\, d\lambda\, d\mu\, d\varphi \\
&= \left(\frac{R}{2}\right)^3 2\pi \left[\int_{-1}^{1} d\mu \int_{1}^{\infty} e^{-R\lambda}\lambda^2\, d\lambda - \int_{1}^{\infty} e^{-R\lambda}\, d\lambda \int_{-1}^{1} \mu^2\, d\mu \right] \\
&= \pi(1 + R + \tfrac{1}{3}R^2)e^{-R}.
\end{aligned}
$$

3.4.3 Prolate Spheroidal Coordinates

The transformation (3.108) can be expressed in a more compact form with still another change of variables. Taking advantage of the identities

$$\sin^2\theta = 1 - \cos^2\theta, \qquad \sinh^2\eta = 1 + \cosh^2\eta,$$

we can set

$$\lambda = \cosh\eta, \quad \mu = \cos\theta. \tag{3.118}$$

With this set of variables, the transformation (3.108) becomes

$$
\begin{aligned}
x &= c\sinh\eta \sin\theta \cos\varphi, \\
y &= c\sinh\eta \sin\theta \sin\varphi, \\
z &= c\cosh\eta \cos\theta.
\end{aligned}
\tag{3.119}
$$

The set of coordinates (η, θ, φ) is known as the prolate spheroidal coordinate system. The range of η is $0 \le \eta < \infty$, the range of θ is $0 \le \theta \le \pi$. The scale factors for this system is

$$h_\eta = c\left(\sinh^2\eta + \sin^2\theta\right)^{1/2}, \tag{3.120}$$

$$h_\theta = h_\eta, \quad h_\varphi = c\sinh\eta \sin\theta. \tag{3.121}$$

The volume element in this system is

$$dV = c^3 \left(\sinh^3\eta \sin\theta + \sin^3\theta \sinh\eta\right) d\eta\, d\theta\, d\varphi.$$

Note that $h_\eta h_\theta h_\varphi \ne h_\lambda h_\varphi h_\mu$, since $d\lambda\, d\mu$ is not equal to $d\eta\, d\theta$.

3.5 Multiple Integrals

So far we have seen how to find surface and volume elements for a multiple integral in an orthogonal coordinate system. In this section, we will show that following the same line of reasoning, this method can also be used for any change of variables in multiple integrals, regardless whether the new coordinates are orthogonal or not.

3.5.1 Jacobian for Double Integral

Consider the double integral in the Cartesian coordinates $\iint_S f(x, y)\mathrm{d}a$ where the area element $\mathrm{d}a$ is of course just $\mathrm{d}x\,\mathrm{d}y$. Very often the variables of integration (x, y) are not the most convenient for evaluating the integral. It is desirable to define double integrals in terms of a general pair of curvilinear coordinates.

Let the curvilinear coordinates be (u, v), and there be a one-to-one transformation between (x, y) and (u, v):

$$x = x(u, v), \quad y = y(u, v). \tag{3.122}$$

The position vector from the origin to a point inside S is

$$\mathbf{r} = x(u, v)\,\mathbf{i} + y(u, v)\,\mathbf{j}. \tag{3.123}$$

Therefore, $\mathbf{r}$ can also be considered as a function of the curvilinear coordinates, that is $\mathbf{r} = \mathbf{r}(u, v)$. Thus,

$$\mathrm{d}\mathbf{r} = \frac{\partial \mathbf{r}}{\partial u}\mathrm{d}u + \frac{\partial \mathbf{r}}{\partial v}\mathrm{d}v. \tag{3.124}$$

Now $(\partial \mathbf{r}/\partial u)\,\mathrm{d}u$ is an infinitesimal vector along the line where v in $\mathbf{r}(u, v)$ is kept constant and $(\partial \mathbf{r}/\partial v)\,\mathrm{d}v$ is an infinitesimal vector along the line where u is kept constant. While they may not be orthogonal, the area of the parallelogram formed by these two vectors is still given by their cross product,

$$\mathrm{d}a = \left| \frac{\partial \mathbf{r}}{\partial u}\mathrm{d}u \times \frac{\partial \mathbf{r}}{\partial v}\mathrm{d}v \right| = \left| \frac{\partial \mathbf{r}}{\partial u} \times \frac{\partial \mathbf{r}}{\partial v} \right| \mathrm{d}u\,\mathrm{d}v. \tag{3.125}$$

It follows from (3.123),

$$\frac{\partial \mathbf{r}}{\partial u} = \frac{\partial x}{\partial u}\mathbf{i} + \frac{\partial y}{\partial u}\mathbf{j}, \tag{3.126}$$

$$\frac{\partial \mathbf{r}}{\partial v} = \frac{\partial x}{\partial v}\mathbf{i} + \frac{\partial y}{\partial v}\mathbf{j}. \tag{3.127}$$

Thus the cross product of these two vectors is

$$\frac{\partial \mathbf{r}}{\partial u} \times \frac{\partial \mathbf{r}}{\partial v} = \begin{vmatrix} \mathbf{i} & \mathbf{j} & \mathbf{k} \\ \dfrac{\partial x}{\partial u} & \dfrac{\partial y}{\partial u} & 0 \\ \dfrac{\partial x}{\partial v} & \dfrac{\partial y}{\partial v} & 0 \end{vmatrix} = \mathbf{k} \begin{vmatrix} \dfrac{\partial x}{\partial u} & \dfrac{\partial y}{\partial u} \\ \dfrac{\partial x}{\partial v} & \dfrac{\partial y}{\partial v} \end{vmatrix}. \tag{3.128}$$

The last determinant is called *Jacobian determinant* (or simply as Jacobian) written as $\dfrac{\partial(x, y)}{\partial(u, v)}$,

$$J = \frac{\partial\,(x,y)}{\partial\,(u,v)} = \begin{vmatrix} \dfrac{\partial x}{\partial u} & \dfrac{\partial y}{\partial u} \\[2mm] \dfrac{\partial x}{\partial v} & \dfrac{\partial y}{\partial v} \end{vmatrix}. \tag{3.129}$$

It follows from (3.125) that the area element is equal to the absolute value of the Jacobian times $du\ dv$

$$da = \frac{\partial\,(x,y)}{\partial\,(u,v)} du\ dv. \tag{3.130}$$

Therefore the double integral can be written as

$$\iint_S f(x,y)\mathrm{d}x\ \mathrm{d}y = \iint_S f(x(u,v),y(u,v))\frac{\partial\,(x,y)}{\partial\,(u,v)} du\ dv. \tag{3.131}$$

The integrand on the right-hand side is a function of u and v. Now suppose we want to change it to an integral over x and y, we should have

$$\iint_S f(x(u,v),y(u,v))\frac{\partial\,(x,y)}{\partial\,(u,v)} du\ dv = \iint_S f(x,y)\frac{\partial\,(x,y)}{\partial\,(u,v)}\frac{\partial\,(u,v)}{\partial\,(x,y)}\mathrm{d}x\ \mathrm{d}y.$$

The right-hand side of this equation must be identical to the left-hand side of (3.131). Therefore

$$\frac{\partial\,(x,y)}{\partial\,(u,v)}\frac{\partial\,(u,v)}{\partial\,(x,y)} = 1. \tag{3.132}$$

This is a useful relation. Often we need $\dfrac{\partial\,(x,y)}{\partial\,(u,v)}$, but $\dfrac{\partial\,(u,v)}{\partial\,(x,y)}$ is much easier to calculate. In that case, we simply set

$$\frac{\partial\,(x,y)}{\partial\,(u,v)} = \left[\frac{\partial\,(u,v)}{\partial\,(x,y)}\right]^{-1}.$$

Now we must be careful not to assert that $\mathrm{d}x\ \mathrm{d}y$ is equal to $\dfrac{\partial\,(x,y)}{\partial\,(u,v)} du\ dv$. They are equal only in the sense that under the integral sign the area element $\mathrm{d}x\ \mathrm{d}y$ can be changed to $\dfrac{\partial\,(x,y)}{\partial\,(u,v)} du\ dv$, provided the area S covered by (x,y) is the same as covered by (u,v). Locally they cannot be equal.

From (3.122), we have

$$\mathrm{d}x = \frac{\partial x}{\partial u}du + \frac{\partial x}{\partial v}dv, \tag{3.133}$$

$$\mathrm{d}y = \frac{\partial y}{\partial u}du + \frac{\partial y}{\partial v}dv. \tag{3.134}$$

If we multiply $\mathrm{d}x$ by $\mathrm{d}y$, it is certainly not equal to $\dfrac{\partial\,(x,y)}{\partial\,(u,v)} du\ dv$.

Incidentally, the transformation of the differentials can be written as

$$
\begin{pmatrix} \mathrm{d}x \\ \mathrm{d}y \end{pmatrix} = \begin{pmatrix} \dfrac{\partial x}{\partial u} & \dfrac{\partial x}{\partial v} \\ \dfrac{\partial y}{\partial u} & \dfrac{\partial y}{\partial v} \end{pmatrix} \begin{pmatrix} \mathrm{d}u \\ \mathrm{d}v \end{pmatrix} = (J) \begin{pmatrix} \mathrm{d}u \\ \mathrm{d}v \end{pmatrix}, \tag{3.135}
$$

where (J) is known as *Jacobian matrix*. It is the matrix associated with the Jacobian determinant $\dfrac{\partial\,(x,y)}{\partial\,(u,v)}$. Jacobian determinant and Jacobian matrix are named after the German mathematician Carl Jacobi (1804–1851). Both are very useful, but we must not get confused by the two.

3.5.2 Jacobians for Multiple Integrals

The definition of the triple integral $\iiint_V f\,(x,y,z)\,\mathrm{d}V$ over a given region V is entirely analogous to the definition of a double integral. If x, y, z are rectangular coordinates, then $\mathrm{d}V = \mathrm{d}x\,\mathrm{d}y\,\mathrm{d}z$. Just as in double integral, often the triple integral is much easier to evaluate with a set of curvilinear coordinates u_1, u_2, u_3. Again, let

$$
x = x\,(u_1, u_2, u_3)\,, \quad y = y\,(u_1, u_2, u_3)\,, \quad z = z\,(u_1, u_2, u_3)\,, \tag{3.136}
$$

and the position vector be

$$
\mathbf{r} = x\,(u_1, u_2, u_3)\,\mathbf{i} + y\,(u_1, u_2, u_3)\,\mathbf{j} + z\,(u_1, u_2, u_3)\,\mathbf{k}, \tag{3.137}
$$

then

$$
\mathrm{d}\mathbf{r} = \frac{\partial \mathbf{r}}{\partial u_1}\mathrm{d}u_1 + \frac{\partial \mathbf{r}}{\partial u_2}\mathrm{d}u_2 + \frac{\partial \mathbf{r}}{\partial u_3}\mathrm{d}u_3. \tag{3.138}
$$

The partial derivative $(\partial \mathbf{r}/\partial u_1)$ is the rate of variation of $\mathbf{r}$ with u_2 and u_3 held fixed. Therefore $(\partial \mathbf{r}/\partial u_1)\,\mathrm{d}u_1$ is an infinitesimal vector along the u_1 coordinate curve. Similarly, $(\partial \mathbf{r}/\partial u_2)\,\mathrm{d}u_2$ and $(\partial \mathbf{r}/\partial u_3)\,\mathrm{d}u_3$ are, respectively, infinitesimal vectors along the u_2 and u_3 coordinate curves. Regardless whether they are orthogonal or not, the volume of parallelepiped formed by these three vectors is equal to the scalar triple product of them

$$
\mathrm{d}V = \frac{\partial \mathbf{r}}{\partial u_1}\mathrm{d}u_1 \cdot \left(\frac{\partial \mathbf{r}}{\partial u_2}\mathrm{d}u_2 \times \frac{\partial \mathbf{r}}{\partial u_3}\mathrm{d}u_3 \right). \tag{3.139}
$$

It follows from (3.137) that

$$
\frac{\partial \mathbf{r}}{\partial u_1} = \frac{\partial x}{\partial u_1}\mathbf{i} + \frac{\partial y}{\partial u_1}\mathbf{j} + \frac{\partial z}{\partial u_1}\mathbf{k}. \tag{3.140}
$$

With similar expressions for $\partial \mathbf{r}/\partial u_2$ and $\partial \mathbf{r}/\partial u_3$, the scalar triple product can be written as

$$dV = \begin{vmatrix} \dfrac{\partial x}{\partial u_1} & \dfrac{\partial y}{\partial u_1} & \dfrac{\partial z}{\partial u_1} \\[2mm] \dfrac{\partial x}{\partial u_2} & \dfrac{\partial y}{\partial u_2} & \dfrac{\partial z}{\partial u_2} \\[2mm] \dfrac{\partial x}{\partial u_3} & \dfrac{\partial y}{\partial u_3} & \dfrac{\partial z}{\partial u_3} \end{vmatrix} \, du_1 \, du_2 \, du_3. \tag{3.141}$$

Again the determinant is known as the Jacobian determinant, written as

$$\frac{\partial(x,y,z)}{\partial(u_1,u_2,u_3)} = \begin{vmatrix} \dfrac{\partial x}{\partial u_1} & \dfrac{\partial y}{\partial u_1} & \dfrac{\partial z}{\partial u_1} \\[2mm] \dfrac{\partial x}{\partial u_2} & \dfrac{\partial y}{\partial u_2} & \dfrac{\partial z}{\partial u_2} \\[2mm] \dfrac{\partial x}{\partial u_3} & \dfrac{\partial y}{\partial u_3} & \dfrac{\partial z}{\partial u_3} \end{vmatrix}. \tag{3.142}$$

Thus, if the region covered by x, y, z and by u_1, u_2, u_3 is the same, then

$$\iiint_V f(x,y,z) \, dx \, dy \, dz = \iiint_V F(u_1,u_2,u_3) \frac{\partial(x,y,z)}{\partial(u_1,u_2,u_3)} du_1 \, du_2 \, du_3, \tag{3.143}$$

where $F(u_1,u_2,u_3) = f\left(x(u_1,u_2,u_3), y(u_1,u_2,u_3), z(u_1,u_2,u_3)\right).$

We mention in passing that it can shown by induction that a multiple integral of n variables can be similarly transformed, that is

$$\iint \cdots \int_V f(x_1, x_2, \ldots, x_n) dx_1 dx_2 \cdots dx_n$$
$$= \iint \cdots \int_V F(u_1, u_2, \ldots, u_3) \frac{\partial(x_1, x_2, \ldots, x_n)}{\partial(u_1, u_2, \ldots, u_n)} du_1 du_2 \cdots du_n. \tag{3.144}$$

Example 3.5.1. Evaluate the integral

$$I = \iint x^2 y^2 dx \, dy$$

over the interior of the ellipse

$$\frac{x^2}{a^2} + \frac{y^2}{b^2} = 1.$$

Solution 3.5.1. Parametrically, the coordinates of a point on the ellipse can be written as

$$x = a\cos\theta, \quad y = b\sin\theta,$$

since

$$\frac{x^2}{a^2} + \frac{y^2}{b^2} = \frac{a^2\cos^2\theta}{a^2} + \frac{b^2\sin^2\theta}{b^2} = 1.$$

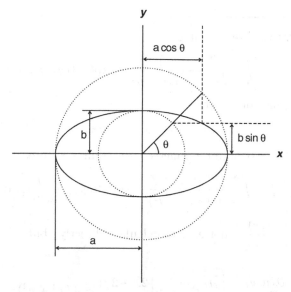

Fig. 3.8. Parametric form of an ellipse. Parametrically an ellipse can be written as $x = a\cos\theta$, $y = b\sin\theta$

This is shown in Fig. 3.8. Any point inside the ellipse can be expressed as

$$x = \gamma a \cos\theta, \quad y = \gamma b \sin\theta,$$

with $\gamma < 1$. Therefore, we can take γ and θ as curvilinear coordinates. (Note that the ellipse of $\gamma =$ constant, and the straight line of $\theta =$ constant are not orthogonal unless $a = b$.) Thus, the integral can be written as

$$I = \iint (\gamma a \cos\theta)^2 (\gamma b \sin\theta)^2 \frac{\partial(x,y)}{\partial(\gamma,\theta)} \mathrm{d}\gamma \, \mathrm{d}\theta$$

where the Jacobian is given by

$$\frac{\partial(x,y)}{\partial(\gamma,\theta)} = \begin{vmatrix} \dfrac{\partial x}{\partial \gamma} & \dfrac{\partial y}{\partial \gamma} \\ \dfrac{\partial x}{\partial \theta} & \dfrac{\partial y}{\partial \theta} \end{vmatrix} = \begin{vmatrix} a\cos\theta & b\sin\theta \\ -\gamma a \sin\theta & \gamma b \cos\theta \end{vmatrix} = \gamma ab.$$

Therefore

$$I = \int_0^1 \int_0^{2\pi} (\gamma a \cos\theta)^2 (\gamma b \sin\theta)^2 \gamma ab \, \mathrm{d}\gamma \, \mathrm{d}\theta$$

$$= a^3 b^3 \int_0^1 \gamma^5 \mathrm{d}\gamma \int_0^{2\pi} \cos^2\theta \sin^2\theta \, \mathrm{d}\theta = a^3 b^3 \left(\frac{1}{6}\right)\frac{\pi}{4} = \frac{\pi}{24} a^3 b^3.$$

Example 3.5.2. Evaluate the integral

$$I = \int_0^\infty \int_0^\infty \frac{x^2 + y^2}{1 + (x^2 - y^2)^2} \exp(-2xy) dx \, dy$$

by making a change of variable

$$u = x^2 - y^2, \quad v = 2xy.$$

Solution 3.5.2. First note the range of u is from $-\infty$ to ∞,

$$I = \int_0^\infty \int_{-\infty}^\infty \frac{x^2 + y^2}{1 + (x^2 - y^2)^2} \exp(-2xy) \frac{\partial (x,y)}{\partial (u,v)} du \, dv.$$

The Jacobian $\dfrac{\partial (x,y)}{\partial (u,v)}$ is not easy to calculate directly, but

$$\frac{\partial (u,v)}{\partial (x,y)} = \begin{vmatrix} \dfrac{\partial u}{\partial x} & \dfrac{\partial u}{\partial y} \\ \dfrac{\partial v}{\partial x} & \dfrac{\partial v}{\partial y} \end{vmatrix} = \begin{vmatrix} 2x & -2y \\ 2y & 2x \end{vmatrix} = 4\left(x^2 + y^2\right).$$

Therefore,

$$\frac{\partial (x,y)}{\partial (u,v)} = \left[\frac{\partial (u,v)}{\partial (x,y)}\right]^{-1} = \frac{1}{4(x^2 + y^2)}.$$

Thus

$$I = \int_0^\infty \int_{-\infty}^\infty \frac{x^2 + y^2}{1 + (x^2 - y^2)^2} \exp(-2xy) \frac{1}{4(x^2 + y^2)} du \, dv$$

$$= \frac{1}{4} \int_0^\infty \int_{-\infty}^\infty \frac{1}{1 + (x^2 - y^2)^2} \exp(-2xy) du \, dv$$

$$= \frac{1}{4} \int_0^\infty \int_{-\infty}^\infty \frac{1}{1 + u^2} \exp(-v) du \, dv = \frac{1}{4} \int_0^\infty \exp(-v) dv \times 2 \int_0^\infty \frac{1}{1 + u^2} du$$

$$= \frac{1}{4} \left[-\exp(-v)\right]_0^\infty 2 \left[\tan^{-1} u\right]_0^\infty = \frac{\pi}{4}.$$

Exercises

1. Express the vector $\mathbf{v} = 2x\mathbf{i} - z\mathbf{j} + y\mathbf{k}$ in cylindrical coordinates.
   Ans. $\mathbf{v} = \left(2\rho \cos^2 \varphi - z \sin \varphi\right) \mathbf{e}_\rho - \left(2\rho \cos \varphi \sin \varphi + z \cos \varphi\right) \mathbf{e}_\varphi + \rho \sin \varphi \mathbf{e}_z$.

2. Find the curl of $\mathbf{A}$ where $\mathbf{A} = \mathbf{e}_z \ln (1/\rho)$ in cylindrical coordinates.
   Ans. $\nabla \times \mathbf{A} = \mathbf{e}_\varphi \frac{1}{\rho}$. (The magnetic vector potential of a long wire carrying

a current I in the z direction is $\mathbf{A} = \mathbf{e}_z \frac{\mu I}{2\pi} \ln (1/\rho)$. The magnetic field is given by $\mathbf{B} = \nabla \times \mathbf{A} = \mathbf{e}_\varphi \frac{\mu I}{2\pi} \ln (1/\rho)$.)

3. Show that $\ln \rho$ satisfies the Laplace's equation $(\nabla^2 \ln \rho = 0)$, (a) use cylindrical coordinates, (b) use spherical coordinates $(\rho = r \sin \theta)$, (c) use Cartesian coordinates $\left(\rho = (x^2 + y^2)^{1/2} \right)$.

4. Show that $1/r$ satisfies the Laplace's equation $(\nabla^2(1/r) = 0)$ for $r \neq 0$, (a) use cylindrical coordinates $\left(r = (\rho^2 + z^2)^{1/2} \right)$, (b) use spherical coordinates, (c) use Cartesian coordinates $\left(r = (x^2 + y^2 + z^2)^{1/2} \right)$.

5. (a) Show that in cylindrical coordinates

$$\frac{d\mathbf{r}}{dt} = \mathbf{e}_\rho \frac{d\rho}{dt} + \mathbf{e}_\varphi \rho \frac{d\varphi}{dt} + \mathbf{e}_z \frac{dz}{dt},$$

$$\frac{ds}{dt} = \left[\left(\frac{d\rho}{dt} \right)^2 + \left(\rho \frac{d\varphi}{dt} \right)^2 + \left(\frac{dz}{dt} \right)^2 \right]^{1/2},$$

where ds is the differential arc length.
(b) Find the length of the spiral described parametrically by $\rho = a$, $\varphi = t$, $z = bt$ from $t = 0$ to $t = 5$.
Ans. $5(a^2 + b^2)^{1/2}$.

6. With the vector field $\mathbf{A}$ given by $\mathbf{A} = \rho \mathbf{e}_\rho + \mathbf{e}_z$ in cylindrical coordinates, (a) show that $\nabla \times \mathbf{A} = 0$. (b) Find a scalar potential Φ, such that $\nabla \Phi = \mathbf{A}$. Ans. $\frac{1}{2}\rho^2 + z$

7. Use the infinitesimal volume element ΔV of Fig. 3.2 and the definition of the divergence

$$\nabla \cdot \mathbf{F} = \frac{1}{\Delta V} \oiint_S \mathbf{F} \cdot \mathbf{n} \, da$$

to derive the expression of the divergence in the cylindrical coordinate system.
Hint: Find the surface elements of the six sides of ΔV, then add pairwise the surface integrals of opposite sides. For example,

$$\iint_{\text{left}} \mathbf{F} \cdot \mathbf{n} \, da + \iint_{\text{right}} \mathbf{F} \cdot \mathbf{n} \, da = -F_\rho(\rho, \varphi, z)\rho \, d\varphi \, dz$$

$$+ F_\rho(\rho + d\rho, \varphi, z)(\rho + d\rho)d\varphi \, dz = \frac{\partial}{\partial \rho} (\rho F_\rho) \, d\rho \, d\varphi \, dz.$$

With the othertwo pairs, the result is seen identical to (3.23).

8. A particle is moving in space. Show that the spherical coordinate components of its velocity and acceleration are given by

$$v_r = \dot{r}, \quad v_\theta = r\dot{\theta}, \quad v_\varphi = r\sin\theta\dot{\varphi},$$

$$a_r = \ddot{r} - r\dot{\theta}^2 - r\sin^2\theta\dot{\varphi}^2,$$

$$a_\theta = r\ddot{\theta} + 2\dot{r}\dot{\theta} - r\cos\theta\sin\theta\dot{\varphi}^2,$$

$$a_\varphi = r\sin\theta\ddot{\varphi} + 2\dot{r}\sin\theta\dot{\varphi} + 2r\cos\theta\dot{\theta}\dot{\varphi}.$$

9. Starting with the expression of $\nabla\Phi$ in spherical system, express $\mathbf{e}_r, \mathbf{e}_\theta, \mathbf{e}_\varphi$ in terms of $\mathbf{i}, \mathbf{j}, \mathbf{k}$, then equate it with $\nabla\Phi$ in rectangular coordinates. In this way, verify that

$$\frac{\partial\Phi}{\partial x} = \sin\theta\cos\varphi\frac{\partial\Phi}{\partial r} + \cos\theta\cos\varphi\frac{1}{r}\frac{\partial\Phi}{\partial\theta} - \frac{\sin\varphi}{r\sin\theta}\frac{\partial\Phi}{\partial\varphi},$$

$$\frac{\partial\Phi}{\partial y} = \sin\theta\sin\varphi\frac{\partial\Phi}{\partial r} + \cos\theta\sin\varphi\frac{1}{r}\frac{\partial\Phi}{\partial\theta} + \frac{\cos\varphi}{r\sin\theta}\frac{\partial\Phi}{\partial\varphi},$$

$$\frac{\partial\Phi}{\partial z} = \cos\theta\frac{\partial\Phi}{\partial r} - \sin\theta\frac{1}{r}\frac{\partial\Phi}{\partial\theta}.$$

10. Use the infinitesimal volume element ΔV of Fig. 3.4 and the definition of the divergence

$$\nabla\cdot\mathbf{F} = \frac{1}{\Delta V}\oiint_S \mathbf{F}\cdot\mathbf{n}\,da$$

to derive the expression of the divergence in the spherical coordinate system.

11. Find the expression of the Laplacian ∇^2 in spherical coordinates by directly transforming $\nabla^2 = \partial^2/\partial x^2 + \partial^2/\partial y^2 + \partial^2/\partial z^2$ into spherical coordinates using the results of the last problem.

12. Show that the following three forms of $\nabla^2\Phi(r)$ are equivalent:

(a) $\dfrac{1}{r^2}\dfrac{d}{dr}\left[r^2\dfrac{d}{dr}\Phi(r)\right]$, (b) $\dfrac{d^2}{dr^2}\Phi(r) + \dfrac{2}{r}\dfrac{d}{dr}\Phi(r)$, (c) $\dfrac{1}{r}\dfrac{d^2}{dr^2}[r\Phi(r)]$.

13. (a) Show that the vector field

$$\mathbf{F} = \left(A - \frac{B}{r^3}\right)\cos\theta\,\mathbf{e}_r - \left(A + \frac{B}{2r^3}\right)\sin\theta\,\mathbf{e}_\theta$$

is irrotational $(\nabla\times\mathbf{F} = \mathbf{0})$.

(b) Find a scalar potential Φ such that $\nabla\Phi = \mathbf{F}$.
(c) Show that Φ satisfies the Laplace equation $\nabla^2\Phi = 0$.

Ans. $\Phi = \left(Ar + \dfrac{B}{2r^2}\right)\cos\theta$.

14. Use spherical coordinates to evaluate the following integrals over a sphere of radius R centered at the origin,

$$(a) \iiint dV, \quad (b) \iiint x^2 \, dV, \quad (c) \iiint y^2 \, dV, \quad (d) \iiint r^2 \, dV.$$

Ans. $\dfrac{4\pi}{3}R^3, \dfrac{4\pi}{15}R^3, \dfrac{4\pi}{15}R^3, \dfrac{4\pi}{5}R^3.$

15. Let

$$\mathbf{L} = -\,\mathrm{i} \left(\mathbf{e}_\varphi \frac{\partial}{\partial \theta} - \mathbf{e}_\theta \frac{1}{\sin\theta} \frac{\partial}{\partial \varphi} \right),$$

show that
(a) $\mathbf{e}_z \cdot \mathbf{L} = -\mathrm{i}\frac{\partial}{\partial\varphi}$,
(b) $\mathbf{L} \cdot \mathbf{L} = -\left[\frac{1}{\sin\theta} \frac{\partial}{\partial\theta} \left(\sin\theta \frac{\partial}{\partial\theta} \right) + \frac{1}{\sin^2\theta} \frac{\partial^2}{\partial\varphi^2} \right].$
(These are quantum mechanical L_z, L^2 angular momentum operators with $\hbar = 1$.)

16. Find the area of the Earth's surface which lies further north than the 45°N latitude. Assume the Earth is a sphere of radius R.
Ans. $\pi R^2 (2 - \sqrt{2})$, which is only about 15% of the total surface area of the earth $(4\pi R^2)$.

17. Use spherical coordinates to verify the divergence theorem

$$\iiint_V \boldsymbol{\nabla} \cdot \mathbf{F} \, dV = \oiint_S \mathbf{F} \cdot \mathbf{n} \, da$$

with

$$\mathbf{F} = r^2 \cos\theta \mathbf{e}_r + r^2 \cos\varphi \mathbf{e}_\theta - r^2 \cos\theta \sin\varphi \mathbf{e}_\varphi$$

over a sphere of radius R.
Ans. Both sides equal to 0.

18. Use elliptical coordinates to evaluate the following integral over all space

$$I = \iiint \frac{1}{r_2} \exp\left(-2r_1\right) dV,$$

where r_1 and r_2 are the distances from two fixed points which are separated by a distance R. (This integral happens to be the so-called Coulomb integral for the H_2 molecule.)
Ans. $\dfrac{\pi}{R} \left[\dfrac{1}{R} - \exp\left(-2R\right)\left(1 + \dfrac{1}{R}\right) \right].$

Hint: $r_2 = \frac{1}{2}R(\lambda + \mu)$, $r_1 = \frac{1}{2}R(\lambda - \mu)$.

19. Parabolic coordinates (u, v, w) are related to Cartesian coordinates (x, y, z) by the relation $x = 2uv$, $y = u^2 - v^2$, $z = w$. (a) Find the scale factors h_u, h_v, h_w. (b) Show that the (u, v, w) coordinate system is orthogonal. Ans. $2\left(u^2 + v^2\right)^{1/2}$, $2\left(u^2 + v^2\right)^{1/2}$, 1.

20. Show that in terms of prolate spheroidal coordinates, the Laplace equation $\left(\nabla^2 \Phi = 0\right)$ is given by

$$\frac{1}{\left(\sinh^2 \eta + \sin^2 \theta\right)} \left[\frac{\partial^2}{\partial \eta^2} \Phi + \coth \eta \frac{\partial}{\partial \eta} \Phi + \frac{\partial^2}{\partial \theta^2} \Phi + \cot \theta \frac{\partial}{\partial \theta} \Phi\right]$$

$$+ \frac{1}{\sinh^2 \eta \sin^2 \theta} \frac{\partial^2}{\partial \varphi^2} \Phi = 0.$$

21. An orthogonal coordinate system (u_1, u_2, u_3) is related to Cartesian coordinates (x, y, z) by

$$x = x\left(u_1, u_2, u_3\right), \quad y = y\left(u_1, u_2, u_3\right), \quad z = z\left(u_1, u_2, u_3\right).$$

Show that

(a) $\dfrac{\partial \mathbf{r}}{\partial u_1} \cdot \dfrac{\partial \mathbf{r}}{\partial u_2} \times \dfrac{\partial \mathbf{r}}{\partial u_3} = h_1 h_2 h_3$,   (b) $\nabla u_1 \cdot \nabla u_2 \times \nabla u_3 = \dfrac{1}{h_1} \dfrac{1}{h_2} \dfrac{1}{h_3}$,

(c) $\dfrac{\partial \mathbf{r}}{\partial u_1} \cdot \dfrac{\partial \mathbf{r}}{\partial u_2} \times \dfrac{\partial \mathbf{r}}{\partial u_3} = \dfrac{\partial(x, y, z)}{\partial(u_1, u_2, u_3)}$,   (d) $\nabla u_1 \cdot \nabla u_2 \times \nabla u_3 = \dfrac{\partial(u_1, u_2, u_3)}{\partial(x, y, z)}$.

22. Use the transformation $x + y = u$, $x - y = v$ to evaluate the double integral

$$I = \iint (x^2 + y^2) dx\, dy$$

within a square whose vertices are $(0, 0)$, $(1, 1)$, $(2, 0)$, $(1, -1)$. Ans. $8/3$.

Hint: Recall the Jacobian is the absolute value of the determinant

$\begin{vmatrix} \dfrac{\partial x}{\partial u} & \dfrac{\partial y}{\partial u} \\ \dfrac{\partial x}{\partial v} & \dfrac{\partial y}{\partial v} \end{vmatrix}$. Draw the square and show that the four sides of the square

are $v = 0$, $u = 2$, $v = 2$, $u = 0$.

4

Vector Transformation and Cartesian Tensors

The universal validity of physical laws is best expressed in terms of mathematical quantities that are independent of any reference frame. Yet physical problems governed by these laws can be solved, in most cases, only if the relevant quantities are resolved into their components in some coordinate system. For example, if we consider a block sliding on an inclined plane, the motion of the block is of course governed by Newton's second law of dynamics $\mathbf{F} = m\mathbf{a}$, in which no coordinates appear. However, to get the actual values of velocity, acceleration, etc. of the block, we have to set a coordinate system. We will obtain the correct answer no matter how we orient the axes, although some orientations are more convenient than others. It is possible to take the x-axis horizontal or along the incline. The components of $\mathbf{F}$ and $\mathbf{a}$ in the x and y directions are of course different in these two cases, but they will combine to give the same correct results. In other words, if the coordinate system is rotated, the components of a vector will of course change. But they must change in a specific way in order for the vector equation to remain valid. For this reason, the vector field is best defined in terms of the behavior of its components under axes rotation.

When the coordinate system is rotated, the transformation of the components of the position vector $\mathbf{r}$ can be expressed in terms of a rotation matrix. We will use this rotation matrix to define all other vectors. The properties of this rotation matrix will be used to analyze a variety of ways of combining the components of two or more vectors. This approach to vector analysis can be easily generalized to vectors of dimensions higher than three. It also naturally leads to tensor analysis.

Many physical quantities are neither vectors nor scalars. For example, the electric current density $\mathbf{J}$ flowing in a material is linearly related to the electric field $\mathbf{E}$ that drives it. If the material is isotropic, the three components of $\mathbf{J}$ and $\mathbf{E}$ are related by the same constant σ in the Ohm's law

$$J_i = \sigma E_i,$$

where σ is known as the conductivity. However, if the material is anisotropic (nonisotropic), the direction of the current is different from the direction of the field. In that case, the Ohm's law is described by

$$J_i = \sum_{j=1}^{3} \sigma_{ij} E_j,$$

where σ_{ij} is the conductivity tensor. It is a tensor of rank two because it has two subscripts i and j, each of them runs from 1 to 3. All together it has nine components. The defining property of a tensor is that, when the coordinate axes are rotated, its components must change according to certain transformation rules, analogous to the vector transformation. In fact, a vector, with one subscript attached to its components, is a tensor of rank one. A tensor of nth rank has n subscripts. In this chapter, the mathematics of tensors will be restricted to Cartesian coordinate systems, therefore the name Cartesian tensors.

4.1 Transformation Properties of Vectors

4.1.1 Transformation of Position Vector

The coordinate frame we use to describe positions in space is of course entirely arbitrary, but there is a specific transformation rule for converting vector components from one frame to another. For simplicity, we will first consider a simple case. Suppose the rectangular coordinate system is rotated counterclockwise about the z-axis through an angle θ. The point P is at the position (x, y, z) before the rotation. After the rotation, the same position becomes (x', y', z'), as shown in Fig. 4.1. Therefore the position vector $\mathbf{r}$ expressed in the original system is

$$\mathbf{r} = x\mathbf{i} + y\mathbf{j} + z\mathbf{k}, \tag{4.1}$$

and expressed in the rotated system is

$$\mathbf{r} = x'\mathbf{i}' + y'\mathbf{j}' + z'\mathbf{k}', \tag{4.2}$$

where $(\mathbf{i}, \mathbf{j}, \mathbf{k})$ and $(\mathbf{i}', \mathbf{j}', \mathbf{k}')$ are the unit vectors along the three axes of the original and the rotated coordinates, respectively. The relationship between primed and unprimed systems can be easily found, since

$$x' = \mathbf{i}' \cdot \mathbf{r} = \mathbf{i}' \cdot (x\mathbf{i} + y\mathbf{j} + z\mathbf{k}) = (\mathbf{i}' \cdot \mathbf{i})\, x + (\mathbf{i}' \cdot \mathbf{j})\, y + (\mathbf{i}' \cdot \mathbf{k})z$$

$$= x\cos\theta + y\cos\left(\frac{\pi}{2} - \theta\right) = x\cos\theta + y\sin\theta \tag{4.3}$$

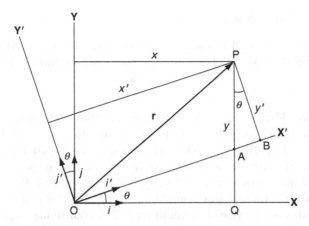

Fig. 4.1. The coordinate system is rotated around z-axis. The primed quantities are those in the rotated system and the unprimed quantities are those in the original system

and

$$y' = \mathbf{j'} \cdot \mathbf{r} = \mathbf{j'} \cdot (x\mathbf{i}+y\mathbf{j}+z\mathbf{k}) = (\mathbf{j'} \cdot \mathbf{i})\,x + (\mathbf{j'} \cdot \mathbf{j})\,y + (\mathbf{i'} \cdot \mathbf{k})z$$
$$= x\cos\left(\frac{\pi}{2}+\theta\right) + y\cos\theta = -x\sin\theta + y\cos\theta. \tag{4.4}$$

Since $\mathbf{k} = \mathbf{k'}$,

$$z' = z.$$

Of course, these relations are also geometrical statements of the rotation. It is seen in Fig. 4.1

$$x' = \text{OA} + \text{AB} = \frac{\text{OQ}}{\cos\theta} + PA\sin\theta = \frac{\text{OQ}}{\cos\theta} + (\text{PQ}-\text{AQ})\sin\theta$$
$$= \frac{x}{\cos\theta} + (y - x\tan\theta)\sin\theta = x\left(\frac{1}{\cos\theta} - \frac{\sin^2\theta}{\cos\theta}\right) + y\sin\theta$$
$$= x\cos\theta + y\sin\theta,$$

$$y' = \text{PA}\cos\theta = (\text{PQ}-\text{AQ})\cos\theta = (y - x\tan\theta)\cos\theta$$
$$= y\cos\theta - x\sin\theta,$$

which are identical to (4.3) and (4.4).

With matrix, these relations can be expressed as

$$\begin{pmatrix} x' \\ y' \\ z' \end{pmatrix} = \begin{pmatrix} \cos\theta & \sin\theta & 0 \\ -\sin\theta & \cos\theta & 0 \\ 0 & 0 & 1 \end{pmatrix} \begin{pmatrix} x \\ y \\ z \end{pmatrix}. \tag{4.5}$$

The 3×3 matrix is known as rotation matrix.

4.1.2 Vector Equations

Vector equations are used to express physical laws which should be independent of the reference frame. For example, Newton's second law of dynamics

$$\mathbf{F} = m\mathbf{a}, \tag{4.6}$$

relates the force $\mathbf{F}$ on the particle of mass m and the acceleration $\mathbf{a}$ of the particle. No coordinates appear explicitly in the equation, as it should be since the law is universal. However, often we find it easier to set up a coordinate system and work with the individual components. In any particular coordinate system, each vector is represented by three components. When we change the refrence frame, these components will change. But they must change in a specific way in order for (4.6) to remain true. The coordinates will be changed by either a translation and/or a rotation of the axes. A translation changes the origin of the coordinate system, resulting in some additive constants in the components of $\mathbf{r}$. Since the derivative of a constant is zero, the translation will not affect vectors $\mathbf{F}$ and $\mathbf{a}$. Therefore the important changes are due to the rotation of the axes.

First we note that if (4.6) holds in one coordinate system it holds in all, for the equation may be written as

$$\mathbf{F} - m\mathbf{a} = \mathbf{0}, \tag{4.7}$$

and under axes rotation, the zero vector will obviously remain zero in the new system. In terms of its components in Cartesian coordinate system, (4.7) can be written as

$$(F_x - ma_x)\mathbf{i} + (F_y - ma_y)\mathbf{j} + (F_z - ma_z)\mathbf{k} = \mathbf{0}, \tag{4.8}$$

which leads to

$$F_x = ma_x, \quad F_y = ma_y, \quad F_z = ma_z. \tag{4.9}$$

Now if the system is rotated counterclockwise about z-axis through an angle θ as indicated in Fig. 4.1, (4.7) becomes

$$(F'_{x'} - ma'_{x'})\mathbf{i}' + (F'_{y'} - ma'_{y'})\mathbf{j}' + (F'_{z'} - ma'_{z'})\mathbf{k}' = \mathbf{0}, \tag{4.10}$$

where by definition

$$a'_{x'} = \frac{d^2}{dt^2}x' = \frac{d^2}{dt^2}(x\cos\theta + y\sin\theta)$$
$$= a_x\cos\theta + a_y\sin\theta, \tag{4.11}$$

$$a'_{y'} = \frac{d^2}{dt^2}y' = \frac{d^2}{dt^2}(-x\sin\theta + y\cos\theta)$$
$$= -a_x\sin\theta + a_y\cos\theta, \tag{4.12}$$

$$a'_{z'} = \frac{d^2}{dt^2}z' = \frac{d^2}{dt^2}z = a_z. \tag{4.13}$$

Each component of (4.10) must be identically equal to zero. This gives

$$F'_{x'} = ma'_{x'} = m(a_x \cos\theta + a_y \sin\theta),$$
$$F'_{y'} = ma'_{y'} = m(-a_x \sin\theta + a_y \cos\theta),$$
$$F'_{z'} = ma'_{z'} = ma_z.$$

Using (4.9), we have

$$F'_{x'} = F_x \cos\theta + F_y \sin\theta,$$

$$F'_{y'} = -F_x \sin\theta + F_y \cos\theta,$$

$$F'_{z'} = F_z.$$

Written in the matrix form, these relations are expressed as

$$\begin{pmatrix} F'_{x'} \\ F'_{y'} \\ F'_{z'} \end{pmatrix} = \begin{pmatrix} \cos\theta & \sin\theta & 0 \\ -\sin\theta & \cos\theta & 0 \\ 0 & 0 & 1 \end{pmatrix} \begin{pmatrix} F_x \\ F_y \\ F_z \end{pmatrix}. \tag{4.14}$$

Comparing (4.5) and (4.14), we see that the rotation matrix is exactly the same. In other words, the components of the vector $\mathbf{F}$ transform in the same way as those of the position vector $\mathbf{r}$.

In physical applications, it means that in order for a quantity to be considered as a vector, the measured values of its components in a rotated system must be related in this way to those in the original system.

The orientation between two coordinate systems is of course not limited to a single rotation around a particular axis. If we know the relative orientation of the systems, we can follow the procedure of (4.3) to establish the relation

$$\begin{pmatrix} x' \\ y' \\ z' \end{pmatrix} = \begin{pmatrix} (\mathbf{i'} \cdot \mathbf{i}) & (\mathbf{i'} \cdot \mathbf{j}) & (\mathbf{i'} \cdot \mathbf{k}) \\ (\mathbf{j'} \cdot \mathbf{i}) & (\mathbf{j'} \cdot \mathbf{j}) & (\mathbf{j'} \cdot \mathbf{k}) \\ (\mathbf{k'} \cdot \mathbf{i}) & (\mathbf{k'} \cdot \mathbf{j}) & (\mathbf{k'} \cdot \mathbf{k}) \end{pmatrix} \begin{pmatrix} x \\ y \\ z \end{pmatrix}.$$

In Sect. 4.1.3 we consider the actual rotation that will bring $(\mathbf{i}, \mathbf{j}, \mathbf{k})$ into $(\mathbf{i'}, \mathbf{j'}, \mathbf{k'})$.

4.1.3 Euler Angles

Often we need to express the transformation matrix in terms of concrete rotations which bring the coordinate axes into a specified orientation. In general, the rotation can be regarded as a combination of three rotations, performed successively, about three different directions in space. The most useful description of this kind is in terms of Euler's angles α, β, γ, which we now define.

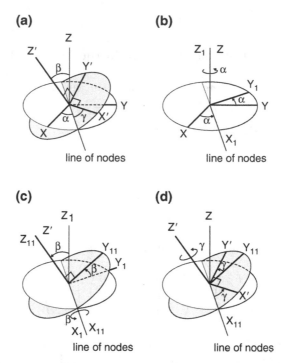

Fig. 4.2. Euler angles. (a) Relative orientation of two rectangular coordinate systems XYZ ang $X'Y'Z'$ with a common origin is specified by three Euler angles α, β, γ. The line of nodes is the intersection of XY and X'Y' planes. The transformation matrix is the product of the matrices representing the following three rotations. (b) First rotate α along the Z axis, bring X axis to coincide with the line of nodes. (c) Rotate β along the line of nodes. (d) Finally rotate γ along Z' axis

The two-coordinate systems are shown in Fig. 4.2a. Let XYZ be the axes of the initial system, $X'Y'Z'$ be the axes of the final system. The intersection of XY plane and X'Y' plane is known as the line of nodes. The relative orientation of the two systems is specified by the three angles α, β, γ. As shown in Fig. 4.2a, α is the angle between X axis and the line of nodes, β is the angle between Z and Z' axes, and γ is the angle between the line of nodes and X' axis.

The transformation matrix from XYZ to X'Y'Z' can be obtained by writing it as the product of the separate rotations, each of which has a relative simple rotation matrix. First rotate the initial axes XYZ, by an angle α counterclockwise about the Z axis, bring the X axis to coincide with the line of nodes. The resultant coordinate system is labeled the X_1, Y_1, Z_1 axes, as shown in Fig. 4.2b. In the second stage the intermediate axes are rotated about X_1 axis counterclockwise by an angle β to produce another intermediate X_{11}, Y_{11}, Z_{11} axes, as shown in Fig. 4.2c. Finally the X_{11}, Y_{11}, Z_{11} axes are rotated counterclockwise by an angle γ about the Z_{11} axis to produce the desired $X'Y'Z'$ axes, as shown in Fig. 4.2d.

After the first rotation, the coordinates (x, y, z) of $\mathbf{r}$ in the initial system becomes (x_1, y_1, z_1) in the $X_1 \, Y_1 \, Z_1$ system. They are related by a rotation matrix,

$$
\begin{pmatrix} x_1 \\ y_1 \\ z_1 \end{pmatrix} = \begin{pmatrix} \cos\alpha & \sin\alpha & 0 \\ -\sin\alpha & \cos\alpha & 0 \\ 0 & 0 & 1 \end{pmatrix} \begin{pmatrix} x \\ y \\ z \end{pmatrix}. \tag{4.15}
$$

The second rotation is about the X_1 axis. After the rotation, (x_1, y_1, z_1) becomes (x_{11}, y_{11}, z_{11}) with the relation

$$
\begin{pmatrix} x_{11} \\ y_{11} \\ z_{11} \end{pmatrix} = \begin{pmatrix} 1 & 0 & 0 \\ 0 & \cos\beta & \sin\beta \\ 0 & -\sin\beta & \cos\beta \end{pmatrix} \begin{pmatrix} x_1 \\ y_1 \\ z_1 \end{pmatrix}. \tag{4.16}
$$

After the final rotation about Z_{11} axis, the coordinates of $\mathbf{r}$ becomes (x', y', z') which is given by

$$
\begin{pmatrix} x' \\ y' \\ z' \end{pmatrix} = \begin{pmatrix} \cos\gamma & \sin\gamma & 0 \\ -\sin\gamma & \cos\gamma & 0 \\ 0 & 0 & 1 \end{pmatrix} \begin{pmatrix} x_{11} \\ y_{11} \\ z_{11} \end{pmatrix}. \tag{4.17}
$$

It is clear from (4.15) to (4.17) that

$$
\begin{pmatrix} x' \\ y' \\ z' \end{pmatrix} = (A) \begin{pmatrix} x \\ y \\ z \end{pmatrix}, \tag{4.18}
$$

where

$$
(A) = \begin{pmatrix} \cos\gamma & \sin\gamma & 0 \\ -\sin\gamma & \cos\gamma & 0 \\ 0 & 0 & 1 \end{pmatrix} \begin{pmatrix} 1 & 0 & 0 \\ 0 & \cos\beta & \sin\beta \\ 0 & -\sin\beta & \cos\beta \end{pmatrix} \begin{pmatrix} \cos\alpha & \sin\alpha & 0 \\ -\sin\alpha & \cos\alpha & 0 \\ 0 & 0 & 1 \end{pmatrix}. \tag{4.19}
$$

Hence the 3×3 matrix (A) is the rotation matrix of the complete transformation. Multiplying the three matrices out, one can readily find the elements of A,

$$
(A) = \begin{pmatrix} \cos\gamma\cos\alpha - \sin\gamma\cos\beta\sin\alpha \\ -\sin\gamma\cos\alpha - \cos\gamma\cos\beta\sin\alpha \\ \sin\beta\sin\alpha \end{pmatrix.
$$

$$
\left.\begin{matrix} \cos\gamma\sin\alpha + \sin\gamma\cos\beta\cos\alpha & \sin\gamma\sin\beta \\ -\sin\gamma\sin\alpha + \cos\gamma\cos\beta\cos\alpha & \cos\gamma\sin\beta \\ -\sin\beta\cos\alpha & \cos\beta \end{matrix}\right) \tag{4.20}
$$

It is not difficult to verify that the product of matrix (A) and its transpose $\left(A^{\mathrm{T}}\right)$ is the identity matrix (I), (The transpose of (A) is the matrix (A) with row and column interchanged.)

$$(A)\left(A^{\mathrm{T}}\right) = (I).$$

Therefore the inverse $(A)^{-1}$ is given by the transpose $(A)^{\mathrm{T}}$,

$$\begin{pmatrix} x \\ y \\ z \end{pmatrix} = \left(A^{\mathrm{T}}\right) \begin{pmatrix} x' \\ y' \\ z' \end{pmatrix}.$$

These are general properties of a rotation matrix which we shall prove in Sect. 4.1.4

It should be noted that different authors define Euler angles in slightly different ways, because the sequence of rotations used to define the final orientation of the coordinate systems is to some extent arbitrary. We have adopted the definition used in most textbooks on classical mechanics.

4.1.4 Properties of Rotation Matrices

To study the general properties of vector space, it is convenient to use a more systematic notation. Let (x, y, z) be (x_1, x_2, x_3); $(\mathbf{i}, \mathbf{j}, \mathbf{k})$ be $(\mathbf{e}_1, \mathbf{e}_2, \mathbf{e}_3)$, and (V_x, V_y, V_z) be (V_1, V_2, V_3). The quantities in the rotated system are similarly labeled as prime quantities. One of the advantages of the new notation is that it permits us to use the summation symbol Σ to write the equations in a more compact form. The orthogonality of $(\mathbf{i}, \mathbf{j}, \mathbf{k})$ is expressed as

$$(\mathbf{e}_i \cdot \mathbf{e}_j) = (\mathbf{e}'_i \cdot \mathbf{e}'_j) = \delta_{ij},$$

where the symbol δ_{ij}, known as the Kronecker delta, is defined as

$$\delta_{ij} = \begin{cases} 1 & i = j \\ 0 & i \neq j \end{cases}.$$

In general, the same position vector $\mathbf{r}$, expressed in two different coordinate systems can be written as

$$\mathbf{r} = \sum_{j=1}^{3} x'_j \mathbf{e}'_j = \sum_{j=1}^{3} x_j \mathbf{e}_j. \tag{4.21}$$

Taking the dot product $\mathbf{e}'_i \cdot \mathbf{r}$, we have

$$\mathbf{e}'_i \cdot \sum_{j=1}^{3} x'_j \mathbf{e}'_j = \sum_{j=1}^{3} (\mathbf{e}'_i \cdot \mathbf{e}'_j) x'_j = \sum_{j=1}^{3} \delta_{ij} x'_j = x'_i. \tag{4.22}$$

The same dot product from (4.21) gives

$$\mathbf{e}'_i \cdot \sum_{j=1}^{3} x_j \mathbf{e}_j = \sum_{j=1}^{3} (\mathbf{e}'_i \cdot \mathbf{e}_j) x_j = \sum_{j=1}^{3} a_{ij} x_j. \qquad (4.23)$$

It follows from (4.22) and (4.23) that

$$x'_i = \sum_{j=1}^{3} (\mathbf{e}'_i \cdot \mathbf{e}_j) x_j = \sum_{j=1}^{3} a_{ij} x_j, \qquad (4.24)$$

where

$$a_{ij} = (\mathbf{e}'_i \cdot \mathbf{e}_j) \qquad (4.25)$$

is the direction cosine between $\mathbf{e}'_i$ and $\mathbf{e}_j$. Note that i in (4.24) remains as a parameter which gives rise to three separate equations when it is set to $1, 2$, and 3. In matrix notation, (4.24) is written as

$$\begin{pmatrix} x'_1 \\ x'_2 \\ x'_3 \end{pmatrix} = \begin{pmatrix} a_{11} & a_{12} & a_{13} \\ a_{21} & a_{22} & a_{23} \\ a_{31} & a_{32} & a_{33} \end{pmatrix} \begin{pmatrix} x_1 \\ x_2 \\ x_3 \end{pmatrix}. \qquad (4.26)$$

If, instead of $\mathbf{e}'_i \cdot \mathbf{r}$, we take $\mathbf{e}_i \cdot \mathbf{r}$ and follow the same procedure, we will obtain

$$x_i = \sum_{j=1}^{3} (\mathbf{e}_i \cdot \mathbf{e}'_j) x'_j.$$

Since $(\mathbf{e}_i \cdot \mathbf{e}'_j)$ is the cosine of the angle between $\mathbf{e}_i$ and $\mathbf{e}'_j$ which can be expressed just as well as $(\mathbf{e}'_j \cdot \mathbf{e}_i)$, and by definition of (4.25) $(\mathbf{e}'_j \cdot \mathbf{e}_i) = a_{ji}$, therefore

$$x_i = \sum_{j=1}^{3} a_{ji} x'_j, \qquad (4.27)$$

or

$$\begin{pmatrix} x_1 \\ x_2 \\ x_3 \end{pmatrix} = \begin{pmatrix} a_{11} & a_{21} & a_{31} \\ a_{12} & a_{22} & a_{32} \\ a_{13} & a_{23} & a_{33} \end{pmatrix} \begin{pmatrix} x'_1 \\ x'_2 \\ x'_3 \end{pmatrix}. \qquad (4.28)$$

Comparing (4.26) and (4.28) we see that the inverse of the rotation matrix is equal to its transpose

$$(a_{ij})^{-1} = (a_{ji}) = (a_{ij})^{\mathrm{T}}. \qquad (4.29)$$

Any transformation that satisfies this condition is known as an orthogonal transformation.

Renaming the indices i and j, we can write (4.27) as

$$x_j = \sum_{i=1}^{3} a_{ij} x'_i. \qquad (4.30)$$

It is thus clear from (4.24) and the last equation that

$$a_{ij} = \frac{\partial x_i'}{\partial x_j} = \frac{\partial x_j}{\partial x_i'}. \tag{4.31}$$

We emphasize this relation is true only in the Cartesian coordinate system.

The nine elements of the rotation matrix are not independent of each other. One way to derive the relationships between them is to note that if the two coordinate systems have the same origin then the length of the position vector should be the same in both systems. This requires

$$\mathbf{r} \cdot \mathbf{r} = \sum_{i=1}^{3} x_i'^2 = \sum_{j=1}^{3} x_j^2. \tag{4.32}$$

Using (4.24), we have

$$\sum_{i=1}^{3} x_i'^2 = \sum_{i=1}^{3} x_i' x_i' = \sum_{i=1}^{3} \left(\sum_{j=1}^{3} a_{ij} x_j \right) \left(\sum_{k=1}^{3} a_{ik} x_k \right)$$

$$= \sum_{j=1}^{3} \sum_{k=1}^{3} x_j x_k \sum_{i=1}^{3} a_{ij} a_{ik} = \sum_{j=1}^{3} x_j^2.$$

This can be true for all points if and only if

$$\sum_{i=1}^{3} a_{ij} a_{ik} = \delta_{jk}. \tag{4.33}$$

This relation is known as the orthogonality condition. Any matrix whose elements satisfy this condition is called an orthogonal matrix. The rotation matrix is an orthogonal matrix. With all possible values of i and j, (4.33) consists of a set of six equations. This set of equations is equivalent to

$$\sum_{i=1}^{3} a_{ji} a_{ki} = \delta_{jk} \tag{4.34}$$

which can be obtained in the same way from (4.32), but starting from right to left with the transformation of (4.27).

Example 4.1.1. Show that the determinant of an orthogonal transformation is equal to either $+1$ or -1.

Solution 4.1.1. Let the matrix of the transformation be (A). Since (A) $(A^{-1}) = (I)$, the determinant of the identity matrix is of course equal to one, $|AA^{-1}| = 1$. For an orthogonal transformation $A^{-1} = A^{\mathrm{T}}$, so $|AA^{\mathrm{T}}| = 1$. Since $|AA^{\mathrm{T}}| = |A||A^{\mathrm{T}}|$ and $|A| = |A^{\mathrm{T}}|$, it follows that $|A|^2 = 1$. Therefore $|A| = \pm 1$.

Example 4.1.2. Show that the determinant of a rotation matrix is equal to +1.

Solution 4.1.2. Express e'_i in terms of $\{e_k\}$: $e'_i = \sum_{i=1}^{3} b_{ik}e_k$.

$$(e'_i \cdot e_j) = \sum_{i=1}^{3} b_{ik}(e_k \cdot e_j) = \sum_{i=1}^{3} b_{ik}\delta_{kj} = b_{ij}.$$

But $(e'_i \cdot e_j) = a_{ij}$, therefore $b_{ij} = a_{ij}$. So

$$e'_1 = a_{11}e_1 + a_{12}e_2 + a_{13}e_3,$$
$$e'_2 = a_{21}e_1 + a_{22}e_2 + a_{23}e_3,$$
$$e'_3 = a_{31}e_1 + a_{32}e_2 + a_{33}e_3.$$

As we have shown in 1.2.7, the scalar triple product is equal to the determinant of the components

$$e'_1 \cdot (e'_2 \times e'_3) = \begin{vmatrix} a_{11} & a_{12} & a_{13} \\ a_{21} & a_{22} & a_{23} \\ a_{31} & a_{32} & a_{33} \end{vmatrix},$$

which is the rotation matrix. On the other hand,

$$e'_1 \cdot (e'_2 \times e'_3) = e'_1 \cdot e'_1 = +1.$$

Therefore

$$\begin{vmatrix} a_{11} & a_{12} & a_{13} \\ a_{21} & a_{22} & a_{23} \\ a_{31} & a_{32} & a_{33} \end{vmatrix} = +1.$$

4.1.5 Definition of a Scalar and a Vector in Terms of Transformation Properties

Now we come to the refined algebraic definition of a scalar and a vector.

Under a rotation of axes, the coordinates of the position vector in the original system x_i transform to x'_i in the rotated system according to

$$x'_i = \sum_j a_{ij}x_j \tag{4.35}$$

with

$$\sum_i a_{ij}a_{ik} = \delta_{jk}. \tag{4.36}$$

If under such a transformation, a quantity φ is unaffected, then φ is called a scalar. This means if φ is a scalar, then

$$\varphi(x_1, x_2, x_3) = \varphi'(x_1', x_2', x_3').$$ (4.37)

Note that after the coordinates are transformed, the functional form may be changed (therefore φ'), but as long as (x_1, x_2, x_3) and (x_1', x_2', x_3') represent the same point, their value is the same.

If a set of quantities (A_1, A_2, A_3) in the original system is transformed into (A_1', A_2', A_3') in the rotated system according to

$$A_i' = \sum_j a_{ij} A_j,$$ (4.38)

then the quantity $\mathbf{A} = (A_1, A_2, A_3)$ is called a vector. Since $(a_{ij})^{-1} = (a_{ji})$, (4.38) is equivalent to

$$A_i = \sum_j a_{ji} A_j'.$$ (4.39)

This definition is capable of generalization and ensures that vector equations are independent of coordinate system.

Example 4.1.3. Suppose $\mathbf{A}$ and $\mathbf{B}$ are vectors. Show that the dot product $\mathbf{A} \cdot \mathbf{B}$ is a scalar.

Solution 4.1.3. Since $\mathbf{A}$ and $\mathbf{B}$ are vectors, under a rotation their components transform according to

$$A_i' = \sum_j a_{ij} A_j; \quad B_i' = \sum_j a_{ij} B_j.$$

To show the dot product

$$\mathbf{A} \cdot \mathbf{B} = \sum_i A_i B_i$$

is a scalar, we must show that its value in the rotated system is the same as its value in the original system.

$$(\mathbf{A} \cdot \mathbf{B})' = \sum_i A_i' B_i' = \sum_i \left(\sum_j a_{ij} A_j \right) \left(\sum_k a_{ik} A_k \right)$$

$$= \sum_j \sum_k \left(\sum_i a_{ij} a_{ik} \right) A_j A_k = \sum_j \sum_k \delta_{jk} A_j B_k = \sum_j A_j B_j = \mathbf{A} \cdot \mathbf{B}.$$

So $\mathbf{A} \cdot \mathbf{B}$ is a scalar.

Example 4.1.4. Show that if (A_1, A_2, A_3) is such that $\sum_i A_i B_i$ is a scalar for every vector **B**, then (A_1, A_2, A_3) is a vector.

Solution 4.1.4. Since $\sum_i A_i B_i$ is a scalar and **B** is a vector,

$$\sum_i A_i B_i = \sum_i A'_i B'_i = \sum_i A'_i \sum_j a_{ij} B_j.$$

Now both i and j are running indices, we can rename i as j, and j as i. So

$$\sum_i A_i B_i = \sum_j A'_j \sum_i a_{ji} B_i = \sum_i \sum_j a_{ji} A'_j B_i.$$

It follows

$$\sum_i \left(A_i - \sum_j a_{ji} A'_j \right) B_i = 0.$$

Since this identity holds for every **B**, we must have

$$A_i = \sum_j a_{ji} A'_j.$$

Therefore A_1, A_2, A_3 are components of a vector.

Example 4.1.5. Show that, in Cartesian coordinates, the gradient of a scalar function $\nabla \varphi$ is a vector function.

Solution 4.1.5. As a scalar it must have the same value at a given point in space, independent of the orientation of the coordinate system

$$\varphi'(x'_1, x'_2, x'_3) = \varphi(x_1, x_2, x_3). \tag{4.40}$$

Differentiating with respect to x'_i and using the chain rule, we have

$$\frac{\partial}{\partial x'_i} \varphi'(x'_1, x'_2, x'_3) = \frac{\partial}{\partial x'_i} \varphi(x_1, x_2, x_3) = \sum_j \frac{\partial \varphi}{\partial x_j} \frac{\partial x_j}{\partial x'_i}. \tag{4.41}$$

It follows from (4.31) that in Cartesian coordinates

$$\frac{\partial x_j}{\partial x'_i} = a_{ij},$$

therefore

$$\frac{\partial \varphi'}{\partial x'_i} = \sum_j a_{ij} \frac{\partial \varphi}{\partial x_j}. \tag{4.42}$$

Now the components of $\boldsymbol{\nabla}\varphi$ are

$$\left(\frac{\partial\varphi}{\partial x_1}, \frac{\partial\varphi}{\partial x_2}, \frac{\partial\varphi}{\partial x_3}\right).$$

They transform under a rotation of coordinates in exactly the same way as the components of a vector, therefore $\boldsymbol{\nabla}\varphi$ is a vector function.

A vector whose components are just numbers is a constant vector. All constant vectors behave like a position vector. When the axes are rotated, the components change into a new set of numbers in accordance with the transformation rule. Therefore, any set of three numbers can be considered as a constant vector.

For vector fields, the components are functions of (x_1, x_2, x_3) themselves. Under a rotation, not only (x_1, x_2, x_3) will change to (x_1', x_2', x_3'), the appearances of the component functions may also change. This leads to some complications.

Mathematically the transformation rules place little restriction on what we can call a vector. We can make any set of three functions the components of a vector field by simply defining, in a rotated system, the corresponding functions obtained from the correct transformation rules as the components of the vector in that system.

However, if we are discussing a physical entity, we are not free to define its components in various systems. They are determined by physical facts. As we stated earlier, all properly formulated physical laws must be independent of the coordinate system. In other words, the appearance of the equation describing physical laws must be the same in all coordinate systems. If vector functions maintain the same appearances in rotated systems, equations written in terms of them will be automatically invariant under rotation. Therefore we include in the definition of a vector field, an additional condition that the transformed components must look the same as the original components.

For example, many authors describe

$$\begin{pmatrix} V_1 \\ V_2 \end{pmatrix} = \begin{pmatrix} x_2 \\ x_1 \end{pmatrix} \tag{4.43}$$

as a vector field in a two-dimensional space (see, for example, E.M. Purcell, "Electricity and Magnetism", McGraw-Hill Book Co. (1965), page 36; D.A. McQuarrie, "Mathematical Methods for Scientists and Engineers," University Science Books, (2003), page 301), still many others would say that (4.43) cannot be called a vector field (see, for example, G. Arfken, "Mathematical Methods for Physicists", Academic Press, (1968), page 8; P.C. Mathews, "Vector Calculus", Springer, Berlin Heidelberg New York (2002), page 118).

If we consider (4.43) as a vector, then the components of this vector in the system where the axes are rotated by an angle θ, are given by

$$\begin{pmatrix} V_1' \\ V_2' \end{pmatrix} = \begin{pmatrix} \cos\theta & \sin\theta \\ -\sin\theta & \cos\theta \end{pmatrix} \begin{pmatrix} V_1 \\ V_2 \end{pmatrix} = \begin{pmatrix} \cos\theta & \sin\theta \\ -\sin\theta & \cos\theta \end{pmatrix} \begin{pmatrix} x_2 \\ x_1 \end{pmatrix}.$$

Furthermore, the coordinates have to change according to

$$\begin{pmatrix} x_1 \\ x_2 \end{pmatrix} = \begin{pmatrix} \cos\theta & -\sin\theta \\ \sin\theta & \cos\theta \end{pmatrix} \begin{pmatrix} x_1' \\ x_2' \end{pmatrix}.$$

One can readily show that

$$\begin{pmatrix} V_1' \\ V_2' \end{pmatrix} = \begin{pmatrix} 2x_1'\sin\theta\cos\theta + x_2'(\cos^2\theta - \sin^2\theta) \\ x_1'(\cos^2\theta - \sin^2\theta) - 2x_2'\sin\theta\cos\theta \end{pmatrix}. \tag{4.44}$$

Mathematically one can certainly define (4.44) as the components of the vector in the rotated system, but they do not look like (4.43).

On the other hand, consider a slightly different expression

$$\begin{pmatrix} V_1 \\ V_2 \end{pmatrix} = \begin{pmatrix} x_2 \\ -x_1 \end{pmatrix}. \tag{4.45}$$

With the same transformation rules, we obtain

$$\begin{pmatrix} V_1' \\ V_2' \end{pmatrix} = \begin{pmatrix} x_2'(\cos^2\theta + \sin^2\theta) \\ -x_1'(\cos^2\theta + \sin^2\theta) \end{pmatrix} = \begin{pmatrix} x_2' \\ -x_1' \end{pmatrix}, \tag{4.46}$$

which has the same form as (4.45). In this sense, we say that (4.45) is invariant under rotations

Under our definition, (4.45) is a vector and (4.43) is not.

4.2 Cartesian Tensors

4.2.1 Definition

The definition of a vector can be extended to define a more general class of objects called tensors, which may have more than one subscript.

If in the rectangular coordinate system of three-dimensional space, under a rotation of coordinates

$$x_i' = \sum_{j=1}^{3} a_{ij} x_j,$$

the 3^N quantities $T_{i_1,i_2,\cdots,i_N}$ (where each of $i_1, i_2, \cdots, i_N$ runs independently from 1 to 3) transform according to the rule

$$T'_{i_1,i_2,\cdots,i_N} = \sum_{j_1=1}^{3} \sum_{j_2=1}^{3} \cdots \sum_{j_N=1}^{3} a_{i_1 j_1} a_{i_2 j_2} \cdots a_{i_N j_N} T_{j_1,j_2,\cdots,j_N}, \tag{4.47}$$

then $T_{i_1,i_2,\cdots,i_N}$ are the components of a Nth rank Cartesian tensor. Since we are going to restrict our discussion to Cartesian tensors, unless explicitly otherwise specified, we shall drop the word Cartesian from here on.

The rank of a tensor is the number of free subscripts. Tensor of zeroth rank has only one ($3^0 = 1$) component. So it can be thought as a scalar. A tensor of first rank has three components ($3^1 = 3$). The rule of transformation of these components under a rotation is the same as the rule for a vector. So a vector is a tensor of rank one.

The most useful other case is the tensor of second rank. It has nine components ($3^2 = 9$), T_{ij} which obey the transformation rule

$$T'_{ij} = \sum_{l=1}^{3} \sum_{m=1}^{3} a_{il} a_{jm} T_{lm}. \tag{4.48}$$

The components of a second rank tensor may be conveniently expressed as a 3×3 matrix:

$$T_{ij} = \begin{pmatrix} T_{11} & T_{12} & T_{13} \\ T_{21} & T_{22} & T_{23} \\ T_{31} & T_{32} & T_{33} \end{pmatrix}.$$

However, this does not mean that any 3×3 matrix forms a tensor. The essential condition is that its components satisfy the transformation rule.

As a matter of terminology, a second-rank tensor in three-dimensional space is a collection of nine components T_{ij}. However, very often T_{ij} is referred to as "tensor" instead of "tensor components" for simplicity. In other words, T_{ij} is used to mean the totality of the components as well as the individual component. The context will make its meaning clear. Another often used symbol for tensors is a double bar over a letter, such as $\overline{\overline{T}}$.

Example 4.2.1. Show that in a two-dimensional space, the following quantity is a second-rank tensor

$$T_{ij} = \begin{pmatrix} x_1 x_2 & -x_1^2 \\ x_2^2 & -x_1 x_2 \end{pmatrix}.$$

Solution 4.2.1. In a two-dimensional space, a second rank tensor has four ($2^2 = 4$) components. If it is a tensor, in a rotated system it must look like

$$T'_{ij} = \begin{pmatrix} x'_1 x'_2 & -x'^2_1 \\ x'^2_2 & -x'_1 x'_2 \end{pmatrix},$$

where

$$\begin{pmatrix} x_1' \\ x_2' \end{pmatrix} = \begin{pmatrix} a_{11} \; a_{12} \\ a_{21} \; a_{22} \end{pmatrix} \begin{pmatrix} x_1 \\ x_2 \end{pmatrix} = \begin{pmatrix} \cos\theta \; \sin\theta \\ -\sin\theta \; \cos\theta \end{pmatrix} \begin{pmatrix} x_1 \\ x_2 \end{pmatrix}.$$

Now we must check if each component satisfies the transformation rule.

$$T_{11}' = x_1' x_2' = (\cos\theta x_1 + \sin\theta x_2)(-\sin\theta x_1 + \cos\theta x_2)$$
$$= -\cos\theta\sin\theta x_1^2 + \cos^2\theta x_1 x_2 - \sin^2\theta x_2 x_1 + \sin\theta\cos\theta x_2^2.$$

This is to be compared with

$$T_{11}' = \sum_{l=1}^{2} \sum_{m=1}^{2} a_{1l} a_{1m} T_{lm}$$
$$= a_{11} a_{11} T_{11} + a_{11} a_{12} T_{12} + a_{12} a_{11} T_{21} + a_{12} a_{12} T_{22}$$
$$= \cos^2\theta x_1 x_2 - \cos\theta\sin\theta x_1^2 + \sin\theta\cos\theta x_2^2 - \sin^2\theta x_1 x_2.$$

It is seen that these two expressions are identical. The same process will show that other components will also satisfy the transformation rule. Therefore T_{ij} is a second rank tensor in the two-dimensional space.

This transformation property is not to be taken for granted. In the above example, if one algebraic sign is changed, the transformation rule will not be satisfied. For example if T_{22} is changed to $x_1 x_2$,

$$T_{ij} = \begin{pmatrix} x_1 x_2 \; -x_1^2 \\ x_2^2 \; x_1 x_2 \end{pmatrix}, \tag{4.49}$$

then

$$T_{11}' \neq \sum_{l=1}^{2} \sum_{m=1}^{2} a_{1l} a_{1m} T_{lm}.$$

Therefore (4.49) is not a tensor.

4.2.2 Kronecker and Levi-Civita Tensors

Kronecker delta tensor. The Kronecker delta which we have already encountered,

$$\delta_{ij} = \begin{cases} 1 & i = j \\ 0 & i \neq j \end{cases},$$

is a second rank tensor. To prove this, consider the transformation

$$\delta_{ij}' = \sum_{l=1}^{3} \sum_{m=1}^{3} a_{il} a_{jm} \delta_{lm} = \sum_{l=1}^{3} a_{il} a_{jl} = \delta_{ij}. \tag{4.50}$$

Therefore

$$\delta'_{ij} = \begin{cases} 1 & i = j \\ 0 & i \neq j \end{cases}.$$

Thus δ_{ij} obeys the tensor transformation rule and is invariant under rotation. In addition, it has a special property. The numerical values of its components are the same in all coordinate systems. A tensor with this property is known as an isotropic tensor.

Since

$$\sum_k D_{ik}\delta_{jk} = D_{ij},$$

the Kronecker delta tensor is also known as the *substitution tensor*. It is also called a unit tensor, because of its matrix representation:

$$\delta_{ij} = \begin{pmatrix} 1 & 0 & 0 \\ 0 & 1 & 0 \\ 0 & 0 & 1 \end{pmatrix}.$$

Levi-Civita Tensor. The Levi-Civita symbol ε_{ijk},

$$\varepsilon_{ijk} = \begin{cases} 1 & \text{if } (i,j,k) \text{ is an even permutation of } (1,2,3) \\ -1 & \text{if } (i,j,k) \text{ is an odd permutation of } (1,2,3) , \\ 0 & \text{if any index is repeated} \end{cases}$$

which we used for the definition of a third-order determinant, is a third rank isotropic tensor. This is also known as the *alternating tensor*. To prove this, recall the definition of the third-order determinant

$$\sum_{l=1}^{3}\sum_{m=1}^{3}\sum_{n=1}^{3} a_{1l}a_{2m}a_{3n}\varepsilon_{lmn} = \begin{vmatrix} a_{11} & a_{12} & a_{13} \\ a_{21} & a_{22} & a_{23} \\ a_{31} & a_{32} & a_{33} \end{vmatrix}. \tag{4.51}$$

Now if the row indices $(1, 2, 3)$ are replaced by (i, j, k), we have

$$\sum_{l=1}^{3}\sum_{m=1}^{3}\sum_{n=1}^{3} a_{il}a_{jm}a_{kn}\varepsilon_{lmn} = \begin{vmatrix} a_{i1} & a_{i2} & a_{i3} \\ a_{j1} & a_{j2} & a_{j3} \\ a_{k1} & a_{k2} & a_{k3} \end{vmatrix}. \tag{4.52}$$

This relation can be demonstrated by writing out the nonvanishing terms of both sides. It can also be proved by noting the following. First for $i = 1$, $j = 2$, $k = 3$, it reduces to (4.51). Now consider the effect of interchanging i and j. The left-hand side changes sign because

$$\sum_{l=1}^{3}\sum_{m=1}^{3}\sum_{n=1}^{3} a_{jl}a_{im}a_{kn}\varepsilon_{lmn} = \sum_{l=1}^{3}\sum_{m=1}^{3}\sum_{n=1}^{3} a_{jm}a_{il}a_{kn}\varepsilon_{m\,ln}$$

$$= -\sum_{l=1}^{3}\sum_{m=1}^{3}\sum_{n=1}^{3} a_{il}a_{jm}a_{kn}\varepsilon_{lmn}.$$

The right-hand side also changes sign because it is an interchange of two rows of the determinant. If two indices of i, j, k are the same, then both sides are equal to zero. The left-hand side is zero, since the quantity is equal to its negative. The right-hand side is equal to zero since two rows of the determinant are identical. This suffices to prove the result since all permutations of i, j, k can be achieved by a sequence of interchanges.

It follows from the properties of determinants and the definition of ε_{ijk} that

$$\begin{vmatrix} a_{i1} & a_{i2} & a_{i3} \\ a_{j1} & a_{j2} & a_{j3} \\ a_{k1} & a_{k2} & a_{k3} \end{vmatrix} = \varepsilon_{ijk} \begin{vmatrix} a_{11} & a_{12} & a_{13} \\ a_{21} & a_{22} & a_{23} \\ a_{31} & a_{32} & a_{33} \end{vmatrix}, \tag{4.53}$$

and (4.52) becomes

$$\sum_{l=1}^{3} \sum_{m=1}^{3} \sum_{n=1}^{3} a_{il} a_{jm} a_{kn} \varepsilon_{lmn} = \varepsilon_{ijk} \begin{vmatrix} a_{11} & a_{12} & a_{13} \\ a_{21} & a_{22} & a_{23} \\ a_{31} & a_{32} & a_{33} \end{vmatrix}. \tag{4.54}$$

This relation is true for any determinant. Now if a_{ij} are elements of a rotation matrix,

$$\begin{vmatrix} a_{11} & a_{12} & a_{13} \\ a_{21} & a_{22} & a_{23} \\ a_{31} & a_{32} & a_{33} \end{vmatrix} = 1,$$

as shown in example 4.1.2.

To decide if ε_{ijk} is a tensor, we should look at its value in a rotated system. The tensor transformation rules require

$$\varepsilon'_{ijk} = \sum_{l=1}^{3} \sum_{m=1}^{3} \sum_{n=1}^{3} a_{il} a_{jm} a_{kn} \varepsilon_{lmn}.$$

But

$$\sum_{l=1}^{3} \sum_{m=1}^{3} \sum_{n=1}^{3} a_{il} a_{jm} a_{kn} \varepsilon_{lmn} = \varepsilon_{ijk} \begin{vmatrix} a_{11} & a_{12} & a_{13} \\ a_{21} & a_{22} & a_{23} \\ a_{31} & a_{32} & a_{33} \end{vmatrix} = \varepsilon_{ijk}.$$

Hence

$$\varepsilon'_{ijk} = \varepsilon_{ijk}. \tag{4.55}$$

Therefore, ε_{ijk} is indeed a third-rank isotropic tensor.

Relation between δ_{ij} and ε_{ijk}. There is an interesting and important relation between Kronecker delta and Levi-Civita tensors

$$\sum_{i=1}^{3} \varepsilon_{ijk} \varepsilon_{ilm} = \delta_{jl} \delta_{km} - \delta_{jm} \delta_{kl}. \tag{4.56}$$

After summed over i, there are four free subscripts j, k, l, m. Therefore (4.56) represents 81 ($3^4 = 81$) equations. Yet it is not difficult to prove (4.56), when we make the following observations.

1. If either $j = k$ or $l = m$, both sides of (4.56) are equal to zero. If $j = k$, the left-hand side is zero, because $\varepsilon_{ikk} = 0$. The right-hand side is also equal to zero because $\delta_{kl}\delta_{km} - \delta_{km}\delta_{kl} = 0$. The same result is obtained for $l = m$. Therefore we only need to check the cases for which $j \neq k$ and $l \neq m$.

2. For the left-hand side not to vanish, i, j, k have to be different. Therefore given $j \neq k$, i is fixed. Consider ε_{ilm}, since i is fixed and i, l, m have to be different for nonvanishing ε_{ilm}, therefore, $l = j$, $m = k$ or $l = k$, $m = j$ are the only two nonvanishing options.

3. For $l = j$ and $m = k$, $\varepsilon_{ijk} = \varepsilon_{ilm}$. Thus ε_{ijk} and ε_{ilm} must have the same sign. (Either both equal to -1, or both equal to $+1$). Therefore on the left-hand side of (4.56), $\varepsilon_{ijk}\varepsilon_{ilm} = +1$. On the right-hand side of (4.56), it is also equal to $+1$, since $l \neq m$,

$$\delta_{jl}\delta_{km} - \delta_{jm}\delta_{kl} = \delta_{ll}\delta_{mm} - \delta_{lm}\delta_{ml} = 1 - 0 = 1.$$

4. For $l = k$ and $m = j$, $\varepsilon_{ijk} = \varepsilon_{iml} = -\varepsilon_{ilm}$. Thus ε_{ijk} and ε_{ilm} have opposite sign. (one equal to -1 and the other $+1$, or vice versa). Therefore on the left-hand side of (4.56), $\varepsilon_{ijk}\varepsilon_{ilm} = -1$. On the right-hand side of (4.56), it is also equal to -1, since $l \neq m$,

$$\delta_{jl}\delta_{km} - \delta_{jm}\delta_{kl} = \delta_{ml}\delta_{lm} - \delta_{mm}\delta_{ll} = 0 - 1 = -1.$$

This covers all 81 cases. In each case the left-hand side is equal to the right-hand side. Therefore (4.56) is established.

4.2.3 Outer Product

If $S_{i_1 i_2 \cdots i_N}$ is a tensor of rank N and $T_{j_1 j_2 \cdots j_M}$ is a tensor of Mth rank, then $S_{i_1 i_2 \cdots i_N} T_{j_1 j_2 \cdots j_M}$ is a tensor of rank $(N + M)$.

This is known as the *outer product* theorem. (Outer product is also known as direct product.) This theorem can be easily demonstrated. First it certainly has 3^{N+M} components. Under a rotation

$$S'_{i_1 i_2 \cdots i_N} = \sum_{k_1 \cdots k_N} a_{i_1 k_1} \cdots a_{i_N k_N} S_{k_1 \cdots k_N},$$

$$T'_{j_1 j_2 \cdots j_M} = \sum_{l_1 \cdots l_N} a_{j_1 l_1} \cdots a_{j_M l_M} T_{l_1 \cdots l_M},$$

where we have written $\displaystyle\sum_{j_1=1}^{3} \sum_{j_2=1}^{3} \cdots \cdot \sum_{j_N=1}^{3}$ as $\displaystyle\sum_{j_1 \cdots j_N}$,

$$(S_{i_1 i_2 \cdots i_N} T_{j_1 j_2 \cdots j_M})' = S'_{i_1 i_2 \cdots i_N} T'_{j_1 j_2 \cdots j_M}$$

$$= \sum_{k_1 \cdots k_N} \sum_{l_1 \cdots l_N} a_{i_1 k_1} \cdots a_{i_N k_N} a_{j_1 l_1} \cdots a_{j_M l_M} S_{k_1 \cdots k_N} T_{l_1 \cdots l_M}, \qquad (4.57)$$

which is how a $(M + N)$th rank tensor should transform.

For example, the outer product of two vectors is a second rank tensor. Let (A_1, A_2, A_3) and (B_1, B_2, B_3) be vectors, so they are first rank tensors. Their outer product $A_i B_j$ is a second rank tensor. Its nine components can be displayed as a matrix

$$A_i B_j = \begin{pmatrix} A_1 B_1 & A_1 B_2 & A_1 B_3 \\ A_2 B_1 & A_2 B_2 & A_2 B_3 \\ A_3 B_1 & A_3 B_2 & A_3 B_3 \end{pmatrix}.$$

Since **A** and **B** are vectors,

$$A_i' = \sum_{k=1}^{3} a_{ik} A_k; \quad B_j' = \sum_{l=1}^{3} a_{jl} B_l,$$

it follows

$$A_i' B_j' = \sum_{k=1}^{3} \sum_{l=1}^{3} a_{ik} a_{jl} A_k B_l,$$

which shows $A_i B_j$ is a second rank tensor, in agreement with the outer product theorem.

We mention in passing, the second rank tensor formed by two vectors **A** and **B** is sometimes denoted as **AB** (without anything between them). When written in this way, it is called a dyad. A linear combination of dyads is a dyadic. Since everything that can be done with vectors and dyadics can also be done by tensors and matrices, but not the other way around, we will not discuss dyadics any further.

Example 4.2.2. Use the outer product theorem to show that the expression in example 4.2.1

$$T_{ij} = \begin{pmatrix} x_1 x_2 & -x_1^2 \\ x_2^2 & -x_1 x_2 \end{pmatrix}$$

is a second rank tensor in a two-dimensional space.

Solution 4.2.2. A two-dimensional position vector is given by

$$\begin{pmatrix} A_1 \\ A_2 \end{pmatrix} = \begin{pmatrix} x_1 \\ x_2 \end{pmatrix}.$$

We have also shown in (4.46) that

$$\begin{pmatrix} B_1 \\ B_2 \end{pmatrix} = \begin{pmatrix} x_2 \\ -x_1 \end{pmatrix}$$

is a vector in the two-dimensional space. The outer product of these two vectors is a second rank tensor

$$A_i B_j = \begin{pmatrix} x_1 x_2 & -x_1^2 \\ x_2^2 & -x_1 x_2 \end{pmatrix} = T_{ij}.$$

4.2.4 Contraction

We can lower the rank of any tensor through the following theorem.

If $T_{i_1 i_2 i_3 \cdots i_N}$ is a tensor of Nth rank, then

$$S_{i_3 \cdots i_N} = \sum_{i_1} \sum_{i_2} \delta_{i_1 i_2} T_{i_1 i_2 i_3 \cdots i_N}$$

is a tensor of rank $(N-2)$.

To prove this theorem, we first note that $S_{i_3 \cdots i_N}$ has 3^{N-2} components. Next we have to show that

$$S'_{i_3 \cdots i_N} = \sum_{i_1} \sum_{i_2} \delta'_{i_1 i_2} T'_{i_1 i_2 i_3 \cdots i_N} \tag{4.58}$$

satisfies the tensor transformation rule.

With

$$\delta'_{i_1 i_2} = \delta_{i_1 i_2},$$

$$T'_{i_1, i_2, \cdots i_N} = \sum_{j_1 \cdots j_N} a_{i_1 j_1} a_{i_2 j_2} \cdots a_{i_N j_N} T_{j_1, j_2, \cdots, j_N},$$

(4.58) becomes

$$S'_{i_3 \cdots i_N} = \sum_{i_1 i_2} \delta_{i_1 i_2} \sum_{j_1 \cdots j_N} a_{i_1 j_1} a_{i_2 j_2} \cdots a_{i_N j_N} T_{j_1, j_2, \cdots, j_N}$$

$$= \sum_{j_1 \cdots j_N} \left(\sum_{i_1 i_2} \delta_{i_1 i_2} a_{i_1 j_1} a_{i_2 j_2} \right) a_{i_3 j_3} \cdots a_{i_N j_N} T_{j_1, j_2, \cdots, j_N}.$$

Now

$$\sum_{i_1 i_2} \delta_{i_1 i_2} a_{i_1 j_1} a_{i_2 j_2} = \sum_{i_1} a_{i_1 j_1} a_{i_1 j_2} = \delta_{j_1 j_2},$$

so

$$S'_{i_3 \cdots i_N} = \sum_{j_1 \cdots j_N} \delta_{j_1 j_2} a_{i_3 j_3} \cdots a_{i_N j_N} T_{j_1, j_2, \cdots, j_N}$$

$$= \sum_{j_3 \cdots j_N} \left(\sum_{j_1 j_2} \delta_{j_1 j_2} T_{j_1, j_2, \cdots, j_N} \right) a_{i_3 j_3} \cdots a_{i_N j_N}$$

$$= \sum_{j_3 \cdots j_N} a_{i_3 j_3} \cdots a_{i_N j_N} S_{j_3 \cdots j_N}. \tag{4.59}$$

Therefore $S_{i_3 \cdots i_N}$ is a $(N-2)$th rank tensor.

The process of multiplying by $\delta_{i_1 i_2}$ and summing over i_1 and i_2 is called *contraction*. For example, we have shown that $A_i B_j$, the outer product of two vectors $\mathbf{A}$ and $\mathbf{B}$, is a second rank tensor. The contraction of this second-rank tensor is a zeroth rank tensor, namely a scalar.

$$\sum_{ij} \delta_{ij} A_i B_j = \sum_i A_i B_i = \mathbf{A} \cdot \mathbf{B}.$$

Indeed, this zeroth rank tensor is the dot product of $\mathbf{A}$ and $\mathbf{B}$.

Simply stated, a new tensor of rank $(N-2)$ will be obtained if two of the indices of a Nth rank tensor are set equal to each other and summed over. (The German word for contraction is *verjüngung*, which can be translated as rejuvenation.) If the rank of the tensor is 3 or higher, we can contract over any two indices. In general we get different $(N-2)$th rank tensors if we contract over different pairs of indices. For example, in the third rank tensor $T_{i_1 i_2 i_3} = A_{i_1} B_{i_2} C_{i_3}$, if i_1 and i_2 are contracted, we obtain a vector

$$\sum_i T_{iii_3} = \sum_i A_i B_i C_{i_3} = (\mathbf{A} \cdot \mathbf{B}) \, C_{i_3},$$

(remember C_{i_3} could represent a particular component of $\mathbf{C}$, it also could represent the totality of the components, namely the vector $\mathbf{C}$ itself). On the other hand, if i_2 and i_3 are contracted, we obtain another vector

$$\sum_i T_{i_1 ii} = \sum_i A_{i_1} B_i C_i = A_{i_1} (\mathbf{B} \cdot \mathbf{C}).$$

So contracting the first two indices we get a scalar times vector $\mathbf{C}$, and contracting the second and third indices we get another scalar times vector $\mathbf{A}$.

Contraction is one of the most important operations in tensor analysis and is worth remembering.

4.2.5 Summation Convention

The summation convention (invented by Einstein) gives tensor analysis much of its appeal. We note that in the definition of tensors (4.47), all indices over which the summation is to be carried out are repeated in the expression. Moreover, the range of the index (such as from 1 to 3) is already known from the context of the discussion. Therefore, without loss of information, we may drop the summation sign with the understanding that the repeated subscripts imply that the term is to be summed over its range. The repeated subscript is referred to as a "dummy" subscript. It must appear no more than twice in a term. The choice of dummy subscript does not matter. Replace one dummy index by another is one of the most useful tricks in tensor analysis one should learn. For example, the dot product $\mathbf{A} \cdot \mathbf{B}$ can be equally represented either by $A_i B_i$ or $A_k B_k$, since both of them mean the same thing, namely

$$A_k B_k = \sum_{i=1}^{3} A_i B_i = \sum_{j=1}^{3} A_j B_j = A_1 B_1 + A_2 B_2 + A_3 B_3 = \mathbf{A} \cdot \mathbf{B}.$$

Example 4.2.3. Express the expression $A_i B_j C_i$ with summation convention in terms of ordinary vector notation.

Solution 4.2.3.

$$A_i B_j C_i = \left(\sum_{i=1}^{3} A_i C_i \right) B_j = (\mathbf{A} \cdot \mathbf{C}) B_j.$$

Note that it is the subscripts that indicate which vector is dotted with which, not the ordering of the components of the vectors. The ordering is immaterial. So $(\mathbf{A} \cdot \mathbf{C}) B_j = A_k C_k B_j$ is equally valid. The letter j is a free subscript, it can be replaced by any letter other than the dummy subscript. However, if the term is used in a equation, then the free subscript of every term in the equation must be represented by the same letter.

From now on when we write down a quantity with N subscripts, if all N subscripts are different, it will be assumed that it is a good Nth rank tensor. If any two of them are the same, it is a contracted tensor of rank $N - 2$.

Example 4.2.4. (a) What is the rank of the tensor $\varepsilon_{ijk} A_l B_m$? (b) what is the rank of the tensor $\varepsilon_{ijk} A_j B_k$? (c) Express $\varepsilon_{ijk} A_j B_k$ in terms of ordinary vector notation.

Solution 4.2.4. (a) Since ε_{ijk} is a third rank tensor and $A_l B_m$ is a tensor of rank 2, so by the outer product theorem $\varepsilon_{ijk} A_l B_m$ is a tensor of rank 5. (b) $\varepsilon_{ijk} A_j B_k$ is twice contracted, so it is a first rank $(5 - 4 = 1)$ tensor. (c)

$$\varepsilon_{ijk} A_j B_k = \sum_j \sum_k \varepsilon_{ijk} A_j B_k.$$

If $i = 1$, the only nonzero terms come from $j = 2$ *or* 3, since ε_{ijk} is equal to zero if any two indices are equal. If $j = 2$, k can only equal to 3. If $j = 3$, k has to be 2. So

$$\varepsilon_{1jk} A_j B_k = \varepsilon_{123} A_2 B_3 + \varepsilon_{132} A_3 B_2 = A_2 B_3 - A_3 B_2 = (\mathbf{A} \times \mathbf{B})_1.$$

Similarly,

$$\varepsilon_{2jk} A_j B_k = (\mathbf{A} \times \mathbf{B})_2, \qquad \varepsilon_{3jk} A_j B_k = (\mathbf{A} \times \mathbf{B})_3.$$

Therefore

$$\varepsilon_{ijk} A_j B_k = (\mathbf{A} \times \mathbf{B})_i.$$

Example 4.2.5. Show that

$$\varepsilon_{ijk} A_i B_j C_k = \begin{vmatrix} A_1 & A_2 & A_3 \\ B_1 & B_2 & B_3 \\ C_1 & C_2 & C_3 \end{vmatrix}.$$

Solution 4.2.5.

$$\varepsilon_{ijk} A_i B_j C_k = (\varepsilon_{ijk} A_i B_j) C_k = (\mathbf{A} \times \mathbf{B})_k C_k = (\mathbf{A} \times \mathbf{B}) \cdot \mathbf{C}$$

$$= \begin{vmatrix} C_1 & C_2 & C_3 \\ A_1 & A_2 & A_3 \\ B_1 & B_2 & B_3 \end{vmatrix} = \begin{vmatrix} A_1 & A_2 & A_3 \\ B_1 & B_2 & B_3 \\ C_1 & C_2 & C_3 \end{vmatrix}.$$

With the summation convention, (4.56) is simply

$$\boxed{\varepsilon_{ijk} \varepsilon_{ilm} = \delta_{jl} \delta_{km} - \delta_{jm} \delta_{kl}.} \tag{4.60}$$

Many vector identities can be quickly and elegantly proved with this equation.

Example 4.2.6. Show that

$$\mathbf{A} \times (\mathbf{B} \times \mathbf{C}) = (\mathbf{A} \cdot \mathbf{C}) \mathbf{B} - \mathbf{C} (\mathbf{A} \cdot \mathbf{B})$$

Solution 4.2.6.

$$(\mathbf{B} \times \mathbf{C})_i = \varepsilon_{ijk} B_j C_k$$

$$[\mathbf{A} \times (\mathbf{B} \times \mathbf{C})]_l = \varepsilon_{lmn} A_m (\mathbf{B} \times \mathbf{C})_n = \varepsilon_{lmn} A_m \varepsilon_{njk} B_j C_k = \varepsilon_{nlm} \varepsilon_{njk} A_m B_j C_k$$

$$\varepsilon_{nlm} \varepsilon_{njk} A_m B_j C_k = (\delta_{lj} \delta_{mk} - \delta_{lk} \delta_{mj}) A_m B_j C_k$$

$$= A_k B_l C_k - A_j B_j C_l = (\mathbf{A} \cdot \mathbf{C}) B_l - (\mathbf{A} \cdot \mathbf{B}) C_l.$$

Since the corresponding components agree, the identity is established.

Example 4.2.7. Show that

$$(\mathbf{A} \times \mathbf{B}) \cdot (\mathbf{C} \times \mathbf{D}) = (\mathbf{A} \cdot \mathbf{C}) (\mathbf{B} \cdot \mathbf{D}) - (\mathbf{A} \cdot \mathbf{D}) (\mathbf{B} \cdot \mathbf{C}).$$

Solution 4.2.7.

$$(\mathbf{A} \times \mathbf{B}) \cdot (\mathbf{C} \times \mathbf{D}) = \varepsilon_{kij} A_i B_j \varepsilon_{klm} C_l D_m$$

$$= (\delta_{il} \delta_{jm} - \delta_{im} \delta_{jl}) A_i B_j C_l D_m = A_l B_m C_l D_m - A_m B_l C_l D_m$$

$$= (\mathbf{A} \cdot \mathbf{C}) (\mathbf{B} \cdot \mathbf{D}) - (\mathbf{A} \cdot \mathbf{D}) (\mathbf{B} \cdot \mathbf{C}).$$

4.2.6 Tensor Fields

A tensor field of Nth rank, $T_{i_1 \cdots i_N} (x_1, x_2, x_3)$ is the totality of 3^N functions which for any given point in space (x_1, x_2, x_3) constitute a tensor of Nth rank.

A scalar field is a tensor field of rank zero. We have shown in example 4.1.5 that the gradient of a scalar field is a vector field. There is a corresponding theorem for tensor fields.

If $T_{i_1 \cdots i_N}(x_1, x_2, x_3)$ is a tensor field of rank N, then

$$\frac{\partial}{\partial x_i} T_{i_1 \cdots i_N}(x_1, x_2, x_3)$$

is a tensor field of rank $N + 1$.

The proof of this theorem goes as follows.

$$\left(\frac{\partial}{\partial x_i} T_{i_1 \cdots i_N}(x_1, x_2, x_3)\right)' = \frac{\partial}{\partial x_i'} T'_{i_1 \cdots i_N}(x_1', x_2', x_3')$$

$$= \frac{\partial}{\partial x_i'} \sum_{j_1 \cdots j_N} a_{i_1 j_1} \cdots a_{i_N j_N} T_{j_1 \cdots j_N}(x_1, x_2, x_3).$$

By chain rule and (4.31)

$$\frac{\partial}{\partial x_i'} = \sum_j \frac{\partial x_j}{\partial x_i'} \frac{\partial}{\partial x_j} = \sum_j a_{ij} \frac{\partial}{\partial x_j},$$

so

$$\left(\frac{\partial}{\partial x_i} T_{i_1 \cdots i_N}(x_1, x_2, x_3)\right)' = \sum_j \sum_{j_1 \cdots j_N} a_{ij} a_{i_1 j_1} \cdots a_{i_N j_N} \frac{\partial}{\partial x_j} T_{j_1 \cdots j_N}(x_1, x_2, x_3),$$

$$(4.61)$$

which is how a tensor of rank $N + 1$ transforms. Therefore the theorem is proven.

To simplify the writing, we introduce another useful notation. From now on, the differential operator $\partial / \partial x_i$ is denoted by ∂_i. For example,

$$\frac{\partial}{\partial x_i} \varphi = \partial_i \varphi,$$

which is the ith component of $\nabla \varphi$. Again it can also represent the totality of

$$\left(\frac{\partial \varphi}{\partial x_1}, \frac{\partial \varphi}{\partial x_2}, \frac{\partial \varphi}{\partial x_3}\right).$$

Thus $\partial_i \varphi$ is a vector. (Note it has one subscript.)

Similarly,

$$\nabla \times \mathbf{A} = \begin{vmatrix} \mathbf{e}_1 & \mathbf{e}_2 & \mathbf{e}_3 \\ \partial_1 & \partial_2 & \partial_3 \\ A_1 & A_2 & A_3 \end{vmatrix},$$

$$\nabla \cdot \mathbf{A} = \sum_i \frac{\partial A_i}{\partial x_i} = \sum_i \partial_i A_i = \partial_i A_i,$$

$$(\boldsymbol{\nabla} \times \mathbf{A})_i = \sum_{j,k} \varepsilon_{ijk} \frac{\partial}{\partial x_j} A_k = \varepsilon_{ijk} \partial_j A_k$$

$$\nabla^2 \varphi = \boldsymbol{\nabla} \cdot \boldsymbol{\nabla}\varphi = \partial_i (\boldsymbol{\nabla}\varphi)_i = \partial_i \partial_i \varphi.$$

With these notations, vector field identities can be easily established.

Example 4.2.8. Show that $\boldsymbol{\nabla} \cdot (\boldsymbol{\nabla} \times \mathbf{A}) = 0$.

Solution 4.2.8.

$$\begin{aligned}
\boldsymbol{\nabla} \cdot (\boldsymbol{\nabla} \times \mathbf{A}) &= \partial_i (\boldsymbol{\nabla} \times \mathbf{A})_i = \partial_i \varepsilon_{ijk} \partial_j A_k = \varepsilon_{ijk} \partial_i \partial_j A_k \\
&= \varepsilon_{ijk} \partial_j \partial_i A_k && (\partial_i \partial_j = \partial_j \partial_i) \\
&= -\varepsilon_{jik} \partial_j \partial_i A_k \\
&= -\varepsilon_{ijk} \partial_i \partial_j A_k && (\text{rename } i \text{ and } j) \\
&= -\boldsymbol{\nabla} \cdot (\boldsymbol{\nabla} \times \mathbf{A})
\end{aligned}$$

Thus $2\boldsymbol{\nabla} \cdot (\boldsymbol{\nabla} \times \mathbf{A}) = 0$.   Hence $\boldsymbol{\nabla} \cdot (\boldsymbol{\nabla} \times \mathbf{A}) = 0$.

Example 4.2.9. Show that $\boldsymbol{\nabla} \times (\boldsymbol{\nabla}\varphi) = 0$.

Solution 4.2.9.

$$\begin{aligned}
[\boldsymbol{\nabla} \times \boldsymbol{\nabla}\varphi]_i &= \varepsilon_{ijk} \partial_j (\boldsymbol{\nabla}\varphi)_k = \varepsilon_{ijk} \partial_j \partial_k \varphi = -\varepsilon_{ikj} \partial_j \partial_k \varphi \\
&= -\varepsilon_{ijk} \partial_j \partial_k \varphi && (\text{rename } j \text{ and } k) \\
&= -[\boldsymbol{\nabla} \times \boldsymbol{\nabla}\varphi]_i .
\end{aligned}$$

It follows that $\boldsymbol{\nabla} \times \boldsymbol{\nabla}\varphi = 0$.

Example 4.2.10. Show that $\boldsymbol{\nabla} \times (\boldsymbol{\nabla} \times \mathbf{A}) = \boldsymbol{\nabla} (\boldsymbol{\nabla} \cdot \mathbf{A}) - \nabla^2 \mathbf{A}$.

Solution 4.2.10.

$$\begin{aligned}
[\boldsymbol{\nabla} \times (\boldsymbol{\nabla} \times \mathbf{A})]_i &= \varepsilon_{ijk} \partial_j (\boldsymbol{\nabla} \times \mathbf{A})_k = \varepsilon_{ijk} \partial_j \varepsilon_{klm} \partial_l A_m \\
&= \varepsilon_{kij} \varepsilon_{klm} \partial_j \partial_l A_m \\
&= (\delta_{il} \delta_{jm} - \delta_{im} \delta_{jl}) \partial_j \partial_l A_m \\
&= \partial_m \partial_i A_m - \partial_l \partial_l A_i = \partial_i \partial_m A_m - \partial_l \partial_l A_i \\
&= [\boldsymbol{\nabla} (\boldsymbol{\nabla} \cdot \mathbf{A})]_i - (\nabla^2 \mathbf{A})_i.
\end{aligned}$$

Hence $\boldsymbol{\nabla} \times (\boldsymbol{\nabla} \times \mathbf{A}) = \boldsymbol{\nabla} (\boldsymbol{\nabla} \cdot \mathbf{A}) - \nabla^2 \mathbf{A}$, since corresponding components from both sides agree.

Example 4.2.11. Show that $\boldsymbol{\nabla} \cdot (\mathbf{A} \times \mathbf{B}) = (\boldsymbol{\nabla} \times \mathbf{A}) \cdot \mathbf{B} - \mathbf{A} \cdot (\boldsymbol{\nabla} \times \mathbf{B})$.

Solution 4.2.11.

$$
\begin{aligned}
\boldsymbol{\nabla} \cdot (\mathbf{A} \times \mathbf{B}) &= \partial_i (\mathbf{A} \times \mathbf{B})_i = \partial_i \varepsilon_{ijk} A_j B_k = \varepsilon_{ijk} \partial_i (A_j B_k) \\
&= \varepsilon_{ijk} (\partial_i A_j) B_k + \varepsilon_{ijk} A_j (\partial_i B_k) = (\varepsilon_{ijk} \partial_i A_j) B_k - A_j (\varepsilon_{jik} \partial_i B_k) \\
&= (\boldsymbol{\nabla} \times \mathbf{A})_k B_k - A_j (\boldsymbol{\nabla} \times \mathbf{B})_j = (\boldsymbol{\nabla} \times \mathbf{A}) \cdot \mathbf{B} - \mathbf{A} \cdot (\boldsymbol{\nabla} \times \mathbf{B}).
\end{aligned}
$$

Example 4.2.12. Show that

$$
\boldsymbol{\nabla} \times (\mathbf{A} \times \mathbf{B}) = (\boldsymbol{\nabla} \cdot \mathbf{B}) \mathbf{A} - (\boldsymbol{\nabla} \cdot \mathbf{A}) \mathbf{B} + (\mathbf{B} \cdot \boldsymbol{\nabla}) \mathbf{A} - (\mathbf{A} \cdot \boldsymbol{\nabla}) \mathbf{B}.
$$

Solution 4.2.12.

$$
\begin{aligned}
[\boldsymbol{\nabla} \times (\mathbf{A} \times \mathbf{B})]_i &= \varepsilon_{ijk} \partial_j (\mathbf{A} \times \mathbf{B})_k = \varepsilon_{ijk} \partial_j \varepsilon_{klm} A_l B_m \\
&= \varepsilon_{kij} \varepsilon_{klm} \partial_j (A_l B_m) = \varepsilon_{kij} \varepsilon_{klm} (B_m \partial_j A_l + A_l \partial_j B_m) \\
&= (\delta_{il} \delta_{jm} - \delta_{im} \delta_{jl}) (B_m \partial_j A_l + A_l \partial_j B_m) \\
&= B_m \partial_m A_i - B_i \partial_l A_l + A_i \partial_m B_m - A_l \partial_l B_i \\
&= [(\mathbf{B} \cdot \boldsymbol{\nabla}) \mathbf{A} - (\boldsymbol{\nabla} \cdot \mathbf{A}) \mathbf{B} + (\boldsymbol{\nabla} \cdot \mathbf{B}) \mathbf{A} - (\mathbf{A} \cdot \boldsymbol{\nabla}) \mathbf{B}]_i.
\end{aligned}
$$

Since the corresponding components agree, the two sides of the desired equation must be equal.

4.2.7 Quotient Rule

Another way to determine if a quantity with two subscripts is a second rank tensor is to use the following quotient rule.

If for an arbitrary vector $\mathbf{B}$, the result of summing over j of the product $K_{ij} B_j$ is another vector A

$$
A_i = K_{ij} B_j, \tag{4.62}
$$

and (4.62) holds in all Cartesian coordinate systems, then K_{ij} is a true second rank tensor.

To prove the quotient rule, we examine the components of $\mathbf{A}$ in a rotated system,

$$
A_i' = a_{il} A_l = a_{il} K_{lm} B_m.
$$

Since $\mathbf{B}$ is a vector,

$$
B_m = a_{jm} B_j'.
$$

It follows

$$A'_i = a_{il}K_{lm}B_m = a_{il}K_{lm}a_{jm}B'_j = a_{il}a_{jm}K_{lm}B'_j.$$

But since (4.62) holds for all systems,

$$A'_i = K'_{ij}B'_j.$$

Subtracting the last two equations,

$$\left(K'_{ij} - a_{il}a_{jm}K_{lm}\right)B'_j = 0.$$

Since B'_j is arbitrary,

$$K'_{ij} = a_{il}a_{jm}K_{lm}.$$

Therefore K_{ij} is a second rank tensor.

   With a similar procedure, one can show that if an Mth rank tensor is linearly related to an Nth rank tensor through a quantity T with $M + N$ subscripts, and the relation holds for all systems, then T is a tensor of rank $M + N$.

Example 4.2.13. If $T_{ij}x_ix_j$ is equal to a scalar S, show that T_{ij} is a second rank tensor.

Solution 4.2.13. Since x_ix_j is the outer product of two position vectors, so it is a second rank tensor. The scalar S is a zeroth rank tensor, therefore by quotient rule T_{ij} is a second rank $(2 + 0 = 2)$ tensor. It is instructive to demonstrate this directly by looking at the components in a rotated system.

$$S = T_{ij}x_ix_j = T_{lm}x_lx_m. \quad x_l = a_{il}x'_i; \quad x_m = a_{jm}x'_j.$$

$$S = T_{lm}a_{il}x'_ia_{jm}x'_j = a_{il}a_{jm}T_{lm}x'_lx'_j,$$

$$S' = T'_{ij}x'_ix'_j, \quad S' = S.$$

Therefore

$$(T'_{ij} - a_{il}a_{jm}T_{lm})x'_ix'_j = 0; \quad T'_{ij} = a_{il}a_{jm}T_{lm}.$$

Hence T_{ij} is a second rank tensor.

4.2.8 Symmetry Properties of Tensors

A tensor $S_{ijk...}$ is said to be symmetric in the indices i and j if

$$S_{ijk...} = S_{jik....}$$

A tensor $A_{ijk\ldots}$ is said to be antisymmetric in the indices i and j if

$$A_{ijk\ldots} = -A_{jik\ldots}.$$

For example, the outer product of $\mathbf{r}$ with itself $x_i x_j$ is a second rank symmetrical tensor, the Kronecker delta δ_{ij} is also a second rank symmetrical tensor. On the other hand, the Levi-Civita symbol ε_{ijk} is antisymmetric with respect to any two of its indices, since $\varepsilon_{ijk} = -\varepsilon_{jik}$.

Symmetry is a physical property of tensors. It is invariant to coordinate transformation. For example, if S_{ij} is a symmetrical tensor in certain coordinate system, in a rotated system

$$S'_{lm} = a_{li}a_{mj}S_{ij} = a_{mj}a_{li}S_{ji} = S'_{ml}.$$

Therefore S_{ij} is also a symmetric tensor in a new system. Similar results hold also for antisymmetric tensors.

A symmetric second rank tensor S_{ij} can be written in the form

$$S_{ij} = \begin{pmatrix} S_{11} & S_{12} & S_{13} \\ S_{12} & S_{22} & S_{23} \\ S_{12} & S_{23} & S_{33} \end{pmatrix},$$

while an antisymmetric second rank tensor A_{ij} is of the form

$$A_{ij} = \begin{pmatrix} 0 & A_{12} & A_{13} \\ -A_{12} & 0 & A_{23} \\ -A_{13} & -A_{23} & 0 \end{pmatrix}.$$

Thus a symmetric second rank tensor has six independent components, while an antisymmetric second tensor has only three independent components.

Any second rank tensor T_{ij} can be represented as the sum of a symmetric tensor and an antisymmetric tensor. Given T_{ij}, one can construct

$$S_{ij} = \frac{1}{2}\left(T_{ij} + T_{ji}\right), \quad A_{ij} = \frac{1}{2}(T_{ij} - T_{ji}).$$

Clearly S_{ij} is symmetric and A_{ij} is antisymmetric. Furthermore

$$T_{ij} = S_{ij} + A_{ij}.$$

Therefore any second rank tensor has a symmetric part and an antisymmetric part.

As shown in the theory of matrices, the six independent elements of a symmetric matrix can be represented by a quadratic surface. In the same way, a symmetric second rank tensor can be represented uniquely by an ellipsoid

$$T_{ij}x_i x_j = \pm 1,$$

where the sign is that of the determinant of $|T_{ij}|$.

The three independent components of an antisymmetric second rank tensor can also be represented geometrically by a vector.

4.2.9 Pseudotensors

One of the reasons why tensors are useful is that they make it possible to formulate the laws of physics in a way that is independent of any preferred direction in space. One might also expect these laws to be independent of whether we choose a right-handed system or a left-handed system of axes. However, under a transformation from right-handed axes to left-handed axes not all tensors transform in the same way.

So far our discussion has been restricted to rotations within right-handed systems. The right-handed system is defined by naming the three basis vectors e_1, e_2, e_3 in such a way that the thumb of your right hand is pointing in the direction of e_3 if the other four fingers can curl from e_1 toward e_2 without passing through negative e_2. A right-handed system can be rotated into another right-handed system. The determinant of the rotation matrix is equal to 1 as shown in example 4.1.2.

Now consider the effect of inversion. The three basis vectors e_1, e_2, e_3 are changed to e_1', e_2', e_3' such that

$$e_1' = -e_1, \quad e_2' = -e_2, \quad e_3' = -e_3.$$

This new set of coordinate axes is a left-handed system. This is called a left-handed system because only if you use your left hand, can the thumb point in the positive e_3' direction while other four fingers curl from e_1' toward e_2' without passing through negative e_2'. Note that one cannot rotate a right-handed system into a left-handed system.

If we use the same right hand rule for the definition of vector cross products in all systems, then with right-handed axes

$$(e_1 \times e_2) \cdot e_3 = 1,$$

and with left-handed axis

$$(e_1' \times e_2') \cdot e_3' = -1.$$

The position vector

$$r = x_1 e_1 + x_2 e_2 + x_3 e_3$$

expressed in the inverted system becomes

$$r' = -x_1 e_1' - x_2 e_2' - x_3 e_3'.$$

In other words, it is the same vector $r = r'$, except in the prime system the coefficients become negative since the axes are inverted. Vectors behaving this way when the coordinates are changed from a right-handed system to a left-handed system are called *polar vectors*. They are just regular vectors.

A fundamental difference appears when we encounter the cross product of two polar vectors. The components of $\mathbf{C} = \mathbf{A} \times \mathbf{B}$ are given by

$$C_1 = A_2 B_3 - A_3 B_2,$$

and so on. Now when the coordinates axes are inverted, A_i goes to $-A_i$, B_i changes to $-B_i$ but C_i goes to $+C_i$ since it is the product of two negative terms. It does not behave like polar vectors under inversion. To distinguish, the cross product is called a *pseudo vector*, also known as *Axial vector*.

In addition to inversion, reflection (reversing one axis) and interchanging of two axes also transform a right-handed system into a left-handed system. The transformation matrices of these right-left operations are as follows.

Inversion:

$$\begin{pmatrix} x_1' \\ x_2' \\ x_3' \end{pmatrix} = \begin{pmatrix} -1 & 0 & 0 \\ 0 & -1 & 0 \\ 0 & 0 & -1 \end{pmatrix} \begin{pmatrix} x_1 \\ x_2 \\ x_3 \end{pmatrix}.$$

Reflection with respect to the $x_2 x_3$ plane:

$$\begin{pmatrix} x_1' \\ x_2' \\ x_3' \end{pmatrix} = \begin{pmatrix} -1 & 0 & 0 \\ 0 & 1 & 0 \\ 0 & 0 & 1 \end{pmatrix} \begin{pmatrix} x_1 \\ x_2 \\ x_3 \end{pmatrix}.$$

Interchange of x_1 and x_2 axes:

$$\begin{pmatrix} x_1' \\ x_2' \\ x_3' \end{pmatrix} = \begin{pmatrix} 0 & 1 & 0 \\ 1 & 0 & 0 \\ 0 & 0 & 1 \end{pmatrix} \begin{pmatrix} x_1 \\ x_2 \\ x_3 \end{pmatrix}.$$

These equations can be written in the form

$$x_i' = a_{ij} x_j.$$

It is obvious that the determinants $|a_{ij}|$ of these transformation matrices are all equal to -1. A left-handed system can be rotated into another left-handed system with Euler angles in the same way as in the right-handed systems. Hence, if a matrix (a_{ij}) transforms a right-handed system into a left-handed system, or vice versa, the determinant $|a_{ij}|$ is always equal to -1. Furthermore, we can show, in the same manner as with the rotation matrix, that its elements also satisfy the orthogonality condition:

$$a_{ik} a_{jk} = \delta_{ij}.$$

Therefore it is also an orthogonal transformation.

Thus the orthogonal transformations can be divided into two classes: proper transformation for which the determinant $|a_{ij}|$ is equal to one, and improper transformation for which the determinant is equal to negative one. If the transformation is a rotation, it is a proper transformation. If the transformation changes a right-handed system into a left-handed system, the transformation is improper.

A pseudovector can now be defined as satisfying the transforming rule

$$V_i' = |a_{ij}|\, a_{ij} V_j.$$

If the transformation is proper, polar vectors and pseudovectors transform in the same way. If the transformation is improper, polar vectors transform as a regular vector, but pseudovectors will flip direction.

Pseudotensors are defined in the same way. The components of a Nth rank pseudotensor transform according to the rule:

$$T_{i_1 \cdots i_N}' = |a_{ij}|\, a_{i_1 j_1} \cdots a_{i_N j_N} T_{j_1 \cdots j_N}, \tag{4.63}$$

which is exactly the same as a regular tensor, except for the determinant $|a_{ij}|$.

It follows from this definition that:

1. The outer product of two pseudotensors of rank M and N is a regular tensor of rank $M + N$.
2. The outer product of a pseudotensor of rank M and a regular tensor of rank N gives a pseudotensor of rank $M + N$.
3. The contraction of a pseudotensor of rank N gives a pseudotensor of rank $N - 2$.

A zeroth rank pseudotensor is a pseudoscalar, which changes sign under inversion, whereas a scalar does not. An example of pseudoscalar is the scalar triple product $(\mathbf{A} \times \mathbf{B}) \cdot \mathbf{C}$. The cross product of $\mathbf{A} \times \mathbf{B}$ is a first rank pseudotensor, the polar vector $\mathbf{C}$ is a regular first rank tensor. Hence $(\mathbf{A} \times \mathbf{B})_i \cdot C_i = (\mathbf{A} \times \mathbf{B}) \cdot \mathbf{C}$ is the contraction of a second rank pseudotensor $(\mathbf{A} \times \mathbf{B})_i \cdot C_j$, therefore a pseudoscalar. If $\mathbf{A}, \mathbf{B}, \mathbf{C}$ are the three sides of a parallelepiped, $(\mathbf{A} \times \mathbf{B}) \cdot \mathbf{C}$ is the volume of the parallelepiped. Defined this way, volume is actually a pseudoscalar.

We have shown ε_{ijk} is an isotropic third rank tensor under rotations. We will now show that if ε_{ijk} is to mean the same thing in both right-handed and left-handed systems, then ε_{ijk} must be regarded as third rank pseudotensor. This is because if we want $\varepsilon_{ijk}' = \varepsilon_{ijk}$ under both proper and improper transformations, then ε_{ijk} must be expressed as

$$\varepsilon_{ijk}' = |a_{ij}|\, a_{il} a_{jm} a_{nk} \varepsilon_{lmn}. \tag{4.64}$$

Since

$$a_{il} a_{jm} a_{kn} \varepsilon_{lmn} = \varepsilon_{ijk} |a_{ij}|,$$

as shown in (4.54), it follows from (4.64) that

$$\varepsilon_{ijk}' = |a_{ij}|\, \varepsilon_{ijk} |a_{ij}| = |a_{ij}|^2 \varepsilon_{ijk} = \varepsilon_{ijk}.$$

Therefore ε_{ijk} is a third rank pseudotensor.

Example 4.2.14. Let T_{12}, T_{13}, T_{23} be the three independent components of an antisymmetric tensor, show that $T_{23}, -T_{13}, T_{13}$ can be regarded as the components of a pseudovector.

Solution 4.2.14. Since ε_{ijk} is a third rank pseudotensor, T_{jk} is a second rank tensor, after contracting twice,

$$C_i = \varepsilon_{ijk} T_{jk}$$

the result C_i is a first rank pseudotensor, which is just a pseudovector. Since $T_{ij} = -T_{ji}$, so

$$T_{21} = -T_{12}, \quad T_{31} = -T_{13}, \quad T_{32} = -T_{23}.$$

Now

$$C_1 = \varepsilon_{123} T_{23} + \varepsilon_{132} T_{32} = T_{23} - T_{32} = 2T_{23},$$
$$C_2 = \varepsilon_{213} T_{13} + \varepsilon_{231} T_{31} = -T_{13} + T_{31} = -2T_{13},$$
$$C_3 = \varepsilon_{312} T_{12} + \varepsilon_{321} T_{21} = T_{12} - T_{21} = 2T_{12}.$$

Since (C_1, C_2, C_3) is a pseudovector, $(T_{23}, -T_{13}, T_{12})$ is also a pseudovector.

Example 4.2.15. Use the fact that ε_{ijk} is a third rank pseudotensor to show that $\mathbf{A} \times \mathbf{B}$ is a pseudovector.

Solution 4.2.15. Let $\mathbf{C} = \mathbf{A} \times \mathbf{B}$, so

$$C_i = \varepsilon_{ijk} A_j B_k.$$

Expressed in a new coordinate system where $\mathbf{A}, \mathbf{B}, \mathbf{C}$ are, respectively, transformed to $\mathbf{A}', \mathbf{B}', \mathbf{C}'$, the cross product becomes $\mathbf{C}' = \mathbf{A}' \times \mathbf{B}'$ which can be written in the component form

$$C_l' = \varepsilon_{lmn}' A_m' B_n'.$$

Since ε_{ijk} is a pseudotensor,

$$\varepsilon_{lmn}' = |a_{ij}| \, a_{li} a_{mj} a_{nk} \varepsilon_{ijk},$$

so C_l' becomes

$$C_l' = |a_{ij}| \, a_{li} a_{mj} a_{nk} \varepsilon_{ijk} A_m' B_n'.$$

But

$$A_j = a_{mj} A_m', \quad B_k = a_{nk} B_n',$$

thus

$$C_l' = |a_{ij}| \, a_{li} \varepsilon_{ijk} A_j B_k = |a_{ij}| \, a_{li} C_i.$$

Therefore $\mathbf{C} = \mathbf{A} \times \mathbf{B}$ is a pseudovector.

Since mathematical equations describing physical laws should be independent of coordinate systems, we cannot equate tensors of different rank because they have different transformation properties under rotation. Likewise, in classical physics, we cannot equate pseudotensors to tensors because they transform differently under inversion. However, surprisingly nature under the influence of weak interaction can distinguish a left-handed system from a right-handed system. By introducing a pseudoscalar, the counterintuitive events that violate parity conservation can be described. (See, T.D. Lee in "Thirty Years Since Parity Nonconservation", Birkhäuser, 1988, page 158).

4.3 Some Physical Examples

4.3.1 Moment of Inertia Tensor

One of the most familiar second rank tensors in physics is the moment of inertial tensor. It relates the angular momentum $\mathbf{L}$ and the angular velocity $\boldsymbol{\omega}$ of the rotational motion of a rigid body. The angular momentum $\mathbf{L}$ of a rigid body rotating about a fixed point is given by

$$\mathbf{L} = \int \mathbf{r} \times \mathbf{v} dm,$$

where $\mathbf{r}$ is the position vector from the fixed point to the mass element dm, and $\mathbf{v}$ is the velocity of dm. We have shown

$$\mathbf{v} = \boldsymbol{\omega} \times \mathbf{r},$$

therefore

$$\mathbf{L} = \int \mathbf{r} \times (\boldsymbol{\omega} \times \mathbf{r}) dm$$

$$= \int \left[(\mathbf{r} \cdot \mathbf{r})\boldsymbol{\omega} - (\mathbf{r} \cdot \boldsymbol{\omega})\mathbf{r} \right] dm.$$

Written in tensor notation with the summation convention, the ith component of $\mathbf{L}$ is

$$L_i = \int [r^2 \omega_i - x_j \omega_j x_i] dm.$$

Since

$$r^2 \omega_i = r^2 \omega_j \delta_{ij},$$

$$L_i = \omega_j \int [r^2 \delta_{ij} - x_j x_i] dm = I_{ij} \omega_j,$$

where I_{ij}, known as the moment of inertia tensor, is given by

$$I_{ij} = \int [x_k x_k \delta_{ij} - x_j x_i] dm. \tag{4.65}$$

Since δ_{ij} and $x_i x_j$ are both second rank tensors, I_{ij} is a symmetric second rank tensor. Explicitly the components of this tensor are

$$I_{ij} = \begin{pmatrix} \int (x_2^2 + x_3^2)\mathrm{d}m & -\int x_1 x_2 \mathrm{d}m & -\int x_1 x_3 \mathrm{d}m \\ -\int x_2 x_1 \mathrm{d}m & \int (x_1^2 + x_3^2)\mathrm{d}m & -\int x_2 x_3 \mathrm{d}m \\ -\int x_3 x_1 \mathrm{d}m & -\int x_3 x_2 \mathrm{d}m & \int (x_2^2 + x_1^2)\mathrm{d}m \end{pmatrix}. \tag{4.66}$$

4.3.2 Stress Tensor

The name tensor comes from the tensile force in elasticity theory. Inside a loaded elastic body, there are forces between neighboring parts of the material. Imagine a cut through the body, the material on the right exerts a force $\mathbf{F}$ on the material to the left, and the material on the left exerts an equal and opposite force $-\mathbf{F}$ on the material to the right.

Let us examine the force across a small area $\Delta x_1 \Delta x_3$, shown in Fig. 4.3, in the imaginary plane perpendicular to the x_2 axis. If the area is small enough, we expect the force is proportional to the area. So we can define the stress $\mathbf{P}_2$ as the force per unit area. The subscript 2 indicates that the force is acting on a plane perpendicular to the positive x_2 axis. The components of $\mathbf{P}_2$ along x_1, x_2, x_3 axes are denoted as P_{12}, P_{22}, P_{32}, respectively. Next, we can look at a small area on the imaginary plane perpendicular to the x_1 axis, and define the stress components P_{11}, P_{21}, P_{31}. Finally, imagine the cut is perpendicular to the x_3 axis, so the stress components P_{13}, P_{23}, P_{33} can be similarly defined. If $\mathbf{e}_1, \mathbf{e}_2, \mathbf{e}_3$ are the unit bases vectors, these relations can be expressed as

$$\mathbf{P}_j = P_{ij}\mathbf{e}_i. \tag{4.67}$$

Thus the stress has nine components

$$P_{ij} = \begin{pmatrix} P_{11} & P_{12} & P_{13} \\ P_{21} & P_{22} & P_{23} \\ P_{31} & P_{32} & P_{33} \end{pmatrix}. \tag{4.68}$$

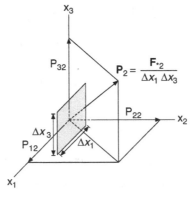

Fig. 4.3. Stress $\mathbf{P}_2$, defined as force per unit area, on a small surface perpendicular to the x_2 axis. Its components along the three axes are, respectively, P_{12}, P_{22}, P_{32}

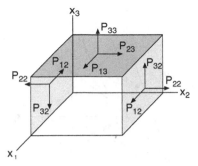

Fig. 4.4. The nine components of a stress tensor at a point can be represented as normal and tangential forces on the surfaces of an infinitesimal cube around the point

The first subscript in P_{ij} indicates the direction of the force component, the second subscript indicates the direction of the normal to the surface on which the force is acted upon.

The physical meaning of P_{ij} is as follows. Imagine an infinitesimal cube around a point inside the material, P_{ij} are the forces per unit area on the faces of this cube, as shown in Fig. 4.4. For clarity, forces are shown only on three surfaces. There are both normal forces P_{ii} (tensions shown but they could be pressures with arrows reversed) and tangential forces P_{ij} ($i \neq j$, shears). Note that, in equilibrium, the forces on the opposite faces must be equal and opposite. Furthermore, (4.67) is symmetric $P_{ij} = P_{ji}$, because of rotational equilibrium. For example, the shearing force on the top surface in the x_2 direction is $P_{23}\Delta x_1 \Delta x_2$. The torque around x_1 axis due to this force is $(P_{23}\Delta x_1 \Delta x_2)\Delta x_3$. The opposite torque due to the shearing force on the right surface is $(P_{32}\Delta x_3 \Delta x_1)\Delta x_2$. Since the net torque around x_1 axis must be zero, therefore

$$(P_{23}\Delta x_1 \Delta x_2)\Delta x_3 = (P_{32}\Delta x_3 \Delta x_1)\Delta x_1,$$

and we have

$$P_{23} = P_{32}.$$

Similar argument will show that in general $P_{ij} = P_{ji}$, therefore (4.68) is symmetric.

Now we are going to show that the nine components of (4.68) is a tensor, known as the *stress tensor*. For this purpose, we construct an infinitesimal tetrahedron with its edges directed along the coordinate axes as shown in Fig. 4.5. Let $\Delta a_1, \Delta a_2, \Delta a_3$ denote the areas of the faces perpendicular to the axes x_1, x_2, x_3, respectively, the forces per unit area on these faces are $-\mathbf{P}_1, -\mathbf{P}_2, -\mathbf{P}_3$, since these faces are directed in the negative direction of the axes. Let Δa_n denote the area of the inclined face with an unit exterior normal $\mathbf{n}$, and $\mathbf{P}_n$ be the force per unit area on this surface. The total force on these

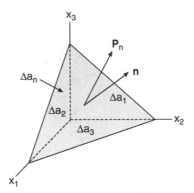

Fig. 4.5. Forces on the surfaces of an infinitesimal tetrahedron. The equilibrium conditions show that the stress must be a second rank tensor

four surfaces must be zero, even if there are body forces, such as gravity. The body forces will be proportional to the volume, whereas all the surface forces are proportional to the area. As dimensions shrink to zero, the body forces have one more infinitesimal, and can always be neglected compared with the surface forces. Therefore

$$\mathbf{P}_n \Delta a_n - \mathbf{P}_1 \Delta a_1 - \mathbf{P}_2 \Delta a_2 - \mathbf{P}_3 \Delta a_3 = \mathbf{0}. \tag{4.69}$$

Since Δa_1 is the area of Δa_n projected on the $x_2 x_3$ plane, therefore

$$\Delta a_1 = (\mathbf{n} \cdot \mathbf{e}_1) \, \Delta a_n.$$

With similar expressions of Δa_2 and Δa_3, we can write (4.69) in the form

$$\mathbf{P}_n \Delta a_n = \mathbf{P}_1 (\mathbf{n} \cdot \mathbf{e}_1) \, \Delta a_n + \mathbf{P}_2 (\mathbf{n} \cdot \mathbf{e}_2) \, \Delta a_n + \mathbf{P}_3 (\mathbf{n} \cdot \mathbf{e}_3) \, \Delta a_n,$$

or

$$\mathbf{P}_n = (\mathbf{n} \cdot \mathbf{e}_k) \, \mathbf{P}_k. \tag{4.70}$$

What we want to find is the stress components in a rotated system with axes $\mathbf{e}_1', \mathbf{e}_2', \mathbf{e}_3'$. Now without loss of generality, we can assume that the jth axis of the rotated system is directed along $\mathbf{n}$, that is

$$\mathbf{n} = \mathbf{e}_j'.$$

Therefore $\mathbf{P}_n$ is $\mathbf{P}_j'$ in the rotated system. Thus (4.70) becomes

$$\mathbf{P}_j' = \left(\mathbf{e}_j' \cdot \mathbf{e}_k \right) \mathbf{P}_k. \tag{4.71}$$

In terms of their components along the respective coordinate axes, as shown in (4.67),

$$\mathbf{P}_j' = P_{ij}' \mathbf{e}_i', \quad \mathbf{P}_k = P_{lk} \mathbf{e}_l$$

the last equation can be written as

$$P'_{ij}\mathbf{e}'_i = \left(\mathbf{e}'_j \cdot \mathbf{e}_k\right)P_{lk}\mathbf{e}_l.$$

Take the dot product with $\mathbf{e}'_i$ on both sides,

$$P'_{ij}(\mathbf{e}'_i \cdot \mathbf{e}'_i) = \left(\mathbf{e}'_j \cdot \mathbf{e}_k\right)P_{lk}(\mathbf{e}_l \cdot \mathbf{e}'_i),$$

we have

$$P'_{ij} = (\mathbf{e}'_i \cdot \mathbf{e}_l)\left(\mathbf{e}'_j \cdot \mathbf{e}_k\right)P_{lk}.$$

Since $(\mathbf{e}'_m \cdot \mathbf{e}_n) = a_{mn}$, we see that

$$P'_{ij} = a_{il}a_{jk}P_{lk}. \tag{4.72}$$

Therefore the array of the stress components (4.68) is indeed a tensor.

4.3.3 Strain Tensor and Hooke's Law

Under imposed forces, an elastic body will deform and exhibit strain. The deformation is characterized by the change of distances between neighboring points. Let P at $\mathbf{r}$ and Q at $\mathbf{r} + \mathbf{\Delta r}$ be two nearby points, as shown in Fig. 4.6. When the body is deformed, point P is displaced by the amount $\mathbf{u}(\mathbf{r})$ to P', and Q by $\mathbf{u}(\mathbf{r} + \mathbf{\Delta r})$ to Q'. If the displacements of these neighboring points are the same, that is if $\mathbf{u}(\mathbf{r}) = \mathbf{u}(\mathbf{r} + \mathbf{\Delta r})$, the relative positions of the points are not changed. That part of the body is unstrained, since the distances PQ and $P'Q'$ will be the same. Therefore the strain is associated with the variation of the displacement vector $\mathbf{u}(\mathbf{r})$. The change of $\mathbf{u}(\mathbf{r})$ can be written as

$$\mathbf{\Delta u} = \mathbf{u}(\mathbf{r} + \mathbf{\Delta r}) - \mathbf{u}(\mathbf{r}).$$

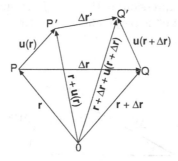

Fig. 4.6. The strain of a deformed elastic body. The body is strained if the relative distance of two nearby points is changed. The strain tensor depends on the variation of the displacement vector $\mathbf{u}$ with respect to the position vector $\mathbf{r}$

Neglecting second and higher terms, the components of $\Delta\mathbf{u}$ can be written as

$$\Delta u_i = u_i\left(x_1 + \Delta x_1, x_2 + \Delta x_2, x_3 + \Delta x_3\right) - u_i\left(x_1, x_2, x_3\right)$$

$$= \frac{\partial u_i}{\partial x_1}\Delta x_1 + \frac{\partial u_i}{\partial x_2}\Delta x_2 + \frac{\partial u_i}{\partial x_3}\Delta x_3 = \frac{\partial u_i}{\partial x_j}\Delta x_j.$$

Since u_i is a vector and $\partial/\partial x_j$ is a vector operator, $\partial u_i/\partial x_j$ is the outer product of two first rank tensors. Therefore $\partial u_i/\partial x_j$ is a second rank tensor. It can be decomposed into a symmetric part and an antisymmetric part

$$\frac{\partial u_i}{\partial x_j} = \frac{1}{2}\left(\frac{\partial u_i}{\partial x_j} + \frac{\partial u_j}{\partial x_i}\right) + \frac{1}{2}\left(\frac{\partial u_i}{\partial x_j} - \frac{\partial u_j}{\partial x_i}\right). \qquad (4.73)$$

We can also divide $\Delta\mathbf{u}$ into two parts

$$\Delta\mathbf{u} = \Delta\mathbf{u}^{\mathrm{s}} + \Delta\mathbf{u}^{\mathrm{a}},$$

where

$$\Delta u_i^{\mathrm{s}} = \frac{1}{2}\left(\frac{\partial u_i}{\partial x_j} + \frac{\partial u_j}{\partial x_i}\right)\Delta x_j,$$

$$\Delta u_i^{\mathrm{a}} = \frac{1}{2}\left(\frac{\partial u_i}{\partial x_j} - \frac{\partial u_j}{\partial x_i}\right)\Delta x_j.$$

The antisymmetric part of (4.73) does not alter the distance between P and Q because of the following. Let the distance $P'Q'$ be $\Delta\mathbf{r}'$. It is clear from Fig. 4.6 that

$$\Delta\mathbf{r}' = \left[\Delta\mathbf{r} + \mathbf{u}\left(\mathbf{r} + \Delta\mathbf{r}\right)\right] - \mathbf{u}\left(\mathbf{r}\right).$$

It follows that

$$\Delta\mathbf{r}' - \Delta\mathbf{r} = \mathbf{u}\left(\mathbf{r} + \Delta\mathbf{r}\right) - \mathbf{u}\left(\mathbf{r}\right) = \Delta\mathbf{u} = \Delta\mathbf{u}^{\mathrm{s}} + \Delta\mathbf{u}^{\mathrm{a}}.$$

Now

$$\Delta\mathbf{u}^a \cdot \Delta\mathbf{r} = \Delta u_i^a \Delta x_i = \frac{1}{2}\left(\frac{\partial u_i}{\partial x_j} - \frac{\partial u_j}{\partial x_i}\right)\Delta x_j \Delta x_i = 0.$$

This is because in this expression, both i and j are summing indices and can be interchanged. Thus $\Delta\mathbf{u}^{\mathrm{a}}$ is perpendicular to $\Delta\mathbf{r}$, and it can be considered as the infinitesimal arc length of a rotation around the tail of $\Delta\mathbf{r}$, as shown in the following sketch.

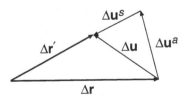

Fig. 4.7. The change of distance between two nearby points in an elastic body. It is determined by the symmetric strain tensor

Therefore $\Delta \mathbf{u}^a$ does not change the length $\Delta \mathbf{r}$. The change of distance between two nearby points of an elastic body is uniquely determined by the symmetric part of (4.73)

$$E_{ij} = \frac{1}{2} \left(\frac{\partial u_i}{\partial x_j} + \frac{\partial u_j}{\partial x_i} \right). \tag{4.74}$$

This quantity is known as the strain tensor. The stain tensor plays an important role in the elasticity theory because it is a measure of the degree of deformation.

Since in one-dimension the elastic force in a spring is given by the Hooke's law $F = -kx$, one might expect that in three-dimensional elastic media the strain is proportional to the stress. For most solid materials with a relative strain of a few percent, this is indeed the case. The linear relationship between the strain tensor and the stress tensor is given by the generalized Hooke's law

$$P_{ij} = c_{ijkl} E_{kl}, \tag{4.75}$$

where c_{ijkl} is known as the elasticity tensor. Since P_{ij} and E_{kl} are both second rank tensors, by the quotient rule, c_{ijkl} must be a fourth rank tensor. While there are 81 components of a fourth rank tensor, but various symmetry considerations will show that the number of independent components in a general crystalline body is only 21. If the body is isotropic, the elastic constants are further reduced to only two. While we are not going into these details which are the subjects of books on elasticity, we only want to show that concepts of tensor are crucial in describing these physical quantities.

Exercises

1. Find the rotation matrix for
   (a) a rotation of $\pi/2$ about z axis,
   (b) a rotation of π about x axis.

   Ans. $\begin{pmatrix} 0 & 1 & 0 \\ -1 & 0 & 0 \\ 0 & 0 & 1 \end{pmatrix}$; $\begin{pmatrix} 1 & 0 & 0 \\ 0 & -1 & 0 \\ 0 & 0 & -1 \end{pmatrix}$.

2. With the transformation matrix (A) given by (4.20), show that
   $$(A)\left(A^{\mathrm{T}}\right) = (I),$$
   where (I) is the identity matrix.

3. With the transformation matrix (A) given by (4.20), explicitly verify that
   $$\sum_{i=1}^{3} a_{ij} a_{ik} = \delta_{jk},$$
   for (a) $j = 1$, $k = 1$, (b) $j = 1$, $k = 2$, (c) $j = 1$, $k = 3$.

4. With the transformation matrix (A) given by (4.20), show explicitly that the determinant A is equal to 1.

5. Show that there is no nontrivial isotropic first rank tensor.
   Hint: (1) Assume there is an isotropic first rank tensor (A_1, A_2, A_3). Under a rotation, $A_1' = A_1$, $A_2' = A_2$, $A_3' = A_3$, since it is isotropic. (2) Consider a rotation of $\pi/2$ about x_3 axis, and show that $A_1 = 0$, $A_2 = 0$. (3) A rotation about x_1 axis will show that $A_3 = 0$. (4) Therefore only the zero vector is a first rank isotropic tensor.

6. Let

$$T_{ij} = \begin{pmatrix} 1 & 0 & 2 \\ 0 & 2 & 1 \\ 1 & 2 & 3 \end{pmatrix}, \quad A_i = \begin{pmatrix} 3 \\ 2 \\ 1 \end{pmatrix}.$$

   Find the following contractions
   (a) $B_i = T_{ij} A_j$,
   (b) $C_j = T_{ij} A_i$,
   (c) $S = T_{ij} A_i A_j$.
   Ans. (a) $B_i = (5, 5, 10)$, (b) $C_i = (4, 6, 11)$, (c) $S = 35$.

7. Let A_{ij} and B_{ij} be two second rank tensors, and let

$$C_{ij} = A_{ij} + B_{ij}.$$

   Show that C_{ij} is also a second rank tensor.

8. The equation of an ellipsoid centered at the origin is of the form

$$A_{ij} x_i x_j = 1.$$

   Show that A_{ij} is a second rank tensor.
   Hint: In a rotated system, the equation of the surface is $A_{ij}' x_i' x_j' = 1$.

9. Show that

$$A_i = \begin{pmatrix} -x_2 \\ x_1 \end{pmatrix}$$

   is a two-dimensional vector.

10. Show that the following 2×2 matrices represent second rank tensors in two-dimensional space:

   (a) $\begin{pmatrix} -x_1 x_2 & x_1^2 \\ -x_2^2 & x_1 x_2 \end{pmatrix}$,   (b) $\begin{pmatrix} x_2^2 & -x_1 x_2 \\ -x_1 x_2 & x_1^2 \end{pmatrix}$,

   (c) $\begin{pmatrix} -x_1 x_2 & -x_2^2 \\ x_1^2 & x_1 x_2 \end{pmatrix}$,   (d) $\begin{pmatrix} x_1^2 & x_1 x_2 \\ x_1 x_2 & x_2^2 \end{pmatrix}$.

   Hint: Show that they are various outer products of the position vector and the vector in the last question.

11. Explicitly show that

$$\sum_{l=1}^{3}\sum_{m=1}^{3}\sum_{n=1}^{3} a_{il}a_{jm}a_{kn}\varepsilon_{lmn} = \begin{vmatrix} a_{i1} & a_{i2} & a_{i3} \\ a_{j1} & a_{j2} & a_{j3} \\ a_{k1} & a_{k2} & a_{k3} \end{vmatrix}$$

by (a) writing out all nonzero terms of

$$\sum_{l=1}^{3}\sum_{m=1}^{3}\sum_{n=1}^{3} a_{il}a_{jm}a_{kn}\varepsilon_{lmn},$$

and (b) expand the determinant

$$\begin{vmatrix} a_{i1} & a_{i2} & a_{i3} \\ a_{j1} & a_{j2} & a_{j3} \\ a_{k1} & a_{k2} & a_{k3} \end{vmatrix}$$

over the elements of the first row.

12. Show that ε_{ijk} can be written as

$$\varepsilon_{ijk} = \begin{vmatrix} \delta_{i1} & \delta_{i2} & \delta_{i3} \\ \delta_{j1} & \delta_{j2} & \delta_{j3} \\ \delta_{k1} & \delta_{k2} & \delta_{k3} \end{vmatrix}.$$

Hint: Show that the determinant has all the properties ε_{ijk} has.

13. Show that
(a) $\sum_{ij} \varepsilon_{ijk}\delta_{ij} = 0$;
(b) $\sum_{jk} \varepsilon_{ijk}\varepsilon_{ljk} = 2\delta_{il}$;
(c) $\sum_{ijk} \varepsilon_{ijk}\varepsilon_{ijk} = 6$.

14. The following equations are written with summation convention, verify them
(a) $\delta_{ij}\delta_{jk}\delta_{ki} = 3$,
(b) $\varepsilon_{ijk}\varepsilon_{klm}\varepsilon_{mni} = \varepsilon_{jnl}$.
Hint: (a) Recall δ_{ij} is a substitution tensor, (b) Use (4.60).

15. With the summation convention, show that
(a) $A_i\delta_{ij} = A_j$
(b) $B_j\delta_{ij} = B_i$
(c) $\delta_{1j}\delta_{j1} = 1$
(d) $\delta_{ij}\delta_{ji} = \delta_{ii} = 3$
(e) $\delta_{ij}\delta_{jl} = \delta_{il}$

16. With subscripts and summation convention, show that
(a) $\partial_i x_j = \delta_{ij}$
(b) $\partial_i (x_j x_j)^{1/2} = \dfrac{1}{(x_j x_j)^{1/2}} x_i.$
Hint: (a) x_1, x_2, x_3 are independent variables. (b) $\partial_i (x_j x_j) = 2x_j \partial_i x_j$.

17. The following equations are written with summation convention
    (a) $\partial_i x_i = 3$, (b) $\partial_i (x_j x_j)^{1/2} = (x_j x_j)^{-1/2} x_i$, translate them into ordinary vector notation.
    Ans. (a) $\nabla \cdot \mathbf{r} = 3$, (b) $\nabla r = \mathbf{r}/r$.

18. The following expressions are written with summation convention
    (a) $V_i A_j B_i e_j$;   (b) $c A_i B_j \delta_{ij}$;
    (c) $A_l B_j \varepsilon_{ijk} \delta_{li} e_k$;   (d) $\varepsilon_{ijk} \varepsilon_{lmk} A_i B_j C_l D_m$,
    translate them into ordinary vector notation.
    Ans. (a) $(\mathbf{V} \cdot \mathbf{B}) \mathbf{A}$, (b) $c \mathbf{A} \cdot \mathbf{B}$, (c) $\mathbf{A} \times \mathbf{B}$, (d) $(\mathbf{A} \times \mathbf{B}) \cdot (\mathbf{C} \times \mathbf{D})$.

19. Use the Levi-Civita tensor technique to prove the following identities:
    (a) $\mathbf{A} \times \mathbf{B} = -\mathbf{B} \times \mathbf{A}$,
    (b) $\mathbf{A} \cdot (\mathbf{B} \times \mathbf{C}) = (\mathbf{A} \times \mathbf{B}) \cdot \mathbf{C}$.

20. Use the Levi-Civita tensor technique to prove the following identity

$$\nabla \times (\phi \mathbf{A}) = \phi (\nabla \times \mathbf{A}) + (\nabla \phi) \times \mathbf{A}.$$

21. Let

$$T_{ij} = \begin{pmatrix} 1 & 2 & 3 \\ 0 & 4 & 5 \\ 0 & 0 & 6 \end{pmatrix}.$$

Find the symmetric part S_{ij} and the antisymmetric part A_{ij} of the tensor T_{ij}.

Ans. $S_{ij} = \begin{pmatrix} 1 & 1 & 1.5 \\ 1 & 4 & 2.5 \\ 1.5 & 2.5 & 6 \end{pmatrix}$, $A_{ij} = \begin{pmatrix} 0 & 1 & 1.5 \\ -1 & 0 & 2.5 \\ -1.5 & -2.5 & 0 \end{pmatrix}$.

22. If S_{ij} is a symmetric tensor and A_{ij} is an antisymmetric tensor, show that

$$S_{ij} A_{ij} = 0.$$

23. Let φ be a scalar, V_i be a pseudovector, T_{ij} be a second rank tensor, and let

$$A_{ijk} = \varepsilon_{ijk} \varphi, \quad B_{ij} = \varepsilon_{ijk} V_k, \quad C_i = \varepsilon_{ijk} T_{jk}.$$

Show that A_{ijk} is a third rank pseudotensor, B_{ij} is a second rank tensor, and C_i is a pseudovector.

24. Find the strain tensor for an isotropic elastic material when it is subjected to
    (a) A stretching deformation $\mathbf{u} = (0, 0, \alpha x_3)$;
    (b) A shearing deformation $\mathbf{u} = (\beta x_3, 0, 0)$.

Ans. (a) $E_{ij} = \begin{pmatrix} 0 & 0 & 0 \\ 0 & 0 & 0 \\ 0 & 0 & \alpha \end{pmatrix}$; (b) $E_{ij} = \begin{pmatrix} 0 & 0 & \beta/2 \\ 0 & 0 & 0 \\ \beta/2 & 0 & 0 \end{pmatrix}$.

Part II

Differential Equations and Laplace Transforms

5

Ordinary Differential Equations

The laws of physics that govern important and significant problems in engineering and sciences are most often expressed in the form of differential equations. A differential equation is an equation involving derivatives of an unknown function that depends upon one or more independent variables. If the unknown function depends on only one independent variable, then the equation is called an ordinary differential equation.

In this chapter, after a review of the standard methods for solving first-order differential equations, we will present a comprehensive treatment of linear differential equations with constant coefficients, in terms of which a great many physical problems are formulated. We will use mechanical vibrations and electrical circuits as illustrative examples. Then we will discuss systems of coupled differential equations and their applications.

Series solutions of differential equations will be discussed in the chapter on special functions. Another important method of solving differential equation is the Laplace transformation, which we will discuss in the next chapter.

5.1 First-Order Differential Equations

To solve a differential equation is to find a way to eliminate the derivatives in the equation so that the relation between the dependent and the independent variables can be exhibited. For a first-order differential equation, this can be achieved by carrying out an integration. The simplest type of differential equations is

$$\frac{dy}{dx} = f(x),\tag{5.1}$$

where $f(x)$ is a given function of x. We know from calculus that

$$y(x) = \int_a^x f(x')dx'\tag{5.2}$$

is a solution. Equation (5.1) contains only the first derivative of y, and is called a first-order differential equation. The order of a differential equation is equal to the order of the highest derivative in the equation. The solution (5.2) is known as a general solution, and contains an arbitrary integration constant. If the integral in (5.2) exists, then by definition there is a function $F(x)$, such that

$$\frac{\mathrm{d}}{\mathrm{d}x}F(x) = f(x), \qquad \mathrm{d}F(x) = f(x)\mathrm{d}x$$

and

$$y(x) = \int_a^x \mathrm{d}F(x') = F(x) + F(a).$$

In this sense, we often use the notation of the indefinite integral

$$y(x) = \int f(x)\mathrm{d}x + C$$

where $C = F(a)$ is an arbitrary constant. If we know that y takes the value y_0 when $x = x_0$, then the constant is determined. This condition "$y = y_0$ when $x = x_0$" is called either "initial condition" or "boundary condition." To satisfy both the equation and the boundary condition, we can carry out the following definite integrals:

$$\int_{y_0}^y \mathrm{d}y' = \int_{x_0}^x f(x')\mathrm{d}x',$$

which can be written in the form of

$$y(x) = \int_{x_0}^x f(x')\mathrm{d}x' + y_0.$$

This is known as the specific solution. (The term "particular solution" is often used, however, this may cause confusion, since "particular solution" is also used in the solution of nonhomogeneous equations, which we shall discuss a little later.) In most physical applications, it is the specific solution that is of interest. A physical problem, when formulated in the mathematical language, usually consists of a differential equation and an appropriate number of boundary and/or initial conditions. The problem is solved only after the specific solution is found.

5.1.1 Equations with Separable Variables

If an equation can be written in the form

$$f(x)\mathrm{d}x + g(y)\mathrm{d}y = 0$$

the solution can be immediately obtained in the form of

$$\int f(x)dx + \int g(y)dy = C.$$

This method is called solution by the separation of variables and is one of the most commonly used methods.

For example, the differential equation

$$\frac{dy}{dx} = -\frac{x}{y}$$

can be solved by noting that the equation can be written as

$$y\,dy + x\,dx = 0.$$

Therefore the solution is given by

$$\int y\,dy + \int x\,dx = C$$

or

$$\frac{1}{2}y^2 + \frac{1}{2}x^2 = C.$$

This general solution can be written as

$$y(x) = (C' - x^2)^{1/2}$$

or equivalently as

$$F(x, y) = C'$$

with

$$F(x, y) = x^2 + y^2.$$

Clearly this general solution represents a family of circles with radius $\sqrt{C'}$ centered at the origin. If it is specified that $x = 5$, $y = 0$, is a point on the circle, then the specific solution is

$$x^2 + y^2 = 25.$$

This specific solution can also be obtained from the definite integral

$$\int_0^y y'\,dy' + \int_5^x x'\,dx' = 0,$$

which gives the same result by way of

$$\frac{1}{2}y^2 + \frac{1}{2}x^2 - \frac{1}{2}5^2 = 0.$$

5.1.2 Equations Reducible to Separable Type

Certain equations of the form

$$g(x,y)\mathrm{d}y = f(x,y)\mathrm{d}x \qquad (5.3)$$

that are not separable can be made separable by a change of variable. This can always be done, if the ratio of $f(x,y)/g(x,y)$ is a function of y/x.

Let $u = y/x$ and the function of y/x be $h(u)$, so that

$$\frac{\mathrm{d}y}{\mathrm{d}x} = \frac{f(x,y)}{g(x,y)} = h(u).$$

Since y is function of x, so is u. It follows that $y(x) = xu(x)$ and

$$\frac{\mathrm{d}y}{\mathrm{d}x} = u + x\frac{\mathrm{d}u}{\mathrm{d}x}.$$

Thus the differential equation can be written as

$$u + x\frac{\mathrm{d}u}{\mathrm{d}x} = h(u),$$

or

$$x\frac{\mathrm{d}u}{\mathrm{d}x} = h(u) - u$$

Clearly it is separable

$$\frac{\mathrm{d}u}{h(u) - u} = \frac{\mathrm{d}x}{x}.$$

So we can solve for $u(x)$ and the solution of the original differential equation is simply

$$y = xu(x).$$

For example, if

$$\frac{\mathrm{d}y}{\mathrm{d}x} = \frac{y^2 + xy}{x^2} = \frac{y^2}{x^2} + \frac{y}{x},$$

then with $u = y/x$, $h(u) = u^2 + u$. Since $h(u) - u = u^2$, so

$$\frac{\mathrm{d}u}{u^2} = \frac{\mathrm{d}x}{x}$$

and

$$\int \frac{\mathrm{d}u}{u^2} = \int \frac{\mathrm{d}x}{x},$$

which gives

$$-\frac{1}{u} + C = \ln x.$$

Since $u = y/x$, the solution of the original differential equation is therefore given by

$$\frac{x}{y} + \ln x = C.$$

5.1.3 Exact Differential Equations

Suppose we want to find a differential equation that represents the following family of curves:

$$F(x, y) = C.$$

First let us look at two nearby points (x, y) and $(x + \Delta x, y + \Delta y)$, both on a specific curve of this family. If the curve is characterized by $C = k$, then

$$F(x, y) = k, \quad F(x + \Delta x, y + \Delta y) = k.$$

Clearly, the difference between the two is equal to zero

$$\Delta F = F(x + \Delta x, y + \Delta y) - F(x, y) = 0.$$

This difference can be written in the form of

$$
\begin{aligned}
F(x + \Delta x, y + \Delta y) - F(x, y) &= F(x + \Delta x, y + \Delta y) - F(x, y + \Delta y) \\
&\quad + F(x, y + \Delta y) - F(x, y).
\end{aligned}
\tag{5.4}
$$

With the understanding that Δx and Δy are approaching zero as a limit, we can use the definition of partial derivative

$$F(x + \Delta x, y + \Delta y) - F(x, y + \Delta y) = \frac{\partial F}{\partial x} \Delta x,$$

$$F(x, y + \Delta y) - F(x, y) = \frac{\partial F}{\partial y} \Delta y$$

to write (5.4) as

$$\Delta F = \frac{\partial F}{\partial x} \Delta x + \frac{\partial F}{\partial y} \Delta y$$

or

$$dF = \frac{\partial F}{\partial x} dx + \frac{\partial F}{\partial y} dy.$$

This is known as the total differential. Since $\Delta F = 0$, so we have

$$\frac{\partial F}{\partial x} dx + \frac{\partial F}{\partial y} dy = 0.
\tag{5.5}$$

This is the differential equation representing the family of curves $F(x, y) = C$. In other words, the solution of the differential equation in the form of (5.5) is given by $F(x, y) = C$.

Now let

$$\frac{\partial F}{\partial x} = f(x, y), \quad \frac{\partial F}{\partial y} = g(x, y)
\tag{5.6}$$

so

$$\frac{\partial^2 F}{\partial y \partial x} = \frac{\partial}{\partial y}\frac{\partial F}{\partial x} = \frac{\partial}{\partial y}f(x,y),$$

$$\frac{\partial^2 F}{\partial x \partial y} = \frac{\partial}{\partial x}\frac{\partial F}{\partial y} = \frac{\partial}{\partial x}g(x,y).$$

Since the order of differentiation can be interchanged as long as the function has continuous partial derivatives

$$\frac{\partial^2 F}{\partial y \partial x} = \frac{\partial^2 F}{\partial x \partial y},$$

one has

$$\frac{\partial}{\partial y}f(x,y) = \frac{\partial}{\partial x}g(x,y). \tag{5.7}$$

Any differential equation of the form

$$f(x,y)\mathrm{d}x + g(x,y)\mathrm{d}y = 0$$

that satisfies (5.7) is known as an exact equation. An exact equation can be expressed as $\mathrm{d}F = 0$, where $\mathrm{d}F$ is a total differential and $F(x,y) = C$ is the solution. The function $F(x,y)$ can be obtained by integrating the two equations of (5.6).

For example, the differential equation

$$\frac{\mathrm{d}y}{\mathrm{d}x} + \frac{xy^2}{2+x^2y} = 0$$

can be written in the form

$$(2+x^2y)\mathrm{d}y + xy^2\mathrm{d}x = 0.$$

Since

$$\frac{\partial}{\partial x}(2+x^2y) = 2xy, \qquad \frac{\partial}{\partial y}(xy^2) = 2xy$$

are equal, the differential equation is exact. Therefore, we can find the general solution in the form of

$$F(x,y) = C$$

with

$$\frac{\partial F(x,y)}{\partial y} = 2+x^2y, \qquad \frac{\partial F(x,y)}{\partial x} = xy^2.$$

The first equation yields

$$F(x,y) = 2y + \frac{1}{2}x^2y^2 + p(x).$$

The second equation requires

$$\frac{\partial F(x,y)}{\partial x} = xy^2 + \frac{d}{dx}p(x) = xy^2.$$

Therefore

$$\frac{d}{dx}p(x) = 0, \qquad p(x) = k.$$

Thus the solution is

$$F(x,y) = 2y + \frac{1}{2}x^2y^2 + k = C.$$

Combining the two constants, we can write the solution as

$$2y + \frac{1}{2}x^2y^2 = C'.$$

5.1.4 Integrating Factors

A multiplying factor which will convert a differential equation that is not exact into an exact one is called an integrating factor. For example, the equation

$$y\,dx + (x^2y^3 + x)dy = 0 \qquad (5.8)$$

is not exact. If, however, we multiply it by $(xy)^{-2}$, the resulting equation

$$\frac{1}{x^2y}dx + (y + \frac{1}{xy^2})dy = 0$$

is exact, since

$$\frac{\partial}{\partial y}\left(\frac{1}{x^2y}\right) = \frac{-1}{x^2y^2},$$

$$\frac{\partial}{\partial x}\left(y + \frac{1}{xy^2}\right) = \frac{-1}{x^2y^2}.$$

Hence by definition, $(xy)^{-2}$ is an integrating factor.

By the method of exact differential equation, we have

$$\frac{\partial}{\partial x}F(x,y) = \frac{1}{x^2y},$$

$$F(x,y) = -\frac{1}{xy} + q(y)$$

and

$$\frac{\partial}{\partial y}F(x,y) = \frac{1}{xy^2} + \frac{d}{dy}q(y) = y + \frac{1}{xy^2},$$

$$\frac{d}{dy}q(y) = y, \qquad q(y) = \frac{1}{2}y^2.$$

Therefore the solution of the original equation is

$$-\frac{1}{xy} + \frac{1}{2}y^2 = C.$$

It is sometimes possible to find an integrating factor by inspection. For example, one may rearrange (5.8) into

$$(y\,dx + x\,dy) + x^2 y^3 dy = 0$$

and recognize $y\,dx + x\,dy = d(xy)$. Then it is readily seen that the equation

$$d(xy) + x^2 y^3 dy = 0$$

can be solved by multiplying by a factor of $(xy)^{-2}$, since it will change the equation to

$$\frac{d(xy)}{(xy)^2} + y\,dy = 0,$$

which immediately gives the result of

$$-\frac{1}{xy} + \frac{1}{2}y^2 = C.$$

Theoretically an integrating factor exists for every differential equation of the form $f(x,y)dx + g(x,y)dy = 0$. Unfortunately no general rule is known to find it. For certain special type of differential equations, integrating factors can be found systematically.

We assume that

$$f(x,y)dx + g(x,y)dy = 0$$

is not an exact differential equation. We wish to find an integrating factor μ, so that

$$\mu f(x,y)dx + \mu g(x,y)dy = 0$$

is exact. For this equation to be exact, it must satisfy the condition

$$\frac{\partial}{\partial y}(\mu f) = \frac{\partial}{\partial x}(\mu g)$$

which gives

$$\mu\left(\frac{\partial f}{\partial y} - \frac{\partial g}{\partial x}\right) = \frac{\partial \mu}{\partial x}g - \frac{\partial \mu}{\partial y}f. \qquad (5.9)$$

Now we consider the following possibilities.

The integrating factor μ is a function of x only. In this case, (5.9) becomes

$$\frac{1}{g}\left(\frac{\partial f}{\partial y} - \frac{\partial g}{\partial x}\right) = \frac{1}{\mu}\frac{\partial \mu}{\partial x}.$$

If the left-hand side of this equation is also a function of x only

$$\frac{1}{g}\left(\frac{\partial f}{\partial y} - \frac{\partial g}{\partial x}\right) = G(x),$$

then clearly

$$\frac{\mathrm{d}\mu}{\mu} = G(x)\mathrm{d}x,$$

which gives

$$\ln \mu = \int G(x)\mathrm{d}x$$

or

$$\mu = e^{\int G(x)\mathrm{d}x}.$$

For example, the differential equation

$$(3xy + y^2)\mathrm{d}x + (x^2 + xy)\mathrm{d}y = 0$$

is not exact. Written in the form of $f(x,y)\mathrm{d}x + g(x,y)\mathrm{d}y = 0$, we see that

$$\frac{\partial f}{\partial y} = \frac{\partial}{\partial y}(3xy + y^2) = 3x + 2y,$$

$$\frac{\partial g}{\partial x} = \frac{\partial}{\partial x}(x^2 + xy) = 2x + y$$

are not equal. But

$$\frac{1}{g}\left(\frac{\partial f}{\partial y} - \frac{\partial g}{\partial x}\right) = \frac{3x + 2y - (2x + y)}{x^2 + xy} = \frac{x + y}{x(x + y)} = \frac{1}{x}$$

is a function of x only. Therefore the integrating factor is given by

$$\mu = e^{\int \frac{1}{x}dx} = e^{\ln x} = x.$$

Multiply the original differential equation by x, it becomes

$$(3x^2y + xy^2)\mathrm{d}x + (x^3 + x^2y)\mathrm{d}y = 0.$$

This equation is exact, since

$$\frac{\partial}{\partial y}(3x^2y + xy^2) = \frac{\partial}{\partial x}(x^3 + x^2y).$$

Integrating these two equations

$$\frac{\partial F}{\partial x} = 3x^2y + xy^2, \quad \frac{\partial F}{\partial y} = x^3 + x^2y.$$

we find

$$F(x, y) = x^3y + \frac{1}{2}x^2y^2.$$

Therefore the solution is

$$x^3y + \frac{1}{2}x^2y^2 = C.$$

The integrating factor μ is a function of y only. In this case, (5.9) becomes

$$\frac{1}{f}\left(\frac{\partial f}{\partial y} - \frac{\partial g}{\partial x}\right) = -\frac{1}{\mu}\frac{\partial \mu}{\partial y}.$$

If the left-hand side of this equation is also a function of y only

$$\frac{1}{f}\left(\frac{\partial f}{\partial y} - \frac{\partial g}{\partial x}\right) = -H(y),$$

then clearly

$$\frac{d\mu}{\mu} = H(y)dy,$$

which gives

$$\ln\mu = \int H(y)dy$$

or

$$\mu = e^{\int H(y)dy}.$$

5.2 First-Order Linear Differential Equations

A special type of first-order differential equation of some importance is of the form

$$\frac{dy}{dx} + p(x)y = q(x), \tag{5.10}$$

in which both the dependent variable y and its first derivative y' are of the first degree, and $p(x)$ and $q(x)$ are continuous functions of the independent variable x. This type of equation is called linear differential equation of the first-order. In what follows, we will derive a general solution for this equation.

First, if $q(x) = 0$, we have

$$\frac{dy}{dx} = -p(x)y.$$

In this case

$$y(x) = e^{-\int^x p(x)dx}.$$

For the general case, we introduce a variable coefficient

$$y(x) = f(x)e^{-\int^x p(x)dx}.$$

With this trial solution, (5.10) becomes

$$\frac{df(x)}{dx}e^{-\int^x p(x)dx} - p(x)f(x)e^{-\int^x p(x)dx} + p(x)f(x)e^{-\int^x p(x)dx} = q(x),$$

or

$$\frac{df(x)}{dx}e^{-\int^x p(x)dx} = q(x).$$

Thus

$$f(x) = \int q(x)e^{\int^x p(x)dx}dx + C.$$

Hence the solution of the first-order linear differential equation (5.10) is given by

$$y(x) = f(x)e^{-\int^x p(x)dx}$$

$$= e^{-\int p(x)dx}\int e^{\int p(x)dx}q(x)dx + Ce^{-\int p(x)dx}. \tag{5.11}$$

To use this formula, it is important to remember to put the differential equation in the form of $y' + p(x)y = q(x)$. In other words, the coefficient of the derivative must be one.

This solution enables us to see that

$$\mu(x) = e^{\int p(x)dx}$$

is the integrating factor of the equation. In terms of $\mu(x)$, (5.11) can be written as

$$\mu(x)y = \int \mu(x)q(x)dx + C,$$

which is a solution of the differential equation

$$\frac{d}{dx}[\mu(x)y] = \mu(x)\,q(x). \tag{5.12}$$

Furthermore

$$\frac{d}{dx}[\mu(x)y] = \mu(x)\frac{dy}{dx} + \left[\frac{d}{dx}\mu(x)\right]y,$$

$$\frac{d}{dx}\mu(x) = \frac{d}{dx}e^{\int p(x)dx} = e^{\int p(x)dx}[p(x)] = \mu(x)\,p(x).$$

Hence (5.12) becomes

$$\mu(x)\frac{dy}{dx} + \mu(x)\,p(x)y = \mu(x)\,q(x),$$

which clearly shows that $\mu(x)$ is a integrating factor of the original equation.

Thus, an easier way to make use of the complicated formula of (5.11) is to write it in terms of the integrating factor

$$y(x) = \frac{1}{\mu(x)}\left[\int \mu(x)\,q(x)dx + C\right]$$

with

$$\mu(x) = e^{\int^x p(x)dx}.$$

Example 5.2.1. Find the general solution of the following differential equation:

$$x\frac{dy}{dx} + (1+x)y = e^x.$$

Solution 5.2.1. This is a linear differential equation of first-order

$$\frac{dy}{dx} + \frac{1+x}{x}y = \frac{e^x}{x}.$$

The integrating factor is given by

$$\mu(x) = e^{\int \frac{1+x}{x}dx}.$$

Since

$$\int^x \frac{1+x}{x}dx = \int\left(\frac{1}{x}+1\right)dx = \ln x + x,$$

$$\mu(x) = e^{\ln x + x} = xe^x.$$

It follows that:

$$y = \frac{1}{xe^x}\left[\int xe^x\frac{e^x}{x}dx + C\right]$$

$$= \frac{1}{xe^x}\left[\int e^{2x}dx + C\right] = \frac{1}{xe^x}\left[\frac{e^{2x}}{2} + C\right].$$

Therefore the solution is given by

$$y = \frac{e^x}{2x} + C\frac{e^{-x}}{x}.$$

5.2.1 Bernoulli Equation

The type of differential equations

$$\frac{dy}{dx} + p(x)y = q(x)y^n$$

is known as Bernoulli equations, named after Swiss mathematician James Bernoulli (1654–1705). This is a nonlinear differential equation if $n \neq 0$ or 1. However, it can be transformed into a linear equation by multiplying both sides with a factor $(1-n)y^{-n}$

$$(1-n)y^{-n}\frac{dy}{dx} + (1-n)p(x)y^{1-n} = (1-n)q(x).$$

Since

$$(1-n)y^{-n}\frac{dy}{dx} = \frac{d}{dx}(y^{1-n}),$$

the last equation can be written as

$$\frac{d}{dx}(y^{1-n}) + (1-n)p(x)y^{1-n} = (1-n)q(x),$$

which is a first-order linear equation in terms of y^{1-n}. This equation can be solved for y^{1-n}, from which the solution of the original equation can be obtained.

Example 5.2.2. Find the solution of

$$\frac{dy}{dx} + \frac{1}{x}y = x^2y^3$$

with the condition $y(1) = 1$.

Solution 5.2.2. This a Bernoulli equation of $n = 3$. Multiplying this equation by $(1-3)y^{-3}$, we have

$$-2y^{-3}\frac{dy}{dx} - 2\frac{1}{x}y^{-2} = -2x^2,$$

which can be written as

$$\frac{d}{dx}y^{-2} - \frac{2}{x}y^{-2} = -2x^2.$$

This equation is first-order in y^{-2} and can be solved by multiplying it with an integrating factor μ,

$$\mu = e^{\int(-\frac{2}{x})dx} = e^{-2\ln x} = \frac{1}{x^2}.$$

Thus

$$\frac{1}{x^2}y^{-2} = \int \frac{1}{x^2}(-2x^2)dx + C = -2x + C.$$

At $x = 1$, $y = 1$, therefore

$$1 = -2 + C, \quad C = 3.$$

Hence the specific solution of the original nonlinear linear differential equation is

$$y^{-2} = -2x^3 + 3x^2$$

or written as

$$y(x) = (3x^2 - 2x^3)^{-1/2}.$$

5.3 Linear Differential Equations of Higher Order

A great many physical problems can be formulated in terms of linear differential equations. A second-order differential equation is called linear if it can be written

$$\frac{\mathrm{d}^2}{\mathrm{d}x^2}y(x) + p(x)\frac{\mathrm{d}}{\mathrm{d}x}y(x) + q(x)y(x) = h(x) \tag{5.13}$$

and nonlinear if it cannot be written in this form. To simplify the notation, this equation is also written as

$$y'' + p(x)y' + q(x)y = h(x).$$

The characteristic feature of this equation is that it is linear in the unknown function y and its derivatives. For example: $y'^2 = x$ is not linear because of the term y'^2. The equation $yy' = 1$ is also not linear because of the product yy'. The functions p and q are called coefficients of the equation.

If $h(x) = 0$ for all x considered, the equation becomes

$$y'' + p(x)y' + q(x)y = 0$$

and is called homogeneous. If $h(x) \neq 0$, it is called nonhomogeneous.

Another convenient way of writing a differential equation is based on the so-called operator notation. The symbol of differentiation $\frac{\mathrm{d}}{\mathrm{d}x}$ is replaced by D:

$$\frac{\mathrm{d}y}{\mathrm{d}x} = \mathrm{D}y, \quad \frac{\mathrm{d}^2y}{\mathrm{d}x^2} = \mathrm{D}^2y,$$

and so on. Therefore (5.13) can be written as

$$\mathrm{D}^2y + p(x)\mathrm{D}y + q(x)y = h(x),$$

or

$$[D^2 + p(x)D + q(x)]y = h(x).$$

If we define

$$f(D) = D^2 + p(x)D + q(x),$$

then the equation is simply

$$f(D)y = h(x).$$

A fundamental theorem about homogeneous linear differential equation is the following. If $f(D)$ is second-order, then there are two linearly independent solutions y_1 and y_2. Furthermore, any linear combination of y_1 and y_2 is also a solution. This means that if

$$f(D)y_1 = 0, \quad f(D)y_2 = 0, \tag{5.14}$$

then with any two arbitrary constants c_1 and c_2

$$f(D)(c_1 y_1 + c_2 y_2) = 0.$$

This is very easy to show,

$$f(D)(c_1 y_1 + c_2 y_2) = f(D)c_1 y_1 + f(D)c_2 y_2$$
$$= c_1 f(D)y_1 + c_2 f(D)y_2 = 0.$$

For the last step we have used (5.14). It is important to remember that this theorem does not hold for nonlinear or nonhomogeneous linear differential equations.

To discuss the general solution of the nonhomogeneous differential equation $f(D)y = h(x)$, we define a complementary function y_c and a particular solution y_p. The complementary function is the solution of the corresponding homogeneous equation, that is

$$f(D)y_c = 0.$$

If this is an nth order equation, then y_c will contain n arbitrary constants.

The particular solution is a function when it is substituted into the original nonhomogeneous equation, the result is an identity

$$f(D)y_p(x) = h(x).$$

The particular solution can be found by various methods as we shall discuss in later sections. There is no arbitrary constant in the particular solution.

The most general solution of the nonhomogeneous differential equation is the sum of the complementary function and the particular solution

$$y(x) = y_c(x) + y_p(x). \tag{5.15}$$

It is a solution since

$$f(\mathrm{D}) \left[y_{\mathrm{c}}(x) + y_{\mathrm{p}}(x) \right] = f(\mathrm{D}) y_{\mathrm{c}}(x) + f(\mathrm{D}) y_{\mathrm{p}}(x) = h(x).$$

It is a general solution since the arbitrary constants, necessary for satisfying boundary or initial conditions, are contained in the complementary function.

These general statements are also true for first-order linear equations. For example, we have found that

$$y = \frac{\mathrm{e}^x}{2x} + C \frac{\mathrm{e}^{-x}}{x}$$

is the general solution of

$$x \frac{\mathrm{d}y}{\mathrm{d}x} + (1 + x)y = \mathrm{e}^x.$$

It can be readily verified that

$$\left[x \frac{\mathrm{d}}{\mathrm{d}x} + (1 + x) \right] \frac{\mathrm{e}^{-x}}{x} = 0,$$

$$\left[x \frac{\mathrm{d}}{\mathrm{d}x} + (1 + x) \right] \frac{\mathrm{e}^x}{2x} = \mathrm{e}^x.$$

Therefore e^{-x}/x is the complementary function and $\mathrm{e}^x/(2x)$ is the particular solution.

5.4 Homogeneous Linear Differential Equations with Constant Coefficients

We will now focus our attention on linear homogeneous differential equations with constant coefficients. In searching for the solution of a homogeneous differential equation such as

$$y'' - 5y' + 6y = 0, \tag{5.16}$$

it is natural to try

$$y = e^{mx}$$

where m is a constant, because all its derivatives have the same functional form. Substituting into (5.16) and using the fact that $y' = m e^{mx}$ and $y'' = m^2 e^{mx}$, we have

$$e^{mx}(m^2 - 5m + 6) = 0.$$

This is the condition to be satisfied if $y = e^{mx}$ is to be a solution. Since e^{mx} can never be zero, it is thus necessary that

$$m^2 - 5m + 6 = 0.$$

This purely algebraic equation is known as the characteristic or auxiliary equation of the differential equation. The roots of this equation are $m = 2$ and $m = 3$. Therefore $y_1 = \exp(2x)$ and $y_2 = \exp(3x)$ are two solutions of (5.16). The general solution is then given by a linear combination of these two functions

$$y = c_1 e^{2x} + c_2 e^{3x}. \tag{5.17}$$

In other words, all solutions of (5.16) can be written in this form. For a second-order linear differential equation, the general solution contains two arbitrary constants c_1 and c_2. These constants can be used to satisfy the initial conditions. For example, suppose it is given that at $x = 0$, $y = 0$ and $y' = 2$, then

$$y(0) = c_1 + c_2 = 0,$$
$$y'(0) = 2c_1 + 3c_2 = 2.$$

Thus $c_1 = -2$, $c_2 = 2$. So the specific solution for the differential equation together with the given initial conditions is

$$y(x) = -2e^{2x} + 2e^{3x}.$$

To facilitate further discussion, we will repeat this process in the operator notation. If we define

$$f(D) = D^2 - 5D + 6, \tag{5.18}$$

then (5.16) can be written as $f(D)y = 0$. Substituting $y = e^{mx}$ into this equation, we obtain $f(m)e^{mx} = 0$, where

$$f(m) = m^2 - 5m + 6.$$

The characteristic equation $f(m) = 0$ has two roots; $m = 2, 3$. So the solution is given by (5.17). Although obvious, it is useful to remember that to get the characteristic equation, we need only to change D in the operator function of (5.18) to m and set it to zero.

5.4.1 Characteristic Equation with Distinct Roots

Clearly this line of reasoning can be applied to any homogeneous linear differential equation with constant coefficients, regardless of its order. If $f(D)y = 0$ is a nth-order differential equation, then the characteristic equation $f(m) = 0$ is a nth-order algebraic equation. It has n roots, $m = m_1, m_2, \cdots m_n$. If they are all distinct (different from each other), exactly n independent solutions $\exp(m_1 x), \exp(m_2), \cdots \exp(m_n x)$ of the differential equation are so obtained and the general solution is

$$y = c_1 e^{m_1 x} + c_2 e^{m_2 x} + \cdots + c_n e^{m_n x}.$$

However, if one or more of the roots are repeated, less than n independent solutions are obtained in this way. Fortunately, it is not difficult to find the missing solutions.

5.4.2 Characteristic Equation with Equal Roots

To find the solutions when two or more roots of the characteristic equation
are the same, we first consider the following identities:

$$(D - a)x^n e^{ax} = Dx^n e^{ax} - ax^n e^{ax}$$
$$= (nx^{n-1} e^{ax} + ax^n e^{ax}) - ax^n e^{ax}$$
$$= nx^{n-1} e^{ax}.$$

If we apply $(D - a)$ once more to both side of this equation, we have

$$(D - a)^2 x^n e^{ax} = (D - a)nx^{n-1} e^{ax}$$
$$= n(n - 1)x^{n-2} e^{ax}.$$

It follows that:
$$(D - a)^n x^n e^{ax} = n! e^{ax}.$$

Applying $(D - a)$ once more

$$(D - a)^{n+1} x^n e^{ax} = n!(D - a)e^{ax} = 0.$$

Clearly, if we continue to apply $(D - a)$ to both the sides of the last equation,
they will all be equal to zero. Therefore

$$(D - a)^l x^n e^{ax} = 0 \quad for \ l > n.$$

This means that $\exp(ax)$, $x \exp(ax), \cdots \cdot x^{n-1} \exp(ax)$ are solutions of the
differential equation $(D - a)^n y = 0$. In other words, if the roots of the characteristic equation are repeated n times, and the common root is a, then the
general solution of the differential equation is

$$y = c_1 e^{ax} + c_2 x e^{ax} + \cdots \cdot + c_n x^{n-1} e^{ax}.$$

5.4.3 Characteristic Equation with Complex Roots

If the coefficients of the differential equation are real and the roots of the
characteristic equation have an imaginary part, then from the theory of algebraic equations, we know that the roots must come in conjugate pairs such as
$a \pm ib$. So the general solution corresponding to these two roots is

$$y = c_1 e^{(a+bi)x} + c_2 e^{(a-bi)x}. \tag{5.19}$$

There are two other very useful equivalent forms of (5.19). Since

$$e^{(a \pm bi)x} = e^{ax} e^{\pm ibx} = e^{ax}(\cos bx \pm i \sin bx),$$

we can write (5.19) as

$$y = e^{ax}[c_1 \cos bx + ic_1 \sin bx + c_2 \cos bx - ic_2 \sin bx]$$
$$= e^{ax}[(c_1 + c_2) \cos bx + (ic_1 - ic_2) \sin bx].$$

Since c_1 and c_2 are arbitrary constants, we can replace $c_1 + c_2$ and $ic_1 - ic_2$ by two new arbitrary constants A and B. Therefore

$$y = e^{ax}(A \cos bx + B \sin bx). \tag{5.20}$$

We can write (5.20) in still another form. Recall

$$C \cos(bx - \phi) = C \cos bx \cos \phi + C \sin bx \sin \phi.$$

If we put

$$C \cos \phi = A, \quad C \sin \phi = B,$$

then $C = (A^2 + B^2)^{1/2}$ and $\phi = \tan^{-1}(B/A)$, and (5.20) becomes

$$y = Ce^{ax} \cos(bx - \phi). \tag{5.21}$$

Therefore (5.19–5.21) are all equivalent. They all contain two arbitrary constants. One set of constants can be easily transformed into another set. However, there is seldom any need to do this. In solving actual problems we simply use the form that seems best for the problem at hand, and determine the arbitrary constants in that form from the given conditions.

We summarize in Table 5.1 the relationships between the roots of the characteristic equation and the solution of the differential equation.

Table 5.1. Relationship between the roots of the characteristic equation and the general solution of the differential equation

m	y(x)
0	c_1
0, 0	$c_1 + c_2 x$
0, 0, 0	$c_1 + c_2 x + c_3 x^2$
...	...
a	$c_1 e^{ax}$
a, a	$c_1 e^{ax} + c_2 x e^{ax}$
...	...
$\pm ib$	$c_1 \cos bx + c_2 \sin bx$
$\pm ib, \pm ib$	$(c_1 + c_2 x) \cos bx + (c_3 + c_4 x) \sin bx$
...	...
$a \pm ib$	$e^{ax}(c_1 \cos bx + c_2 \sin bx)$
$a \pm ib, a \pm ib$	$e^{ax}[(c_1 + c_2 x) \cos bx + (c_3 + c_4 x) \sin bx]$
...	...

Example 5.4.1. Find the general solution of the differential equation

$$y''' = 0.$$

Solution 5.4.1. We can write the equation as

$$D^3 y = 0.$$

The characteristic equation is

$$m^3 = 0.$$

The three roots are $0, 0, 0$. Therefore the general solution is given by

$$y = c_1 e^{0x} + c_2 x e^{0x} + c_3 x^2 e^{0x}$$
$$= c_1 + c_2 x + c_3 x^2.$$

This seems to be a trivial example. Obviously the result can be obtained by inspection. Here, we have demonstrated that by using the general method, we can find all the linear independent terms.

Example 5.4.2. Find the general solution of the differential equation

$$y''' - 6y'' + 9y' = 0.$$

Solution 5.4.2. We can write the equation as

$$(D^3 - 6D^2 + 9D)y = 0.$$

The characteristic equation is

$$m^3 - 6m^2 + 9m = m(m^2 - 6m + 9)$$
$$= m(m - 3)^2 = 0.$$

The three roots are $0, 3, 3$. Therefore the general solution is

$$y = c_1 + c_2 e^{3x} + c_3 x e^{3x}.$$

Again, for this third-order differential equation, the general solution has three arbitrary constants. To determine these constants, we need three conditions.

Example 5.4.3. Find the solution of

$$y'' + 9y = 0,$$

satisfying the boundary conditions $y(\pi/2) = 1$ and $y'(\pi/2) = 2$.

Solution 5.4.3. We can write the equation as

$$(D^2 + 9)y = 0.$$

The characteristic equation is

$$m^2 + 9 = 0.$$

The two roots of this equation are $m = \pm 3i$. Therefore the general solution according to (5.20) is

$$y(x) = A \cos 3x + B \sin 3x.$$

We also need y' to determine A and B

$$y'(x) = -3A \sin 3x + 3B \cos 3x.$$

The initial conditions require that

$$y\left(\frac{\pi}{2}\right) = A \cos \frac{3\pi}{2} + B \sin \frac{3\pi}{2}$$
$$= -B = 1,$$

$$y'\left(\frac{\pi}{2}\right) = -3A \sin \frac{3\pi}{2} + 3B \cos \frac{3\pi}{2}$$
$$= 3A = 2.$$

Thus $A = 2/3$, and $B = -1$. Therefore the solution is

$$y = \frac{2}{3} \cos 3x - \sin 3x.$$

We will get the same solution if we use either (5.19) or (5.21) instead of (5.20).

Example 5.4.4. Find the general solution of

$$y'' + y' + y = 0.$$

Solution 5.4.4. The characteristic equation is

$$m^2 + m + 1 = 0.$$

The roots of this equation are

$$m = \frac{1}{2}(-1 \pm \sqrt{1 - 4}) = -\frac{1}{2} \pm i\frac{\sqrt{3}}{2}.$$

Therefore the general solution is

$$y = e^{-x/2}\left(A \cos \frac{\sqrt{3}}{2}x + B \sin \frac{\sqrt{3}}{2}x\right).$$

Example 5.4.5. Find the general solution of the differential equation

$$(D^4 + 8D^2 + 16)y = 0.$$

Solution 5.4.5. The characteristic equation is

$$m^4 + 8m^2 + 16 = (m^2 + 4)^2 = 0.$$

The four roots of this equation are $m = \pm 2i, \pm 2i$. The two independent solutions associated with the roots $\pm 2i$ are $\cos 2x$, $\sin 2x$. The other two independent solutions corresponding to the repeated roots $\pm 2i$ are $x \cos 2x$, $x \sin 2x$. Therefore the general solution is given by

$$y = A \cos 2x + B \sin 2x + x(C \cos 2x + D \sin 2x).$$

5.5 Nonhomogeneous Linear Differential Equations with Constant Coefficients

5.5.1 Method of Undetermined Coefficients

We will use an example to illustrate that the general solution of nonhomogeneous differential equation is given by (5.15), namely the sum of the complementary function and the particular solution. For an equation, such as

$$(D^2 + 5D + 6)y = e^{3x} \tag{5.22}$$

it is not difficult to find the particular function $y_p(x)$. Because of e^{3x} in the right-hand side, we try

$$y_p(x) = ce^{3x}. \tag{5.23}$$

Replace y by $y_p(x)$, (5.22) becomes

$$(D^2 + 5D + 6)ce^{3x} = e^{3x}.$$

Since

$$(D^2 + 5D + 6)ce^{3x} = (9 + 5 \times 3 + 6)ce^{3x} = 30ce^{3x},$$

thus $c = 1/30$. The function $y_p(x)$ of (5.23) with $c = 1/30$ is therefore the particular solution y_p of the nonhomogeneous differential equation, that is

$$y_p = \frac{1}{30}e^{3x}.$$

Other than this particular solution, the nonhomogeneous equation has many more solutions. In fact the general solution of (5.22) should also have two

arbitrary constants. To find the general solution, let us first solve the corresponding homogeneous differential equation

$$(D^2 + 5D + 6)y_c = 0.$$

The general solution of this homogeneous equation is known as the complementary function y_c of the nonhomogeneous equation. Following the rules of
solving homogeneous equation, we find

$$y_c = c_1 e^{-2x} + c_2 e^{-3x}.$$

Thus

$$y = y_c + y_p$$
$$= c_1 e^{-2x} + c_2 e^{-3x} + \frac{1}{30} e^{3x} \qquad (5.24)$$

is the general solution of the nonhomogeneous equation (5.22).

First it is certainly a solution, since

$$(D^2 + 5D + 6)(y_c + y_p) = (D^2 + 5D + 6)y_c + (D^2 + 5D + 6)y_p$$
$$= 0 + e^{3x} = e^{3x}.$$

Furthermore, it has two arbitrary constants c_1 and c_2. This means that every
possible solution of the nonhomogeneous equation (5.22) can be obtained
by assigning suitable values to the arbitrary constants c_1 and c_2 in (5.24).
Clearly this principle is not limited to this particular problem.

As we have already learnt how to solve homogeneous equations, we shall
now discuss some systematical ways of finding particular solutions.

Let us consider another nonhomogeneous equation

$$(D^2 - 5D + 6)y = e^{3x}. \qquad (5.25)$$

Since this equation is only slightly different from (5.22), to find the particular
solution y_p of this equation, we may again try

$$y_p = c e^{3x}.$$

Putting it into (5.25), we find

$$(D^2 - 5D + 6)ce^{3x} = (9 - 15 + 6)ce^{3x} = 0.$$

Obviously no value of c can make it equal to e^{3x}. Clearly, we need a more
general method.

The idea is that if we can transform the nonhomogeneous equation into a
homogeneous equation, then we know what to do. First we note that

$$(D - 3)e^{3x} = 0.$$

Applying $(D-3)$ to both side of (5.25), we have

$$(D-3)(D^2-5D+6)y = (D-3)e^{3x} = 0. \tag{5.26}$$

Then we note that the general solution y_c+y_p of the nonhomogeneous equation (5.25) must also satisfy the newly formed homogeneous equation (5.26), since

$$(D-3)(D^2-5D+6)(y_c+y_p) = (D-3)(0+e^{3x}) = 0, \tag{5.27}$$

where we have used

$$(D^2-5D+6)y_c = 0 \tag{5.28}$$

and

$$(D^2-5D+6)y_p = e^{3x}. \tag{5.29}$$

This means we can obtain the general solution y_c+y_p of the original nonhomogeneous equation (5.25) by assigning certain specific values to some constants in the general solution of the newly formed homogeneous equation (5.26). For example, since the roots of the characteristic equation $(m-3)(m^2-5m+6) = 0$ of the newly formed homogeneous equation are

$$m = 2, 3, 3,$$

the general solution is

$$y = c_1 e^{2x} + c_2 e^{3x} + c_3 x e^{3x}. \tag{5.30}$$

But the complementary function given by (5.28) is

$$y_c = c_1 e^{2x} + c_2 e^{3x},$$

we see that the particular solution y_p can be obtained by assigning an appropriate value to c_3, since the general solution of the original nonhomogeneous equation $y_c + y_p$ is a special solution of the newly formed homogeneous equation. To determine c_3, we substitute $c_3 x e^{3x}$ into (5.29)

$$(D^2-5D+6)c_3 x e^{3x} = e^{3x}.$$

Since

$$(D^2-5D+6)c_3 x e^{3x} = c_3[D(e^{3x}+3xe^{3x})-5(e^{3x}+3xe^{3x})+6xe^{3x}]$$
$$= c_3 e^{3x},$$

clearly $c_3 = 1$. Thus the particular solution is

$$y_p = x e^{3x}.$$

Hence the general solution of the nonhomogeneous differential equation (5.25) is

$$y = c_1 e^{2x} + c_2 e^{3x} + x e^{3x}.$$

Now we summarize the general procedure of solving a nonhomogeneous differential equation

$$f(D)y(x) = h(x).$$

1. The general solution is

$$y = y_c + y_p,$$

   where y_c is the complementary function and y_p is the particular solution, and

$$f(D)y_c = 0, \quad f(D)y_p = h(x).$$

2. Let the roots of $f(m) = 0$ be

$$m = m_1, m_2 \ldots \ldots$$

   Complementary function y_c is given by the linear combination of all the linear independent functions arising from m.
3. To find y_p, we first find another equation such that

$$g(D)h(x) = 0.$$

   Applying $g(D)$ to both sides of $f(D)y(x) = h(x)$, we have

$$g(D)f(D)y = 0.$$

4. The general solution $y_c + y_p$ of the original nonhomogeneous differential equation $f(D)y(x) = h(x)$ is a special solution of the newly formed homogeneous differential equation $g(D)f(D)y = 0$, since

$$g(D)f(D)(y_c + y_p) = g(D)[f(D)y_c + f(D)y_p]$$
$$= g(D)[0 + h(x)] = 0.$$

5. The general solution of $g(D)f(D)y = 0$ is associated with the roots of the characteristic equation $g(m)f(m) = 0$.
   Since the roots of $f(m) = 0$ lead to y_c, the roots of $g(m) = 0$ must be associated with y_p.
6. Let the roots of $g(m) = 0$ be

$$m = m_1', m_2' \ldots \ldots$$

   If there is no duplication between $m_1', m_2', \cdots$ and $m_1, m_2, \cdots \ldots \ldots$, then the particular solution y_p is given by the linear combination of all the linear independent functions arising from $m_1', m_2', \cdots$.
   If there is duplication between $m_1', m_2', \cdots$ and $m_1, m_2, \cdots$, then the functions arising from $m_1', m_2', \cdots$ must be multiplied by the lowest positive integer powers of x which will eliminate all such duplications.

7. Determine the arbitrary constants in the functions arising from $m_1', m_2' \cdots$ from

$$f(D)y_p = h(x).$$

This procedure may seem to be complicated. Once understood, the implementation is actually relative simple. This we illustrate with following examples.

Example 5.5.1. Find the general solution of

$$(D^2 + 5D + 6)y = 3e^{-2x} + e^{3x}.$$

Solution 5.5.1. The characteristic equation is

$$f(m) = m^2 + 5m + 6 = 0$$

its roots are

$$m = -2, -3. \ (m_1 = -2, \ m_2 = -3).$$

Therefore

$$y_c = c_1 e^{-2x} + c_2 e^{-3x}.$$

For

$$g(D)(3e^{-2x} + e^{3x}) = 0,$$

the roots of

$$g(m) = 0$$

must be

$$m = -2, 3 \ (m_1' = -2, \ m_2' = 3).$$

Since m_1' repeats m_1, the term arising from $m_1' = -2$ must be multiplied by x. Therefore

$$y_p = c_3 x e^{-2x} + c_4 e^{3x}.$$

Since

$$(D^2 + 5D + 6)(c_3 x e^{-2x} + c_4 e^{3x}) = c_3 e^{-2x} + 30 c_4 e^{3x},$$

$$(D^2 + 5D + 6)y_p = 3e^{-2x} + e^{3x},$$

we have

$$c_3 = 3; \quad c_4 = \frac{1}{30}.$$

Thus

$$y = c_1 e^{-2x} + c_2 e^{-3x} + 3x e^{-2x} + \frac{1}{30} e^{3x}.$$

Example 5.5.2. Find the general solution of

$$(D^2 + 1)y = x^2.$$

Solution 5.5.2. The roots of the characteristic equation $f(m) = m^2 + 1 = 0$ are

$$m = \pm i,$$

therefore

$$y_c = c_1 \cos x + c_2 \sin x.$$

For $g(D)x^2 = 0$, we can regard x^2 as $c_1 + c_2 x + c_3 x^2$ with $c_1 = 0$, $c_2 = 0$, $c_3 = 1$. The roots of $g(m) = 0$, as seen in Table 5.1, are

$$m = 0, 0, 0.$$

Thus

$$y_p = a + bx + cx^2.$$

Substituting into the original equation

$$(D^2 + 1)(a + bx + cx^2) = x^2,$$

we have

$$2c + a + bx + cx^2 = x^2.$$

Thus

$$2c + a = 0, \quad b = 0, \quad c = 1.$$

It follows that $a = -2$ and

$$y_p = -2 + x^2.$$

Finally

$$y = c_1 \cos x + c_2 \sin x - 2 + x^2.$$

Example 5.5.3. Find the general solution of

$$(D^2 + 4D + 5)y = 3e^{-2x}.$$

Solution 5.5.3. The roots of $f(m) = m^2 + 4m + 5 = 0$ are

$$m = -2 \pm i.$$

Therefore

$$y_c = e^{-2x}(c_1 \cos x + c_2 \sin x).$$

For $g(D)3e^{-2x} = 0$, $g(D) = D + 2$. Obviously, the root of $g(m) = 0$ is

$$m = -2.$$

Thus

$$y_{\mathrm{p}} = A\mathrm{e}^{-2x}.$$

With

$$(\mathrm{D}^2 + 4\mathrm{D} + 5)A\mathrm{e}^{-2x} = 3\mathrm{e}^{-2x}$$

we have

$$A = 3.$$

The general solution is therefore given by

$$y = \mathrm{e}^{-2x}(c_1 \cos x + c_2 \sin x) + 3\mathrm{e}^{-2x}.$$

Example 5.5.4. Find the general solution of

$$(\mathrm{D}^2 - 2\mathrm{D} + 1)y = x\mathrm{e}^x - \mathrm{e}^x.$$

Solution 5.5.4. The roots of $m^2 - 2m + 1 = 0$ are

$$m = 1,1 \quad (m_1 = 1, \ m_2 = 1)$$

Hence

$$y_c = c_1\mathrm{e}^x + c_2 x\mathrm{e}^x.$$

The characteristic equation of the differential equation for which $x\mathrm{e}^x - \mathrm{e}^x$ is the solution

$$g(\mathrm{D})(x\mathrm{e}^x - \mathrm{e}^x) = 0$$

is of course $g(m) = 0$. We can regard $x\mathrm{e}^x - \mathrm{e}^x$ as $a\mathrm{e}^x + bx\mathrm{e}^x$ (with $a = -1$, $b = 1$). We see from Table 5.1 that the roots of $g(m) = 0$ are

$$m = 1,1 \quad (m_1' = 1, \ m_2' = 1).$$

Therefore

$$y_{\mathrm{p}} = Ax^2\mathrm{e}^x + Bx^3\mathrm{e}^x.$$

With

$$(\mathrm{D}^2 - 2\mathrm{D} + 1)(Ax^2\mathrm{e}^x + Bx^3\mathrm{e}^x) = x\mathrm{e}^x - \mathrm{e}^x$$

we find

$$A = -\frac{1}{2}, \qquad B = \frac{1}{6}.$$

The general solution is therefore given by

$$y = c_1\mathrm{e}^x + c_2 x\mathrm{e}^x - \frac{1}{2}x^2\mathrm{e}^x + \frac{1}{6}x^3\mathrm{e}^x.$$

Example 5.5.5. Find the general solution of

$$(D^2 + 1)y = \sin x.$$

Solution 5.5.5. The roots of $m^2 + 1 = 0$ are

$$m = \pm i.$$

Therefore

$$y_c = c_1 \cos x + c_2 \sin x.$$

To find

$$g(D) \sin x = 0,$$

we can regard $\sin x$ as $a \cos x + b \sin x$ with $a = 0$ and $b = 1$. From Table 5.1, we see that the roots of $g(m) = 0$ are

$$m = \pm i \quad (m' = \pm i).$$

Thus

$$y_p = Ax \cos x + Bx \sin x.$$

With

$$(D^2 + 1)(Ax \cos x + Bx \sin x) = \sin x$$

we find

$$A = -\frac{1}{2}, \quad B = 0.$$

Therefore

$$y = c_1 \cos x + c_2 \sin x - \frac{1}{2}x \cos x.$$

5.5.2 Use of Complex Exponentials

In applied problems, the function $h(x)$ is very often a sine or a cosine representing alternating voltage in an electric circuit or a periodic force in a vibrating system. The particular solution y_p can be found more efficiently by replacing sine or cosine by the complex exponential form.

In the last example, we can replace $\sin x$ by e^{ix} and solving the equation

$$(D^2 + 1)Y = e^{ix}. \tag{5.31}$$

The solution Y will also be complex. $Y = Y_R + iY_I$. Since $e^{ix} = \cos x + i \sin x$, the equation is equivalent to

$$(D^2 + 1)Y_R = \cos x,$$
$$(D^2 + 1)Y_I = \sin x.$$

Since the second equation is exactly the same as the original equation, we see that to find y_p, we can solve (5.31) for Y and take its imaginary part. Following the procedure of the last section, we assume

$$Y = Axe^{ix},$$

so

$$DY = Ae^{ix} + iAxe^{ix},$$
$$D^2Y = 2iAe^{ix} - Axe^{ix},$$
$$(D^2 + 1)Y = 2iAe^{ix} - Axe^{ix} + Axe^{ix} = 2iAe^{ix} = e^{ix}.$$

Thus

$$A = \frac{1}{2i}.$$

Taking the imaginary part of

$$Y = \frac{1}{2i}xe^{ix} = -\frac{1}{2}ix(\cos x + i \sin x),$$

we have

$$y_{\mathrm{p}} = -\frac{1}{2}x \cos x,$$

which is, of course, the same as obtained in the last example.

5.5.3 Euler–Cauchy Differential Equations

An equation of the form

$$a_n x^n \frac{d^n y}{dx^n} + a_{n-1}x^{n-1}\frac{d^{n-1}y}{dx^{n-1}} + \cdots a_1 x \frac{dy}{dx} + a_0 y = h(x), \qquad (5.32)$$

where the a_i are constants, is called Euler, or Cauchy, or Euler–Cauchy differential equation. By a change of variable, it can be transformed into an equation with constant coefficients which can then be solved.

If we set

$$x = e^z, \quad z = \ln x,$$

then

$$\frac{dz}{dx} = \frac{1}{x} = e^{-z}.$$

With the notation $D = \dfrac{d}{dz}$, we can write

$$\frac{dy}{dx} = \frac{dy}{dz}\frac{dz}{dx} = e^{-z}Dy,$$

$$\begin{aligned}
\frac{d^2y}{dx^2} &= \frac{d}{dx}\frac{dy}{dx} = \frac{d}{dx}(e^{-z}Dy) = \frac{d}{dz}(e^{-z}Dy)\frac{dz}{dx} \\
&= (-e^{-z}Dy + e^{-z}D^2y)e^{-z} = e^{-2z}D(D-1)y,
\end{aligned}$$

$$\begin{aligned}
\frac{d^3y}{dx^3} &= \frac{d}{dz}\left[e^{-2z}D(D-1)y\right]\frac{dz}{dx} \\
&= \left[-2e^{-2z}D(D-1)y + e^{-2z}D^2(D-1)y\right]e^{-z} \\
&= e^{-3z}D(D-1)(D-2)y.
\end{aligned}$$

Clearly

$$\frac{d^n y}{dx^n} = e^{-nz} D(D-1)(D-2)\cdots(D-n+1)y. \qquad (5.33)$$

Substituting (5.33) into (5.32) and using $x^n = e^{nz}$, we have a differential equation with constants coefficients,

$$a_n D(D-1)(D-2)\cdots(D-n+1)y + \cdots a_1 Dy + a_0 y = h(e^z).$$

If the solution of this equation is denoted

$$y = F(z),$$

then the solution of the original equation is given by

$$y = F(\ln x).$$

The following example will make this procedure clear.

Example 5.5.6. Find the general solution of

$$x^2 \frac{d^2 y}{dx^2} + x\frac{dy}{dx} - y = x\ln x.$$

Solution 5.5.6. With $x = e^z$, this equation becomes

$$[D(D-1) + D - 1]\,y = ze^z.$$

The complementary function comes from

$$m(m-1) + m - 1 = m^2 - 1 = 0, \quad m = 1, \ -1,$$

which gives

$$y_c = c_1 e^z + c_2 e^{-z}.$$

For $g(D)ze^z = 0$, $g(D)$ must be $(D-1)^2$. The characteristic equation is then

$$(m'-1)^2 = 0, \quad m' = 1,\ 1.$$

Therefore the particular solution is of the form

$$y_p = c_3 z e^z + c_4 z^2 e^z.$$

Substituting it back into the differential equation

$$[D(D-1) + D - 1]\,(c_3 z e^z + c_4 z^2 e^z) = ze^z,$$

we find

$$c_3 = -\frac{1}{4}, \quad c_4 = \frac{1}{4}.$$

Therefore

$$y = c_1 e^z + c_2 e^{-z} - \frac{1}{4} z e^z + \frac{1}{4} z^2 e^z.$$

For the general solution of the original equation, we must change z back to x. With $z = \ln x$

$$y(x) = c_1 x + c_2 \frac{1}{x} - \frac{1}{4} x \ln x + \frac{1}{4} x (\ln x)^2,$$

which can be readily verified that this is indeed the general solution with two arbitrary constants.

For a homogeneous Euler–Cauchy equation, the following procedure is perhaps simpler. For example, to solve the equation

$$x^2 \frac{d^2 y}{dx^2} + x \frac{dy}{dx} - y = 0,$$

we can simply start with the solution

$$y(x) = x^m.$$

So

$$x^2 m(m-1) x^{m-2} + x m x^{m-1} - x^m = 0$$

or

$$[m(m-1) + m - 1] x^m = 0.$$

Thus

$$m(m-1) + m - 1 = m^2 - 1 = 0,$$

$$m = 1, \quad m = -1.$$

It follows that:

$$y(x) = c_1 x + c_2 \frac{1}{x}.$$

5.5.4 Variation of Parameters

The method of undetermined coefficients is simple and has important physical applications, but it applies only to constant coefficient equations with special forms of the nonhomogeneous term $h(x)$. In this section, we discuss the method of variation of parameters, which is more general. It applies to equations

$$(D^2 + p(x)D + q(x))y = h(x), \tag{5.34}$$

where p, q, and h are continuous functions of x in some interval. Let

$$y_c = c_1 y_1(x) + c_2 y_2(x)$$

be the solution of the corresponding homogeneous equation

$$(D^2 + p(x)D + q(x))y_c = 0.$$

The method of variation of parameters involves replacing the parameters c_1 and c_2 by functions u and v to be determined so that

$$y_p = u(x)y_1(x) + v(x)y_2(x)$$

is the particular solution of (5.34). Now this expression contains two unknown functions u and v, but the requirement that y_p satisfies (5.34) imposes only one condition on u and v. Therefore, we are free to impose a second arbitrary condition without loss of generality. Further calculation will show that it is convenient to require

$$u'y_1 + v'y_2 = 0. \tag{5.35}$$

Now

$$Dy_p = u'y_1 + uy_1' + v'y_2 + vy_2'.$$

With the imposed condition (5.35), we are left with

$$y_p' = uy_1' + vy_2'.$$

It follows that:

$$D^2 y_p = u'y_1' + uy_1'' + v'y_2' + vy_2''.$$

Substituting them back into the equation

$$(D^2 + p(x)D + q(x))y_p = h(x)$$

and collecting terms, we have

$$u(y_1'' + py_1' + qy_1) + v(y_2'' + py_2' + qy_2) + u'y_1' + v'y_2' = h.$$

Since y_1 and y_2 satisfy the homogeneous equation, the quantities in the parenthesis are equal to zero. Thus

$$u'y_1' + v'y_2' = h.$$

This equation together with the imposed condition (5.35) can be solved for u' and v'

$$u' = -\frac{hy_2}{y_1y_2' - y_2y_1'}, \quad v' = \frac{hy_1}{y_1y_2' - y_2y_1'}. \tag{5.36}$$

The quantity in the denominator, known as the Wronskian W of y_1 and y_2,

$$W = \begin{vmatrix} y_1 & y_2 \\ y_1' & y_2' \end{vmatrix} = y_1y_2' - y_2y_1'$$

will not be equal to zero as long as y_1 and y_2 are linearly independent. Integration of (5.36) will enable us to determine u and v

$$u = -\int \frac{hy_2}{W}dx, \qquad v = \int \frac{hy_1}{W}dx.$$

The particular solution is then

$$y_p = -y_1 \int \frac{hy_2}{W}dx + y_2 \int \frac{hy_1}{W}dx.$$

Example 5.5.7. Use the variation of parameters method to find the general solution of

$$(D^2 + 4D + 4)y = 3xe^{-2x}$$

Solution 5.5.7. This equation can be solved easily by the method of undetermined coefficients. But we want to show that the solution can also be found by the variation of parameters method. Since

$$m^2 + 4m + 4 = (m+2)^2 = 0, \qquad m = -2, \ -2$$

the two independent solutions of the homogeneous equation are

$$y_1 = e^{-2x}, \qquad y_2 = xe^{-2x}.$$

The Wronskian of y_1 and y_2 is

$$W = \begin{vmatrix} e^{-2x} & xe^{-2x} \\ -2e^{-2x} & e^{-2x} - 2xe^{-2x} \end{vmatrix} = e^{-4x}(1-2x) + 2xe^{-4x} = e^{-4x}$$

and u' and v' are given by

$$u' = -\frac{3xe^{-2x}y_2}{W} = -\frac{3xe^{-2x}xe^{-2x}}{e^{-4x}} = -3x^2,$$

$$v' = \frac{3xe^{-2x}y_1}{W} = \frac{3xe^{-2x}e^{-2x}}{e^{-4x}} = 3x.$$

It follows that:

$$u = -\int 3x^2 dx = -x^3 + c_1, \qquad v = \int 3x\,dx = \frac{3}{2}x^2 + c_2.$$

Therefore

$$y = (-x^3 + c_1)e^{-2x} + \left(\frac{3}{2}x^2 + c_2\right)xe^{-2x}$$

$$= \frac{1}{2}x^3 e^{-2x} + c_1 e^{-2x} + c_2 xe^{-2x}.$$

This is the general solution. It is seen that this solution includes the complementary function

$$y_c = c_1 e^{-2x} + c_2 xe^{-2x}$$

and the particular solution

$$y_p = \frac{1}{2}x^3 e^{-2x}.$$

Example 5.5.8. Find the general solution of

$$(D^2 + 2D + 1)y = h(x), \quad h(x) = \frac{2}{x^2}e^{-x}.$$

Solution 5.5.8. This equation cannot be solved by the method of undetermined coefficients, therefore we seek the solution by variation of parameters. The complementary function is obtained from

$$m^2 + 2m + 1 = (m+1)^2 = 0, \quad m = -1, \, -1$$

to be

$$y_c = c_1 y_1 + c_2 y_2 = c_1 e^{-x} + c_2 x e^{-x}.$$

Let

$$y_p = u y_1 + v y_2.$$

From the Wronskian of y_1 and y_2

$$W = \begin{vmatrix} e^{-x} & x e^{-x} \\ -e^{-x} & e^{-x} - x e^{-x} \end{vmatrix} = e^{-2x}(1-x) + x e^{-2x} = e^{-2x},$$

we have

$$u = -\int \frac{y_2 h}{W} = -\int \frac{x e^{-x} 2 x^{-2} e^{-x}}{e^{-2x}} dx = -2 \ln x,$$

$$v = \int \frac{y_1 h}{W} = \int \frac{e^{-x} 2 x^{-2} e^{-x}}{e^{-2x}} dx = -\frac{2}{x}.$$

Thus

$$y_p = -2 e^{-x} \ln x - 2 e^{-x}.$$

Therefore the general solution is

$$y = c_1 e^{-x} + c_2 x e^{-x} - 2 e^{-x} \ln x.$$

Note that we have dropped the term $-2 e^{-x}$, since it is absorbed in $c_1 e^{-x}$.

5.6 Mechanical Vibrations

There are countless applications of differential equations in engineering and physical sciences. As illustrative examples, we will first discuss mechanical vibrations. Any motion that repeats itself after certain time interval is called vibration or oscillation. A simple model is the spring–mass system shown in Fig. 5.1. The block of mass m is constrained to move on a frictionless table and

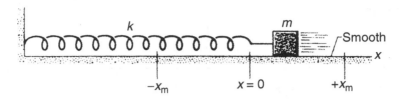

Fig. 5.1. A simple harmonic oscillator. The block moves on a frictionless table. The equilibrium position of the spring is at $x = 0$. At $t = 0$, the block is released from rest at $x = x_m$

is fastened to a spring with spring constant k. The block is pulled a distance x_m from its equilibrium position at $x = 0$ and released from rest. We want to know the subsequent motion of the block.

This system is simple enough for us to demonstrate the following steps in mathematical physics:

- Formulate the physical problem in terms of mathematical language, usually in the form of a differential equation.
- Solve the mathematical equation.
- Understand the physical meaning of the mathematical solution.

5.6.1 Free Vibration

The first step is to observe that the only horizontal force on the block is coming from the spring. According to the Hooke's law, the force is proportional to the displacement but opposite in sign, that is

$$F = -kx.$$

The motion of the block is governed by the Newton's dynamic equation

$$F = ma.$$

Since the acceleration is equal to the second derivative of the displacement

$$a = \frac{\mathrm{d}^2 x}{\mathrm{d}t^2},$$

therefore

$$m\frac{\mathrm{d}^2 x}{\mathrm{d}t^2} = -kx.$$

This is a second-order linear homogeneous differential equation. Since the block is released from rest at $x = x_m$, the velocity of the block, which is the first derivative of the displacement, is zero at $t = 0$. Therefore the initial conditions are

$$x(0) = x_m, \quad v(0) = \left.\frac{\mathrm{d}x}{\mathrm{d}t}\right|_{t=0} = 0.$$

With the differential equation and the initial conditions, the mathematical problem is uniquely defined.

The second step is to solve this equation. Since the coefficients are constants, the solution of the differential equation is of the exponential form, $x = \exp(\alpha t)$ with α determined by the characteristic equation

$$m\alpha^2 = -k.$$

Clearly the roots of this equation are

$$\alpha = \pm\sqrt{-\frac{k}{m}} = \pm i\omega_0, \quad \omega_0 = \sqrt{\frac{k}{m}}.$$

Thus the general solution of the differential equation is given by

$$x(t) = Ae^{i\omega_0 t} + Be^{-i\omega_0 t}.$$

The initial conditions requires that

$$x(0) = A + B = x_m,$$

$$\left.\frac{dx}{dt}\right|_{t=0} = i\omega_0 A - i\omega_0 B = 0.$$

Therefore $A = B = \frac{1}{2}x_m$, and

$$x(t) = x_m \cos \omega_0 t.$$

The third step is to interpret this solution. The cosine function varies between 1 and -1, it repeats itself when its argument is increased by 2π. Therefore the block oscillates between x_m and $-x_m$. The period T of the oscillation is defined as the time required for the motion to repeat itself, this means

$$x(t + T) = x(t).$$

Thus

$$\cos(\omega_0 t + \omega_0 T) = \cos(\omega_0 t).$$

Clearly

$$\omega_0 T = 2\pi.$$

Therefore the period is given by

$$T = \frac{2\pi}{\omega_0} = 2\pi\sqrt{\frac{m}{k}}.$$

The frequency f is defined as the number of oscillations in one second, that is

$$f = \frac{1}{T} = \frac{\omega_0}{2\pi} = \frac{1}{2\pi}\sqrt{\frac{k}{m}}.$$

Thus the block oscillates with a frequency that is prescribed by k and m. Since $\omega_0 = 2\pi f$, ω_0 is called angular frequency. Often ω_0 is referred simply as the natural frequency with the understanding that it is actually the angular frequency.

From the solution of the differential equation, we can derive all attributes of the motion, such as the velocity of the block at any given time. This type of periodic motion is called simple harmonic motion. The mass–spring system is known as a harmonic oscillator.

5.6.2 Free Vibration with Viscous Damping

In practical systems, the amplitude of the oscillation gradually decreases due to friction. This is known as damping. For example, if the system is vibrating in a fluid medium, such as air, water, oil, the resisting force offered by the viscosity of the fluid is generally proportional to the velocity of the vibrating body. Therefore with viscous damping, there is an additional force

$$F_{\mathrm{v}} = -c\frac{\mathrm{d}x}{\mathrm{d}t}$$

where c is the coefficient of viscous damping and the negative sign indicates that the damping force is opposite to the direction of velocity. Thus the equation of motion of the mass–spring system becomes

$$m\frac{\mathrm{d}^2x}{\mathrm{d}t^2} = -kx - c\frac{\mathrm{d}x}{\mathrm{d}t}$$

or

$$m\frac{\mathrm{d}^2x}{\mathrm{d}t^2} + c\frac{\mathrm{d}x}{\mathrm{d}t} + kx = 0.$$

This equation can be written in the form

$$\frac{\mathrm{d}^2x}{\mathrm{d}t^2} + 2\beta\frac{\mathrm{d}x}{\mathrm{d}t} + \omega_0^2 x = 0,$$

where $\beta = c/2m$, $\omega_0^2 = k/m$. With $x = \exp(\alpha t)$, α must satisfy the equation

$$\alpha^2 + 2\beta\alpha + \omega_0^2 = 0.$$

The roots of this equation are

$$\alpha_1 = -\beta + \sqrt{\beta^2 - \omega_0^2}, \qquad \alpha_2 = -\beta - \sqrt{\beta^2 - \omega_0^2}.$$

The solution is therefore given by

$$x(t) = A_1 e^{\alpha_1 t} + A_2 e^{\alpha_2 t}. \tag{5.37}$$

Depending on the strength of damping, this solution takes the following three forms.

Over damping. If $\beta^2 > \omega_0^2$, then the values of α_1 and α_2 are both negative. Thus both terms in x exponentially go to zero as $t \to \infty$. In this case, the damping force represented by β overpowers the restoring force represented by ω_0 and hence prevents oscillation. The system is called overdamped.

Critical damping. In this case $\beta^2 = \omega_0^2$, the characteristic equation has a double root at $\alpha = -\beta$ twice. Hence the solution is of the form

$$x(t) = (A + Bt)e^{-\beta t}.$$

Since $\beta > 0$, both $e^{-\beta t}$ and $te^{-\beta t}$ go to zero as $t \to \infty$. The motion dies out with time and is not qualitatively different from the overdamped motion. In this case the damping force is just as strong as the restoring force, therefore the system is called critically damped.

Under damping. If $\beta^2 < \omega_0^2$, then the roots of the characteristic equation are complex

$$\alpha_1, \alpha_2 = -\beta \pm i\omega,$$

where

$$\omega = \sqrt{\omega_0^2 - \beta^2}.$$

Therefore the solution $x = e^{-\beta t}(Ae^{i\omega t} + Be^{-i\omega t})$ can be written in the form of

$$x(t) = Ce^{-\beta t}\cos(\omega t + \varphi).$$

Because of the cosine term in the solution, the motion is oscillatory. Since the maximum value of cosine is one, the displacement x must lie between the curves $x(t) = \pm Ce^{-\beta t}$. Hence it resembles a cosine curve with decreasing amplitude. In this case, the damping force represented by β is weaker than the restoring force represented by ω and thus cannot prevent oscillation. For this reason, the system is called under damped.

These three cases are illustrated in the following example with specific parameters.

Example 5.6.1. The displacement $x(t)$ of a damped harmonic oscillator satisfies the equation

$$\frac{d^2x}{dt^2} + 2\beta\frac{dx}{dt} + \omega_0^2 x = 0.$$

Let the initial conditions be

$$x(0) = x_0; \qquad v(0) = \frac{dx}{dt}\bigg|_{t=0} = 0.$$

Find x as a function of time t, if $\omega_0 = 4$, and (a) $\beta = 5$, (b) $\beta = 4$, (c) $\beta = 1$. Show a sketch of the solutions of these three cases.

Solution 5.6.1. (a) Since $\beta = 5$ and $\omega_0 = 4$, the motion is overdamped. The roots of the characteristic equation are

$$\alpha_1, \alpha_2 = -5 \pm \sqrt{25 - 16} = -2, \; -8.$$

Thus

$$x(t) = A_1 e^{-2t} + A_2 e^{-8t}.$$

The initial conditions require A_1 and A_2 to satisfy

$$A_1 + A_2 = x_0, \qquad -2A_1 - 8A_2 = 0.$$

Therefore

$$x(t) = x_0 \left(\frac{4}{3} e^{-2t} - \frac{1}{3} e^{-8t} \right).$$

The graph of this function is shown as the dotted line in Fig. 5.2.

(b) Since $\beta = 4$ and $\omega_0 = 4$, the motion is critically damped. The roots of the characteristic equation are $-\beta$ twice

$$\alpha_1, \alpha_2 = -\beta = -4.$$

Thus

$$x(t) = (A + Bt)e^{-4t}.$$

From the initial conditions, we find

$$A = x_0, \qquad B = 4x_0.$$

Therefore

$$x(t) = x_0(1 + 4t)e^{-4t}.$$

The graph of this function is shown as the line of open circles in Fig. 5.2.

(c) For $\beta = 1$ and $\omega_0 = 4$, the motion is under damped. The solution can be written as

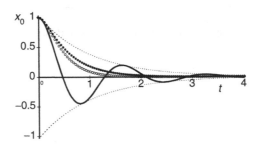

Fig. 5.2. Free vibrations with viscous damping. Initially the block is at x_0 and released from rest. The *dotted line* is for the over damped motion, the *line of open circles* is for the critically damped motion, the *solid line* is for the under damped motion. The damped amplitude is shown as *dashed lines*

$$x(t) = Ce^{-\beta t} \cos(\omega t - \phi),$$

where

$$\omega = \sqrt{\omega_0^2 - \beta^2} = \sqrt{15}.$$

From the initial conditions

$$x(0) = C \cos(-\phi) = x_0,$$
$$\left. \frac{dx}{dt} \right|_{t=0} = -\beta C \cos(-\phi) - \omega C \sin(-\phi) = 0,$$

we find

$$\tan \phi = \frac{\beta}{\omega} = \frac{1}{\sqrt{15}}, \qquad C = \frac{1}{\cos \phi} x_0.$$

Since $\cos \phi = (1 + \tan^2 \phi)^{-1/2}$,

$$C = \frac{\sqrt{\omega^2 + \beta^2}}{\omega} x_0 = \frac{4}{\sqrt{15}} x_0.$$

Therefore

$$x(t) = \frac{4}{\sqrt{15}} x_0 e^{-t} \cos \left(\sqrt{15} t - \tan^{-1} \frac{1}{\sqrt{15}} \right).$$

The graph of this function is shown as the solid line in Fig. 5.2. The damped amplitude $Ce^{-\beta t}$ is also shown as the dashed line.

In the over damped and critically damped cases, if there is a large negative initial velocity, it is possible for the block to overshoot the equilibrium position. In that case, it will come to a temporarily stop in a negative x position. After that, it will return to $x = 0$ in a monotonically decreasing way. Therefore even if it overshoots the equilibrium position, it can do that only once. Hence the motion is not oscillatory.

The under damped motion is oscillatory although its amplitude is approaching zero as time goes to ∞. The damped frequency $\sqrt{\omega_0^2 - \beta^2}$ is always less than the natural frequency ω_0.

5.6.3 Free Vibration with Coulomb Damping

From the first course of physics, we all learned that the friction force of a block sliding on a plane is proportional to the normal force acting on the plane of contact. This friction force acts in a direction opposite to the direction of velocity and is given by

$$F_c = \mu N$$

where N is the normal force and μ is the coefficient of friction. When the motion is damped by this friction force, it is known as Coulomb damping. Charles Augustin Coulomb (1736–1806) first proposed this relationship, but he is much better known for his law of electrostatic force. His name is also remembered through the unit of electric charge. Coulomb damping is also known as constant damping, since the magnitude of the damping force is independent of the displacement and velocity.

However, the sign of the friction force changes with the direction of the velocity, and we need to consider the motion in two directions separately.

When the block moves from right to left, the friction force is pointing toward the right and has a positive sign. With this friction force, the equation of motion is given by

$$m\frac{d^2x}{dt^2} = -kx + \mu N.$$

This equation has a constant nonhomogeneous term. The solution is

$$x(t) = A\cos\omega_0 t + B\sin\omega_0 t + \frac{\mu N}{k}, \tag{5.38}$$

where $\omega_0 = \sqrt{k/m}$, which is the same as the angular frequency of the undamped oscillator.

When the block is moving from left to right, the friction force is pointing toward the left and has a negative sign, and the equation of motion becomes

$$m\frac{d^2x}{dt^2} = -kx - \mu N.$$

The solution of this equation is

$$x(t) = C\cos\omega_0 t + D\sin\omega_0 t - \frac{\mu N}{k}. \tag{5.39}$$

The constants A, B, C, D are determined by the initial conditions. For example, if the block is released from rest at a distance x_0 to the right of the equilibrium position, then $x(0) = x_0$ and the velocity v, which is the first derivative of x, at $t = 0$ is zero. In this case, the motion starts from right to left. Using (5.38), we have

$$v(t) = \frac{dx}{dt} = -\omega A\sin\omega_0 t + \omega B\cos\omega_0 t.$$

Thus the initial conditions are

$$x(0) = A + \frac{\mu N}{k} = x_0, \quad v(0) = \omega_0 B = 0.$$

Therefore $A = x_0 - \mu N/k$, $B = 0$, and

$$x(t) = (x_0 - \frac{\mu N}{k})\cos\omega_0 t + \frac{\mu N}{k}.$$

This equation represents a simple harmonic motion with the equilibrium position shifted from zero to $\mu N/k$. However, this equation is valid only for the first half of the first cycle. When $t = \pi/\omega_0$, the velocity of the block is equal to zero and the block is at its extreme left position x_1 which is

$$x_1 = \left(x_0 - \frac{\mu N}{k}\right)\cos\omega_0\frac{\pi}{\omega_0} + \frac{\mu N}{k} = -x_0 + 2\frac{\mu N}{k}.$$

In the next half cycle, the block moves from left to right, so we have to use (5.39). To determine C and D, we use the fact $x(t = \pi/\omega_0) = x_1$ and $v(t = \pi/\omega_0) = 0$. With these conditions, we have

$$x\left(t = \frac{\pi}{\omega_0}\right) = C\cos\omega_0\frac{\pi}{\omega_0} + D\sin\omega_0\frac{\pi}{\omega_0} - \frac{\mu N}{k} = -C - \frac{\mu N}{k} = x_1,$$

$$v\left(t = \frac{\pi}{\omega_0}\right) = -\omega_0 C\sin\omega_0\frac{\pi}{\omega_0} + \omega_0 D\cos\omega_0\frac{\pi}{\omega_0} = -\omega_0 D = 0.$$

Therefore $C = -x_1 - \frac{\mu N}{k} = x_0 - 3\frac{\mu N}{k}$, $D = 0$, and (5.39) becomes

$$x(t) = \left(x_0 - 3\frac{\mu N}{k}\right)\cos\omega_0 t - \frac{\mu N}{k}.$$

This is also a simple harmonic motion with the equilibrium position shifted to $-\mu N/k$. This equation is valid for $\pi/\omega_0 \le t \le 2\pi/\omega_0$. At the end of this half cycle, the velocity is again equal to zero

$$v\left(t = \frac{2\pi}{\omega_0}\right) = -\omega_0\left(x_0 - 3\frac{\mu N}{k}\right)\sin\omega_0\frac{2\pi}{\omega_0} = 0$$

and the block is at x_2 which is

$$x_2 = \left(x_0 - 3\frac{\mu N}{k}\right)\cos\omega_0\frac{2\pi}{\omega_0} - \frac{\mu N}{k} = x_0 - 4\frac{\mu N}{k}.$$

These become the initial conditions for the third half cycle, and the procedure can be continued until the motion stops. The displacement x as a function of time t of this motion is shown in Fig. 5.3.

It is to be noted that the frequency of a Coulomb damped vibration is the same as that of the free vibration without damping. This is to be contrasted with the viscous damping. Furthermore, if x_n is a local maximum, then

$$x_n - x_{n-2} = -\frac{4\mu N}{k}.$$

This means that in each successive cycle, the amplitude of the motion is reduced by $4\mu N/k$ in a time interval of $2\pi/\omega_0$. Therefore the maxima of the oscillation all fall on a straight line. The slope of this straight enveloping line is

$$-\frac{4\mu N/k}{2\pi/\omega_0} = -\frac{2\mu N\omega_0}{\pi k}.$$

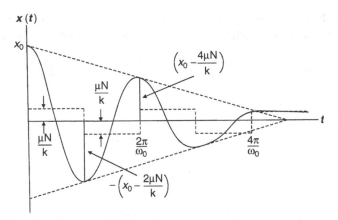

Fig. 5.3. The displacement x of the block as a function of time t in the mass–spring system with Coulomb damping. Initially the block is at $x = x_0$ and is released from rest

Similarly, all local minima must fall on a straight line with a slope of $2\mu N \omega_0 / \pi k$. These characteristics are also shown in Fig. 5.3.

The motion stops when the velocity becomes zero and the restoring force kx of the spring is equal or less than the friction force μN. Thus the number of half cycles n_0 that elapse before the motion ceases can be found from the condition

$$k\left(x_0 - n_0 \frac{2\mu N}{k}\right) \leq \mu N.$$

If $x_0 \leq \mu N/k$, the motion will not even start. For $\mu N/k < x_0 < 2\mu N/k$, the block will stop before it reaches the equilibrium position. The final position of the block is usually different from the equilibrium position and represents a permanent displacement.

5.6.4 Forced Vibration without Damping

A dynamic system is often subjected to some type of external force. In this section, we shall consider the response of a spring-mass system under the external force of the form $F_0 \cos \omega t$. First suppose there is no damping, then the equation of motion is given by

$$m\frac{d^2 x}{dt^2} + kx = F_0 \cos \omega t$$

or

$$\frac{d^2 x}{dt^2} + \omega_0^2 x = \frac{F_0}{m} \cos \omega t,$$

where $\omega_0 = (k/m)^{1/2}$ is the natural frequency of the system. The general solution is the sum of the complementary function x_c and the particular solution x_p. The complementary function satisfies the homogeneous equation

$$\frac{d^2 x_c}{dt^2} + \omega_0^2 x_c = 0$$

and is given by

$$x_c(t) = c_1 \cos \omega_0 t + c_2 \sin \omega_0 t.$$

Since the nonhomogeneous term has a frequency ω, the particular solution takes the form

$$x_p(t) = A \cos \omega t + B \sin \omega t.$$

Substituting it into the equation

$$\frac{d^2 x_p}{dt^2} + \omega_0^2 x_p = \frac{F_0}{m} \cos \omega t$$

we find $B = 0$, and $A = F_0/[m(\omega_0^2 - \omega^2)]$. Thus

$$x_p(t) = \frac{F_0}{m(\omega_0^2 - \omega^2)} \cos \omega t$$

and the general solution, given by $x_c + x_p$

$$x(t) = c_1 \cos \omega_0 t + c_2 \sin \omega_0 t + \frac{F_0}{m(\omega_0^2 - \omega^2)} \cos \omega t$$

is the sum of two periodic motions of different frequencies.

Beats. Suppose the initial conditions are $x(0) = 0$ and $v(0) = 0$, then c_1 and c_2 are found to be

$$c_1 = -\frac{F_0}{m(\omega_0^2 - \omega^2)}, \quad c_2 = 0.$$

Thus the solution is given by

$$x(t) = \frac{F_0}{m(\omega_0^2 - \omega^2)} (\cos \omega t - \cos \omega_0 t)$$

$$= \frac{2F_0}{m(\omega_0^2 - \omega^2)} \sin \frac{\omega_0 - \omega}{2} t \sin \frac{\omega_0 + \omega}{2} t.$$

Let the forcing frequency ω be slightly less than the natural frequency ω_0 so that $\omega_0 - \omega = 2\epsilon$, where ϵ is a small positive quantity. Then $\omega_0 + \omega \approx 2\omega$ and the solution becomes

$$x(t) \approx \frac{F_0}{2m\omega\epsilon} \sin \epsilon t \sin \omega t.$$

Since ϵ is small, the function $\sin \epsilon t$ varies slowly. Thus the factor

$$\frac{F_0}{2m\omega\epsilon} \sin \epsilon t$$

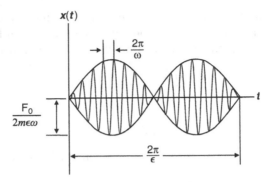

Fig. 5.4. Beats produced by the sum of two waves with approximately the same frequencies

can be regarded as the variable amplitude of the vibration whose period is $2\pi/\omega$. The oscillation of the amplitude has a large period of $2\pi/\epsilon$, which is called the period of beats. This kind of motion is shown in Fig. 5.4.

Resonance. In the case that the frequency of the forcing function is the same as the natural frequency of the system, that is $\omega = \omega_0$, then the particular solution takes the form

$$x_p(t) = At \cos \omega_0 t + Bt \sin \omega_0 t.$$

Substituting it into the original nonhomogeneous equation, we find $A = 0$, $B = F_0/(2m\omega_0)$. Thus the general solution is given by

$$x(t) = c_1 \cos \omega_0 t + c_2 \sin \omega_0 t + \frac{F_0}{2m\omega_0} t \sin \omega_0 t.$$

Because of the presence of the term $t \sin \omega_0 t$, the motion will become unbounded as $t \to \infty$. This is known as resonance. The phenomenon of resonance is characterized by x_p which is shown in Fig. 5.5. If there is damping, the

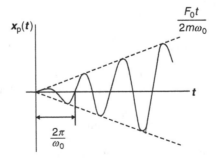

Fig. 5.5. Resonance without damping. When the forcing frequency coincide with the natural frequency, the motion will become unbounded

motion will remain bounded. However, there may still be a large response if the damping is small and ω is close to ω_0.

5.6.5 Forced Vibration with Viscous Damping

With viscous damping, the equation of motion of the spring–mass system under a harmonic forcing function is given by

$$m\frac{\mathrm{d}^2 x}{\mathrm{d}t^2} + c\frac{\mathrm{d}x}{\mathrm{d}t} + kx = F_0\cos\omega t. \tag{5.40}$$

The solution of this equation is again the sum of the complementary function and the particular solution. The complementary function satisfies the homogeneous equation

$$m\frac{\mathrm{d}^2 x_\mathrm{c}}{\mathrm{d}t^2} + c\frac{\mathrm{d}x_\mathrm{c}}{\mathrm{d}t} + kx_\mathrm{c} = 0,$$

which represents free vibrations with damping. As discussed earlier, the free vibration dies out with time under all possible initial conditions. This part of the solution is called transient. The rate at which the transient motion decays depends on the system parameters m, k, c.

The general solution of the equation eventually reduces to the particular solution which represents the steady-state vibration.

The particular solution is expected to have the same frequency as the forcing function, we can write the solution in the following form:

$$x_\mathrm{p}(t) = A\cos(\omega t - \phi). \tag{5.41}$$

Substituting it into the equation of motion, we find

$$A\left[(k - m\omega^2)\cos(\omega t - \phi) - c\omega\sin(\omega t - \phi)\right] = F_0\cos\omega t.$$

Using the trigonometric identities

$$\cos(\omega t - \phi) = \cos\omega t\cos\phi + \sin\omega t\sin\phi,$$
$$\sin(\omega t - \phi) = \sin\omega t\cos\phi - \cos\omega t\sin\phi$$

and equating the coefficients of $\cos\omega t$ and $\sin\omega t$ on both sides of the resulting equation, we obtain

$$A\left[(k - m\omega^2)\cos\phi + c\omega\sin\phi\right] = F_0, \tag{5.42a}$$
$$A\left[(k - m\omega^2)\sin\phi - c\omega\cos\phi\right] = 0. \tag{5.42b}$$

It follows from (5.42b) that:

$$(k - m\omega^2)\sin\phi = c\omega\cos\phi,$$

but $\sin^2\phi = 1 - \cos^2\phi$, so

$$(k - m\omega^2)^2(1 - \cos^2\phi) = (c\omega\cos\phi)^2.$$

Therefore

$$\cos\phi = \frac{k - m\omega^2}{[(k - m\omega^2)^2 + (c\omega)^2]^{1/2}}.$$

It follows that:

$$\sin\phi = \frac{c\omega}{[(k - m\omega^2)^2 + (c\omega)^2]^{1/2}}.$$

Substituting $\cos\phi$ and $\sin\phi$ into (5.42a), we find

$$A = \frac{F_0}{[(k - m\omega^2)^2 + (c\omega)^2]^{1/2}} = \frac{F_0}{[m^2(\omega_0^2 - \omega^2)^2 + c^2\omega^2]^{1/2}}, \tag{5.43}$$

where $\omega_0^2 = k/m$. The particular solution $x_{\mathrm{p}}(t)$ is, therefore, given by

$$x_{\mathrm{p}}(t) = \frac{F_0\cos(\omega t - \phi)}{[m^2(\omega_0^2 - \omega^2)^2 + c^2\omega^2]^{1/2}}, \tag{5.44}$$

where

$$\phi = \tan^{-1}\frac{c\omega}{m(\omega_0^2 - \omega^2)}.$$

Notice that $m^2(\omega_0^2 - \omega^2)^2 + c^2\omega^2$ is never zero, even for $\omega = \omega_0$. Hence with damping, the motion is always bounded. However, if the damping is not strong enough, the amplitude can still get to be very large.

To find the maximum amplitude, we take the derivative of A with respect to ω, and set it to zero. This shows that the frequency that makes $\dfrac{dA}{d\omega} = 0$ must satisfy the equation

$$2m^2(\omega_0^2 - \omega^2) - c^2 = 0.$$

Therefore the maximum amplitude occurs at

$$\omega = \sqrt{\omega_0^2 - \frac{c^2}{2m^2}}. \tag{5.45}$$

Note that for $c^2 > 2m^2\omega_0^2$, no real ω can satisfy this equation. In that case, there will not be any maximum for $\omega \neq 0$. The amplitude is a monotonically decreasing function of the forcing frequency.

However, if $c^2 < 2m^2\omega_0^2$, then there will be a maximum. Substituting (5.45) into the expression of A, we obtain the maximum amplitude

$$A_{\max} = \frac{2mF_0}{c(4m^2\omega_0^2 - c^2)^{1/2}}.$$

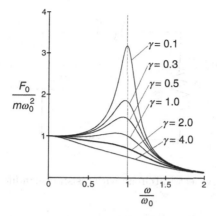

Fig. 5.6. Forced vibration with viscous damping. Amplitude of the steady-state as a function of ω/ω_0, $\gamma = (c/m\omega_0)^2$ represents the strength of damping. For $\gamma \geq 2$, there is no maximum

To see the relation between the amplitude A and the forcing frequency ω, it is convenient to express A of (5.43) as

$$A = \frac{F_0}{m\omega_0^2 \left[\left(1 - (\omega/\omega_0)^2\right)^2 + \gamma(\omega/\omega_0)^2\right]^{1/2}},$$

where $\gamma = c^2/m^2\omega_0^2$. The graphs of A in units of $F_0/m\omega_0^2$ as functions of ω/ω_0 are shown in Fig. 5.6 for several different values of γ. For $\gamma = 0$, it is the forced vibration without damping and the motion is unbound at $\omega = \omega_0$. For a small γ, the amplitude still has a sharp peak at a frequency slightly less than ω_0. As γ gets larger, the peak becomes smaller and wider. When $\gamma \geq 2$, there is no longer any maximum.

In designing structures, we want to include sufficient amount of damping to avoid resonance which can lead to disaster. On the other hand, if we design a device to detect periodic force, we would want to choose m, k, c to satisfy (5.45) so that the response of the device to such a force is maximum.

5.7 Electric Circuits

As a second example of application of theory of linear second-order differential equations with constant coefficients, we consider the simple electric circuit shown in Fig. 5.7.

It consists of three kinds of circuit elements; a resistor with a resistance R measured in ohms, an inductor with an inductance L measured in henries, and a capacitor with capacitance C measured in farads. They are connected in series with a source of electromotive force (emf) that supplies at time t

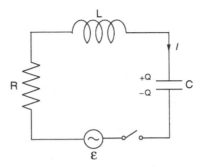

Fig. 5.7. An oscillatory electrical circuit with resistance, inductance, and capacitance

a voltage $V(t)$ measured in volts. The capacitor is a device to store electric charges Q, measured in coulombs. If the switch is closed, there will be current $I(t)$, measured in amperes, flowing in the circuit. In elementary physics, we learned that the voltage drop acrose the resistor is equal to IR, the voltage drop across the inductor is $L\dfrac{\mathrm{d}I}{\mathrm{d}t}$, and the voltage acrose the capacitor is $\dfrac{1}{C}Q$. The sum of these is equal to the applied voltage. Therefore

$$L\frac{\mathrm{d}I}{\mathrm{d}t} + RI + \frac{1}{C}Q = V(t).$$

Furthermore, the rate of increase of the charge Q on the capacitor is, by definition, equal to the current

$$\frac{\mathrm{d}Q}{\mathrm{d}t} = I.$$

With this relation, we obtain the following second-order linear nonhomogeneous equation for Q:

$$L\frac{\mathrm{d}^2Q}{\mathrm{d}t^2} + R\frac{\mathrm{d}Q}{\mathrm{d}t} + \frac{1}{C}Q = V(t).$$

Suppose the circuit is driven by a generator with a pure cosine wave oscillation, $V(t) = V_0 \cos \omega t$, then the equation becomes

$$L\frac{\mathrm{d}^2Q}{\mathrm{d}t^2} + R\frac{\mathrm{d}Q}{\mathrm{d}t} + \frac{1}{C}Q = V_0 \cos \omega t. \tag{5.46}$$

5.7.1 Analog Computation

We see that the equation describing an LRC circuit is exactly the same as (5.40), the equation describing the forced vibration of a spring–mass system with viscous damping. The fact that the same differential equation serves to describe two entirely different physical phenomena is a striking example of the

Table 5.2. The analogy between mechanical and electrical systems

Mechanical Property	Electrical Property
$m\dfrac{d^2x}{dt^2} + c\dfrac{dx}{dt} + kx = F_0 \cos\omega t$	$L\dfrac{d^2Q}{dt^2} + R\dfrac{dQ}{dt} + \dfrac{1}{C}Q = V_0 \cos\omega t$
displacement x	charge Q
velocity $v = \dfrac{dx}{dt}$	current $I = \dfrac{dQ}{dt}$
mass m	inductance L
spring constant k	inverse capacitance $1/C$
damping coefficient c	resistance R
applied force $F_0 \cos\omega t$	applied voltage $V_0 \cos\omega t$
resonant frequency $\omega_0^2 = \dfrac{k}{m}$	resonant frequency $\omega_0^2 = \dfrac{1}{LC}$

unifying role of mathematics in natural sciences. With appropriate substitutions, the solution of (5.40) can be applied to electric circuits. The correspondence between the electrical and mechanical cases are shown in Table 5.2.

The correspondence between mechanical and electrical properties can also be used to construct an electrical model of a given mechanical system. This is a very useful way to predict the performance of a mechanical system, since the electrical elements are inexpensive and electrical measurements are usually very accurate. The method of computing the motion of a mechanical system from an electrical circuit is known as analog computation.

By directly converting $x_p(t)$ of (5.44) into its electrical equivalent, the steady-state solution of (5.46) is found to be

$$Q(t) = \frac{V_0 \cos(\omega t - \phi)}{\left[\left(\frac{1}{C} - \omega^2 L\right)^2 + (\omega R)^2\right]^{1/2}},$$

$$\phi = \tan^{-1}\frac{\omega R}{\frac{1}{C} - \omega^2 L} = \tan^{-1}\frac{R}{\frac{1}{\omega C} - \omega L}.$$

Generally, it is the current that is of primary interests, so we differentiate Q with respect to t to get the steady-state current

$$I(t) = \frac{dQ}{dt} = \frac{-\omega V_0 \sin(\omega t - \phi)}{\left[\left(\frac{1}{C} - \omega^2 L\right)^2 + (\omega R)^2\right]^{1/2}} = \frac{-V_0 \sin(\omega t - \phi)}{\left[\left(\frac{1}{\omega C} - \omega L\right)^2 + R^2\right]^{1/2}}.$$

To see more clearly the phase relation between the current $I(t)$ and the applied voltage $V(t) = \cos\omega t$, we would like to express the current also in terms of a cosine function. This can be done by noting that

$$\tan\phi = \frac{R}{\frac{1}{\omega C} - \omega L}$$

can be expressed geometrically in the following triangle.

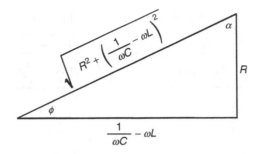

It is clear that $\phi = \frac{\pi}{2} - \alpha$ and

$$\tan \alpha = \frac{\frac{1}{\omega C} - \omega L}{R}.$$

Since $\sin(\omega t - \phi) = \sin(\omega t - \frac{\pi}{2} + \alpha) = -\cos(\omega t + \alpha)$, it follows that:

$$I(t) = \frac{V_0 \cos(\omega t + \alpha)}{\left[\left(\frac{1}{\omega C} - \omega L\right)^2 + R^2\right]^{1/2}}.$$

For reasons that will soon be clear, often $I(t)$ is written in still another form:

$$I(t) = \frac{V_0 \cos(\omega t - \beta)}{\left[\left(\frac{1}{\omega C} - \omega L\right)^2 + R^2\right]^{1/2}}, \qquad (5.47)$$

where $\beta = -\alpha$, and

$$\tan \beta = \tan(-\alpha) = -\tan \alpha = \frac{\omega L - \frac{1}{\omega C}}{R}.$$

5.7.2 Complex Solution and Impedance

The particular solution of (5.46) can be found by the complex exponential method. This method offers some computational and conceptual advantages. We can replace $V_0 \cos \omega t$ by $V_0 e^i \omega$ and solve the equation

$$L\frac{d^2 Q_c}{dt^2} + R\frac{d Q_c}{dt} + \frac{1}{C}Q_c = V_0 e^{i\omega t}. \qquad (5.48)$$

The real part of the solution Q_c is the charge Q, and the real part of I_c, defined as $\frac{d}{dt}Q_c$, is the current I. The time dependence of charges and currents must also be in the form of $e^i \omega t$

$$Q_c = \widehat{Q}e^{i\omega t}, \qquad I_c = \widehat{I}e^{i\omega t},$$

where $\widehat{Q}$ and $\widehat{I}$ are complex but independent of t. Since

$$\frac{dQ_c}{dt} = i\omega Q_c, \qquad \frac{d^2 Q_c}{dt^2} = -\omega^2 Q_c,$$

the differential equation (5.48) becomes the algebraic equation

$$(-\omega^2 L + i\omega R + \frac{1}{C})Q_c = V_0 e^{i\omega t}.$$

Clearly

$$Q_c = \frac{V_0 e^{i\omega t}}{-\omega^2 L + i\omega R + \frac{1}{C}},$$

and

$$I_c = \frac{dQ_c}{dt} = \frac{i\omega V_0 e^{i\omega t}}{-\omega^2 L + i\omega R + \frac{1}{C}} = \frac{V_0 e^{i\omega t}}{R + i(\omega L - \frac{1}{\omega C})}. \qquad (5.49)$$

Writing the denominator in the polar form

$$R + i(\omega L - \frac{1}{\omega C}) = \left[R^2 + (\omega L - \frac{1}{\omega C})^2\right]^{1/2} e^{i\beta},$$

$$\beta = \tan^{-1} \frac{\omega L - \frac{1}{\omega C}}{R},$$

We see that

$$I_c = \frac{V_0 e^{i\omega t}}{\left[R^2 + (\omega L - \frac{1}{\omega C})^2\right]^{1/2} e^{i\beta}} = \frac{V_0 e^{i(\omega t - \beta)}}{\left[R^2 + (\omega L - \frac{1}{\omega C})^2\right]^{1/2}}.$$

The real part of I_c is

$$I = \frac{V_0 \cos(\omega t - \beta)}{\left[R^2 + (\omega L - \frac{1}{\omega C})^2\right]^{1/2}},$$

which is identical to (5.47).

In electrical engineering, it is customary to define $V_0 e^{i\omega t}$ as the complex voltage V_c, and to define

$$Z = R + i\omega L + \frac{1}{i\omega C}$$

as the complex impedance Z. With these notations, (5.49) can be written in the form

$$I_c = \frac{V_c}{Z}.$$

Note that if the circuit element had consisted of the resistance R alone, the impedance would be equal simply to R, so this relation would resembles Ohm's law for a direct current circuit: $V = RI$. Thus the role the impedance plays in an alternating circuit with a sinusoidal voltage is exactly the same as the resistor in a direct current circuit.

It is a simple matter to show that if the circuit element consists only of the inductance L, the impedance is simply $i\omega L$. Similarly, with only capacitance C, the impedance is just $1/(i\omega C)$. Thus we see that when electrical elements are connected in series, the corresponding impedances combine just as simple resistances do.

In a similar way, we can show that when electrical elements are connected in parallel, the corresponding impedances also combine just as simple resistances do. For example, if R, L, C are connected in parallel, the complex current can be found by dividing the complex voltage by the simple impedance Z defined by the relation.

$$\frac{1}{Z} = \frac{1}{R} + \frac{1}{i\omega L} + i\omega C.$$

The real part of the result is the current in this AC circuit. This makes it very easy to determine the steady state behavior of an electrical system.

5.8 Systems of Simultaneous Linear Differential Equations

In many applications, it is necessary to simultaneously consider several dependent variables, each depending on the same independent variable, usually time t. The mathematical model is generally a system of linear differential equations. The elementary approach of solving systems of differential equations is to eliminate the dependent variables one by one through combining pairs of equations, until there is only one equation left containing one dependent variable. This equation will usually be of higher order, and can be solved by the methods we have discussed. Once this equation is solved, the other dependent variables can be found in turn. This method is similar to the solution of systems of simultaneous algebraic equations.

A closely related method is to find the eigenvalues of the matrix formed by the differential equations. This method provides a mathematical framework for the discussion of normal frequencies of the system, which are physically important.

5.8.1 The Reduction of a System to a Single Equation

Let us solve the following system of equations with two dependent variables $x(t)$ and $y(t)$:

$$\frac{dx}{dt} = -2x + y, \tag{5.50a}$$

$$\frac{dy}{dt} = -4x + 3y + 10\cos t \tag{5.50b}$$

with the initial conditions

$$x(0) = 0, \quad y(0) = -1.$$

From the first equation, we have

$$y = \frac{dx}{dt} + 2x, \qquad \frac{dy}{dt} = \frac{d^2x}{dt^2} + 2\frac{dx}{dt}.$$

Substitute them into the second equation

$$\frac{d^2x}{dt^2} + 2\frac{dx}{dt} = -4x + 3(\frac{dx}{dt} + 2x) + 10\cos t$$

or

$$\frac{d^2x}{dt^2} - \frac{dx}{dt} - 2x = 10\cos t. \tag{5.51}$$

This is an ordinary second-order nonhomogeneous differential equation, the complementary function $x_c(t)$ and the particular solution $x_p(t)$ are found to be, respectively, $c_1 e^{-t} + c_2 e^{2t}$ and $-3\cos t - \sin t$. Therefore

$$x = x_c + x_p = c_1 e^{-t} + c_2 e^{2t} - 3\cos t - \sin t.$$

The solution for $y(t)$ is then given by

$$y = \frac{dx}{dt} + 2x = c_1 e^{-t} + 4c_2 e^{2t} - 7\cos t + \sin t.$$

The constants c_1, c_2 are determined by the initial conditions

$$x(0) = c_1 + c_2 - 3 = 0,$$
$$y(0) = c_1 + 4c_2 - 7 = -1,$$

which gives $c_1 = 2$, $c_2 = 1$. Thus

$$x(t) = 2e^{-t} + e^{2t} - 3\cos t - \sin t,$$
$$y(t) = 2e^{-t} + 4e^{2t} - 7\cos t + \sin t.$$

If the number of coupled equations is small (2 or 3), the simplest method of solving the problem is this kind of direct substitution. However, for a larger system, one may prefer the more systematic approach of Sect. 5.8.2.

5.8.2 Cramer's Rule for Simultaneous Differential Equations

We will use the same example of the last section to illustrate this method. First, use the notation D to represent $\frac{d}{dt}$, and write the set of equations (5.50) as

$$(D + 2)x - y = 0, \tag{5.52a}$$
$$4x + (D - 3)y = 10\cos t. \tag{5.52b}$$

Recall that for a system of algebraic equations

$$a_{11}x + a_{12}y = b_1$$
$$a_{21}x + a_{22}y = b_2,$$

the solution can be obtained by the Cramer's rule

$$x = \frac{\begin{vmatrix} b_1 & a_{12} \\ b_2 & a_{22} \end{vmatrix}}{\begin{vmatrix} a_{11} & a_{12} \\ a_{21} & a_{22} \end{vmatrix}}, \quad y = \frac{\begin{vmatrix} a_{11} & b_1 \\ a_{21} & b_2 \end{vmatrix}}{\begin{vmatrix} a_{11} & a_{12} \\ a_{21} & a_{22} \end{vmatrix}}.$$

We can use the same formalism to solve a system of differential equations. That is, $x(t)$ of (5.52) can be written as

$$x = \frac{\begin{vmatrix} 0 & -1 \\ 10\cos t & (D+3) \end{vmatrix}}{\begin{vmatrix} (D+2) & -1 \\ 4 & (D-3) \end{vmatrix}},$$

or

$$\begin{vmatrix} (D+2) & -1 \\ 4 & (D-3) \end{vmatrix} x = \begin{vmatrix} 0 & -1 \\ 10\cos t & (D-3) \end{vmatrix}.$$

Expanding the determinant, we have

$$[(D+2)(D-3) + 4]x = 10\cos t.$$

This means

$$(D^2 - D - 2)x = 10\cos t,$$

which is identical to (5.51) of the last section. Proceeding in exactly the same way as in the last section, we find

$$x(t) = x_c + x_p = c_1 e^{-t} + c_2 e^{2t} - 3\cos t - \sin t.$$

Substituting it into the original differential equation, $y(t)$ is found to be

$$y(t) = c_1 e^{-t} + 4c_2 e^{2t} - 7\cos t + \sin t.$$

An alternative way of finding $y(t)$ is to note that

$$y = \frac{\begin{vmatrix} (D+2) & 0 \\ 4 & 10\cos t \end{vmatrix}}{\begin{vmatrix} (D+2) & -1 \\ 4 & (D-3) \end{vmatrix}}$$

or

$$\begin{vmatrix} (D+2) & -1 \\ 4 & (D-3) \end{vmatrix} y = \begin{vmatrix} (D+2) & 0 \\ 4 & 10\cos t \end{vmatrix}.$$

Expanding the determinant, we have

$$(D^2 - D - 2)y = 20\cos t - 10\sin t.$$

The solution of this equation is

$$y(t) = y_c + y_p = k_1 e^{-t} + k_2 e^{2t} - 7\cos t + \sin t.$$

Note that the complementary functions x_c and y_c satisfy the same homogeneous differential equation

$$(D^2 - D - 2)x_c = 0 \quad (D^2 - D - 2)y_c = 0.$$

Since we have already written $x_c = c_1 e^{-t} + c_2 e^{2t}$, we must avoid using c_1 and c_2 as the constants in y_c. That is, in $y_c = k_1 e^{-t} + k_2 e^{2t}$, k_1 and k_2 are not necessarily equal to c_1 and c_2, because there is no reason that they should be equal. To find the relationship between them, we have to substitute $x(t)$ and $y(t)$ back into one of the original differential equations. For example, substituting them back into $(D+2)x = y$, we have

$$c_1 e^{-t} + 4c_2 e^{2t} - 7\cot + \sin t = k_1 e^{-t} + k_2 e^{2t} - 7\cos t + \sin t.$$

Therefore

$$k_1 = c_1, \quad k_2 = 4c_2.$$

Thus we obtain the same result as before.

It is seen that after the first dependent variable $x(t)$ is found from Cramer's rule, it is simpler to find the second dependent variable $y(t)$ by direct substitution. If we continue to use Cramer's rule to find $y(t)$, we will introduce some additional constants which must be eliminated by substituting both $x(t)$ and $y(t)$ back into the original differential equation.

5.8.3 Simultaneous Equations as an Eigenvalue Problem

A system of simultaneous differential equations can be solved as an eigenvalue problem in matrix theory. We will continue to use the same example to illustrate the procedures of this method. First write the set of equations (5.50) in the following form:

$$-2x + y = x',$$
$$-4x + 3y = y' - 10\cos t.$$

With matrix notation, they become

$$\begin{pmatrix} -2 & 1 \\ -4 & 3 \end{pmatrix} \begin{pmatrix} x \\ y \end{pmatrix} = \begin{pmatrix} x' \\ y' - 10\cos t \end{pmatrix}.$$

Let

$$x = x_c + x_p, \quad y = y_c + y_p.$$

The complementary functions x_c and y_c satisfy the equation

$$\begin{pmatrix} -2 & 1 \\ -4 & 3 \end{pmatrix} \begin{pmatrix} x_c \\ y_c \end{pmatrix} = \begin{pmatrix} x_c' \\ y_c' \end{pmatrix}, \tag{5.53}$$

and the particular solutions x_p and y_p satisfy the equation

$$\begin{pmatrix} -2 & 1 \\ -4 & 3 \end{pmatrix} \begin{pmatrix} x_p \\ y_p \end{pmatrix} = \begin{pmatrix} x_p' \\ y_p' - 10\cos t \end{pmatrix}. \tag{5.54}$$

Since these are linear equations with constant coefficients, we assume

$$x_c = c_1 e^{\lambda t}, \quad y_c = c_2 e^{\lambda t},$$

so

$$x' = \frac{dx_c}{dt} = \lambda c_1 e^{\lambda t}, \quad y_c' = \frac{dy_c}{dt} = \lambda c_1 e^{\lambda t}.$$

It follows that the matrix equation for the complementary functions is given by:

$$\begin{pmatrix} -2 & 1 \\ -4 & 3 \end{pmatrix} \begin{pmatrix} c_1 e^{\lambda t} \\ c_2 e^{\lambda t} \end{pmatrix} = \begin{pmatrix} \lambda c_1 e^{\lambda t} \\ \lambda c_2 e^{\lambda t} \end{pmatrix}.$$

This is an eigenvalue problem

$$\begin{pmatrix} -2 & 1 \\ -4 & 3 \end{pmatrix} \begin{pmatrix} c_1 \\ c_2 \end{pmatrix} = \lambda \begin{pmatrix} c_1 \\ c_2 \end{pmatrix}$$

with eigenvalue λ and eigenvector $\begin{pmatrix} c_1 \\ c_2 \end{pmatrix}$. Therefore, λ must satisfy the secular equation

$$\begin{vmatrix} -2 - \lambda & 1 \\ -4 & 3 - \lambda \end{vmatrix} = 0$$

or

$$(-2 - \lambda)(3 - \lambda) + 4 = 0.$$

The two roots λ_1, λ_2 of this equation are easily found to be

$$\lambda_1 = -1, \quad \lambda_2 = 2.$$

Corresponding to each λ_i, there is an eigenvector $\begin{pmatrix} c_1^i \\ c_2^i \end{pmatrix}$. The coefficients c_1^i and c_2^i are not independent of each other, they must satisfy the equation

$$\begin{pmatrix} -2 & 1 \\ -4 & 3 \end{pmatrix} \begin{pmatrix} c_1^i \\ c_2^i \end{pmatrix} = \lambda_i \begin{pmatrix} c_1^i \\ c_2^i \end{pmatrix}.$$

It follows from this equation that for $\lambda_1 = -1$, $c_2^1 = c_1^1$, and for $\lambda_2 = 2$, $c_2^2 = 4c_1^2$. Therefore, other than some multiplicative constants, the eigenvector for $\lambda = -1$ is $\begin{pmatrix} 1 \\ 1 \end{pmatrix}$, and for $\lambda = 2$ is $\begin{pmatrix} 1 \\ 4 \end{pmatrix}$.

The complementary functions x_c and y_c are given by the linear combinations of the these two sets of solutions,

$$\begin{pmatrix} x_c \\ y_c \end{pmatrix} = c_1 \begin{pmatrix} 1 \\ 1 \end{pmatrix} e^{-t} + c_2 \begin{pmatrix} 1 \\ 4 \end{pmatrix} e^{2t}.$$

For the particular solution, because of the nonhomogeneous term $10\cos t$, we can assume $x_p = A\cos t + B\sin t$ and $y_p = C\cos t + D\sin t$. However, it is less cumbersome to make use of the fact that $10\cos t$ is the real part of $10e^{it}$. We can assume x_p is the real part of $A_c e^{it}$ and y_p is the real part of $B_c e^{it}$, where A_c and B_c are complex numbers. With these assumptions, (5.54) becomes

$$\begin{pmatrix} -2 & 1 \\ -4 & 3 \end{pmatrix} \begin{pmatrix} A_c e^{it} \\ B_c e^{it} \end{pmatrix} = \begin{pmatrix} iA_c e^{it} \\ iB_c e^{it} - 10e^{it} \end{pmatrix}.$$

Thus

$$-2A_c + B_c = iA_c,$$
$$-4A_c + 3B_c = iB_c - 10,$$

which yields

$$A_c = -3 + i, \quad B_c = -7 - i.$$

Therefore

$$x_p = \text{Re}(A_c e^{it}) = -3\cos t - \sin t,$$
$$y_p = \text{Re}(B_c e^{it}) = -7\cos t + \sin t.$$

Finally, we have the general solution

$$\begin{pmatrix} x \\ y \end{pmatrix} = \begin{pmatrix} x_c \\ y_c \end{pmatrix} + \begin{pmatrix} x_p \\ y_p \end{pmatrix} = \begin{pmatrix} c_1 e^{-t} + c_2 e^{2t} - 3\cos t - \sin t \\ c_1 e^{-t} + 4c_2 e^{2t} - 7\cos t + \sin t \end{pmatrix},$$

which is what we had before.

5.8.4 Transformation of an nth Order Equation into a System of n First-Order Equations

We have seen that a system of equations can be reduced to a single equation of higher order. The reverse is also true. Any nth-order differential equation can always be transformed into a simultaneous n first-order equations. Let us use this method to solve the second-order differential equation for the damped harmonic oscillator

$$\frac{d^2x}{dt^2} + \frac{c}{m}\frac{dx}{dt} + \frac{k}{m}x = 0.$$

Let

$$x_1 = x, \qquad x_2 = \frac{dx_1}{dt}.$$

It follows:

$$\frac{dx_2}{dt} = \frac{d^2x_1}{dt^2} = -\frac{c}{m}\frac{dx_1}{dt} - \frac{k}{m}x_1.$$

Thus the second-order equation can be written as a set of two first-order equations:

$$x_2 = \frac{dx_1}{dt},$$

$$-\frac{k}{m}x_1 - \frac{c}{m}x_2 = \frac{dx_2}{dt}.$$

With matrix notation, we have

$$\begin{pmatrix} 0 & 1 \\ -\dfrac{k}{m} & -\dfrac{c}{m} \end{pmatrix} \begin{pmatrix} x_1 \\ x_2 \end{pmatrix} = \begin{pmatrix} x_1' \\ x_2' \end{pmatrix}.$$

Since the coefficients are constants, we can assume

$$\begin{pmatrix} x_1 \\ x_2 \end{pmatrix} = \begin{pmatrix} c_1 \\ c_2 \end{pmatrix} e^{\lambda t},$$

thus we have the eigenvalue problem

$$\begin{pmatrix} 0 & 1 \\ -\dfrac{k}{m} & -\dfrac{c}{m} \end{pmatrix} \begin{pmatrix} c_1 \\ c_2 \end{pmatrix} = \lambda \begin{pmatrix} c_1 \\ c_2 \end{pmatrix}.$$

The eigenvalues λ can be found from the characteristic equation

$$\begin{vmatrix} 0 - \lambda & 1 \\ -\dfrac{k}{m} & -\dfrac{c}{m} - \lambda \end{vmatrix} = 0.$$

The two roots of this equation are

$$\lambda_1, \lambda_2 = \frac{1}{2m}(-c \pm \sqrt{c^2 - 4km}).$$

Thus the general solution of the original problem is given by

$$x(t) = x_1 = c_1 e^{\lambda_1 t} + c_2 e^{\lambda_2 t}.$$

This result is identical to (5.37).

The fact that a linear differential equation of nth-order can be transformed into a system of n coupled first-order equations is of some importance because mathematically one can show that there is an unique solution for a linear first-order system, provided the initial conditions are specified. However, we are not too concerned with uniqueness and existence because in physical applications, the mathematical model, if formulated correctly, must have a solution.

5.8.5 Coupled Oscillators and Normal Modes

The motion of a harmonic oscillator is described by a second-order differential equation. Its solution shows that the motion is characterized by a single natural frequency. A real physical system usually has many different characteristic frequencies. A vibration with any of these frequencies is called a normal mode of the system. The motion of the system is generally a linear combination of these normal modes.

A simple example is the system of two coupled oscillators shown in Fig. 5.8.

The system consists of two identical mass–spring oscillators of mass m and spring constant k. The two masses rest on a frictionless table and are connected by a spring with spring constant K. When the displacements x_A and x_B are zero, the springs are neither stretched or compressed. At any moment, the connecting spring is stretched an amount $x_A - x_B$ and therefore pulls or pushes on A and B with a force whose magnitude is $K(x_A - x_B)$. Thus the magnitude of the restoring force on A is

$$-kx_A - K(x_A - x_B).$$

The force on B must be

$$-kx_B + K(x_A - x_B).$$

Therefore the equations of motion for A and B are

$$-kx_A - K(x_A - x_B) = m\frac{d^2x_A}{dt^2},$$

$$-kx_B + K(x_A - x_B) = m\frac{d^2x_B}{dt^2}.$$

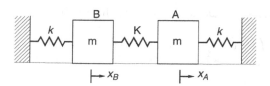

Fig. 5.8. Two coupled oscillators

With matrix notation, these equations can be written as

$$\begin{pmatrix} -\dfrac{k}{m} - \dfrac{K}{m} & \dfrac{K}{m} \\[2mm] \dfrac{K}{m} & -\dfrac{k}{m} - \dfrac{K}{m} \end{pmatrix} \begin{pmatrix} x_A \\ x_B \end{pmatrix} = \begin{pmatrix} \dfrac{d^2 x_A}{dt^2} \\[2mm] \dfrac{d^2 x_B}{dt^2} \end{pmatrix}.$$

To simplify the writing, let $\omega_0^2 = k/m$ and $\omega_1^2 = K/m$. With the assumption $x_A = ae^{\lambda t}$ and $x_B = be^{\lambda t}$, the last equation becomes

$$\begin{pmatrix} -\omega_0^2 - \omega_1^2 & \omega_1^2 \\ \omega_1^2 & -\omega_0^2 - \omega_1^2 \end{pmatrix} \begin{pmatrix} a \\ b \end{pmatrix} e^{\lambda t} = \lambda^2 \begin{pmatrix} a \\ b \end{pmatrix} e^{\lambda t}.$$

This can be regarded as an eigenvalue problem. The secular equation

$$\begin{vmatrix} -\omega_0^2 - \omega_1^2 - \lambda^2 & \omega_1^2 \\ \omega_1^2 & -\omega_0^2 - \omega_1^2 - \lambda^2 \end{vmatrix} = 0$$

shows that λ^2 satisfies the equation

$$(-\omega_0^2 - \omega_1^2 - \lambda^2)^2 - \omega_1^4 = 0$$

or

$$\lambda^2 = -\omega_0^2 - \omega_1^2 \pm \omega_1^2.$$

Thus

$$\lambda^2 = -\omega_0^2, \quad \lambda^2 = -(\omega_0^2 + 2\omega_1^2).$$

The four roots of λ are

$$\lambda_1, \lambda_2 = \pm i\omega_0, \quad \lambda_3, \lambda_4 = \pm i\omega_c,$$

where

$$\omega_c = \sqrt{\omega_0^2 + 2\omega_1^2}.$$

These frequencies, ω_0 and ω_c are known as the normal frequencies of the system. The amplitudes a and b are not independent of each other, since they must satisfy the equation

$$\begin{pmatrix} -\omega_0^2 - \omega_1^2 - \lambda^2 & \omega_1^2 \\ \omega_1^2 & -\omega_0^2 - \omega_1^2 - \lambda^2 \end{pmatrix} \begin{pmatrix} a \\ b \end{pmatrix} = 0.$$

Thus for $\lambda = \lambda_1, \lambda_2 = \pm i\omega_0$, so $\lambda^2 = -\omega_0^2$, the amplitudes a and b must satisfy

$$\begin{pmatrix} -\omega_0^2 - \omega_1^2 + \omega_0^2 & \omega_1^2 \\ \omega_1^2 & -\omega_0^2 - \omega_1^2 + \omega_0^2 \end{pmatrix} \begin{pmatrix} a \\ b \end{pmatrix} = 0.$$

It follows that $a = b$.

Similarly, for $\lambda = \lambda_3, \lambda_4 = \pm i\omega_c$, the relation between a and b is given by

$$\begin{pmatrix} -\omega_0^2 - \omega_1^2 + \omega_0^2 + 2\omega_1^2 & \omega_1^2 \\ \omega_1^2 & -\omega_0^2 - \omega_1^2 + \omega_0^2 + 2\omega_1^2 \end{pmatrix} \begin{pmatrix} a \\ b \end{pmatrix} = 0,$$

which gives $b = -a$.

The displacements x_A and x_B are given by a linear combinations of these four solutions,

$$x_A = a_1 e^{\lambda_1 t} + a_2 e^{\lambda_2 t} + a_3 e^{\lambda_3 t} + a_4 e^{\lambda_4 t},$$
$$x_B = a_1 e^{\lambda_1 t} + a_2 e^{\lambda_2 t} - a_3 e^{\lambda_3 t} - a_4 e^{\lambda_4 t},$$

where we have substituted a_1, a_2 for b_1, b_2, and $-a_3, -a_4$ for b_3, b_4. Since $\lambda_1 = i\omega_0$, $\lambda_2 = -i\omega_0$,

$$a_1 e^{\lambda_1 t} + a_2 e^{\lambda_2 t} = a_1 e^{i\omega_0 t} + a_2 e^{-i\omega_0 t} = C \cos(\omega_0 t + \alpha).$$

Similarly

$$a_3 e^{\lambda_3 t} + a_4 e^{\lambda_4 t} = D \cos(\omega_c t + \beta).$$

Thus x_A and x_B can be written as

$$x_A = C \cos(\omega_0 t + \alpha) + D \cos(\omega_c t + \beta),$$
$$x_B = C \cos(\omega_0 t + \alpha) - D \cos(\omega_c t + \beta).$$

The four constants a_1, a_2, a_3, a_4 (or C, α, D, β) depend on the initial conditions. It is seen that both x_A and x_B are given by some combination of the vibrations of two normal frequencies ω_0 and ω_c.

Suppose the motion is started when we pull both A and B toward the same direction by an equal amount x_0 and then release them from rest. The distance between A and B equals the relaxed length of the coupling spring and therefore it exerts no force on each mass. Thus A and B will oscillate in phase with the same natural frequency ω_0 as if they were not coupled. Mathematically, we see that is indeed the case. With the initial conditions

$$x_A(0) = x_B(0) = x_0 \quad \text{and} \quad \left. \frac{dx_A}{dt} \right|_{t=0} = 0, \quad \left. \frac{dx_B}{dt} \right|_{t=0} = 0,$$

one can easily show that $C = x_0$, $D = 0$, $\alpha = \beta = 0$. Therefore

$$x_A = x_0 \cos \omega_0 t, \qquad x_B = x_0 \cos \omega_0 t.$$

This represents a normal mode of the coupled system. Once the system is vibrating with a normal frequency, it will continue to vibrate with that frequency.

Suppose initially we pull A and B in opposite direction by the same amount x_m and then release them. The symmetry of the arrangement tells us that A and B will be mirror images of each other. They will vibrate with certain frequency, which we might expect it to be ω_c, since ω_c is the only other normal frequency of the system. This is indeed the case. Since with the initial conditions

$$x_A(0) = -x_B(0) = x_m, \quad \text{and} \quad \left.\frac{dx_A}{dt}\right|_{t=0} = 0, \quad \left.\frac{dx_B}{dt}\right|_{t=0} = 0,$$

one can show that $C = 0$, $D = x_m$, $\alpha = \beta = 0$. Therefore

$$x_A = x_m \cos \omega_c t, \quad x_B = -x_m \cos \omega_c t.$$

They oscillate with the same frequency ω_c but they are always 180° out of phase. This constitutes the second normal mode of the system. The general motion is a linear combination of these two modes.

For a real molecule or crystal, there will be many normal modes. Each normal mode corresponds to a certain symmetry of the structure. The fact that these modes can be excited by their corresponding normal frequencies is widely used in scientific applications.

5.9 Other Methods and Resources for Differential Equations

Many readers probably had previously taken a course in ordinary differential equations. Here we just give a review so that even those who did not have previous exposure can gain enough background to continue. The literature of the theory and applications of differential equations is vast. Our discussion is far from complete.

Among the methods we have not yet discussed are the Laplace transform, Fourier analysis and power series solutions.

The Laplace transform is especially useful in solving problems with nonhomogeneous terms of a discontinuous or impulsive nature. In Chap. 6 we will study these problems in detail.

If the nonhomogeneous term is periodic but not sinusoidal, Fourier series method is particularly convenient. We will discuss this method after we study the Fourier series.

In general, a differential equation with variable coefficients cannot be solved by the methods of this chapter. The usual procedure for such equations is to obtain solutions in the form of infinite series. This is known as series method. Some most important equations in physics and engineering lead us to this type equations. The series so obtained can be taken as definitions of new functions. Some important ones are named and tabulated. We shall study this method in the chapter on special functions.

In addition, differential equations can be solved numerically. Sometimes this is the only way to solve the equation. Digital computers have made numerical solutions readily available. There are several computer programs for the integration of ordinary differential equations in "Numerical Recipes" by William H. Press, Brian P. Flannery, Saul A. Teukolsky and William T. Vetterling (Cambridge University Press, 1986). For a discussion of the numerical methods, see R.J. Rice, "Numerical Methods, Software and Analysis" (McGraw-Hill, New York, 1983).

Finally it should be mentioned that a number of commercial computer packages are available to perform algebraic manipulations, including solving differential equations. They are called computer algebraic systems, some prominent ones are Matlab, Maple, Mathematica, MathCad and MuPAD.

This book is written with the software "Scientific WorkPlace", which also provides an interface to MuPAD. (Before version 5, it also came with Maple). Instead of requiring the user to adhere to a rigid syntax, the user can use natural mathematical notations. For example, to solve the differential equation

$$\frac{d^2y}{dx^2} + \frac{dy}{dx} = x + y$$

all you have to do is (1) type this equation in the math-mode, and (2) click on the "Compute" button, and (3) click on the "Solve ODE" button in the pull-down menu, and (4) click on the "Exact" button in the submenu. The program will return with

$$\text{Exact solution is: } C_1 e^{x(\frac{1}{2}\sqrt{5}-\frac{1}{2})} - x + C_2 e^{x(-\frac{1}{2}\sqrt{5}-\frac{1}{2})} - 1.$$

Unfortunately, not every problem can be solved by a computer algebraic system. Sometimes it fails to find the solution. Even worse, for a variety of reasons, the intention of the user is sometimes misinterpreted, and the computer returns with an answer to a wrong problem without the user knowing it. Therefore these systems must be used with caution.

Exercises

1. Find the general solutions of the following separable differential equations:
   (a) $xy' + y + 3 = 0$,
   (b) $2yy' + 4x = 0$.
   Ans. (a) $x(y + 3) = c$, (b) $2x^2 + y^2 = c$.

2. Find the specific solutions of the following separable differential equations:
   (a) $\dfrac{dy}{dx} = \dfrac{x(1 + y^2)}{y(1 + x^2)}$, $y(0) = 1$,

   (b) $ye^{x+y}dy = dx$, $y(0) = 0$.
   Ans. (a) $1 + y^2 = 2(1 + x^2)$, (b) $(1 - y)e^y = e^{-x}$.

3. Change the following equations into separable differential equations and find the general solutions:
   (a) $xyy' = y^2 - x^2$,
   (b) $\dfrac{dy}{dx} = \dfrac{x-y}{x+y}$.

   Ans. (a) $\ln x + \dfrac{y^2}{2x^2} = c$, (b) $y^2 + 2xy - x^2 = c$.

4. Show that the following differential equations are exact and find the general solutions:
   (a) $(2xy - \cos x)dx + (x^2 - 1)dy = 0$,
   (b) $(2x + e^y)dx + xe^y dy = 0$.

   Ans. (a) $x^2 y - \sin x - y = c$, (b) $x^2 + xe^y = c$.

5. Solve the following differential equations by first finding an integrating factor:
   (a) $2(y^3 - 2)dx + 3xy^2 dy = 0$,
   (b) (b) $(y + x^4)dx - xdy = 0$.

   Ans. (a) $\mu = x$, $x^2 y^3 - 2x^2 = c$, (b) $\mu = 1/x^2$, $\dfrac{x^3}{3} - \dfrac{y}{x} = c$.

6. Find the general solutions of the following first-order linear differential equations:
   (a) $y' + y = x$,
   (b) $xy' + (1+x)y = e^{-x}$.

   Ans. (a) $y = x - 1 + ce^{-x}$, (b) $y = e^{-x} + ce^{-x}/x$.

7. Find the specific solutions of the following first-order linear differential equations:
   (a) $y' - y = 1 - x$, $y(0) = 1$.
   (b) $y' + \frac{1}{x}y = 3x^2$, $y(1) = 5$.

   Ans. (a) $y = x + e^x$,   (b) $y = \frac{3}{4}x^3 + \frac{17}{4}x^{-1}$.

8. The RL circuit is described by the equation

$$L\dfrac{di}{dt} + Ri = A\cos t, \quad i(0) = 0$$

   where i is current. Find the current i as a function of time t.
   Ans. $i(t) = \dfrac{AR}{R^2 + L^2}\left[\cos t + \dfrac{L}{R}\sin t - e^{-Rt/L}\right]$.

9. Find the general solutions of the following homogeneous second-order differential equations:
   (a) $y'' - k^2 y = 0$,
   (b) $y'' - (a+b)y' + aby = 0$,
   (c) $y'' + 2ky' + k^2 y = 0$.

Ans. (a) $y(x) = c_1 e^{kx} + c_2 e^{-kx}$,   (b) $y(x) = c_1 e^{ax} + c_2 e^{bx}$,   (c) $y(x) = c_1 e^{-kx} + c_2 x e^{-kx}$.

10. Find the specific solutions of the following homogeneous second-order differential equations:

(a) $y'' - 2ay' + (a^2 + b^2)y = 0$, $y(0) = 0$, $y'(0) = 1$.

(b) $y'' + 4y = \sin x$, $y(0) = 0$, $y'(0) = 0$.

(c) $y''' + y' = e^{2z}$, $y(0) = y'(0) = y''(0) = 0$.

Ans. (a) $y(x) = \frac{1}{b} e^{ax} \sin bx$, (b) $y(x) = \frac{1}{3} \sin x - \frac{1}{6} \sin 2x$,
(c) $y(x) = \frac{1}{10} e^{2x} - \frac{1}{5} \sin x + \frac{2}{5} \cos x - \frac{1}{2}$.

11. Find the general solutions of the following nonhomogeneous differential equations:

(a) $y'' + k^2 y = a$,

(b) $y'' - 4y = x$,

(c) $y'' - 2y' + y = 3x^2 - 12x + 7$.

Ans. (a) $y(x) = c_1 \sin kx + c_2 \cos kx + a/k^2$, (b) $y(x) = c_1 e^{2x} + c_2 e^{-2x} - \frac{1}{4} x$,
(c) $y(x) = (c_1 + c_2 x)e^x + 3x^2 + 1$.

12. Find the general solutions of the following nonhomogeneous differential equations:

(a) $y'' - 3y' + 2y = e^{2x}$,

(b) $y'' - 6y' + 9y = 4e^{3x}$,

(c) $y'' + 9y = \cos(3x)$.

Ans. (a) $y = c_1 e^x + c_2 e^{2x} + x e^{2x}$, (b) $y = c_1 e^{3x} + c_2 x e^{3x} + 2x^2 e^{3x}$,
(c) $y = c_1 \cos(3x) + c_2 \sin(3x) + \frac{1}{6} x \sin(3x)$.

13. Find the specific solutions of the following nonhomogeneous differential equations:

(a) $y'' + y' = x^2 + 2x$, $y(0) = 4$, $y'(0) = -2$.

(b) $y'' - 4y' + 4y = 6 \sin x - 8 \cos x$, $y(0) = 3$, $y'(0) = 4$.

(c) $y'' - 4y = 8e^{2x}$, $y(0) = 4$, $y'(0) = 6$.

Ans. (a) $y = \frac{1}{3} x^3 + 2e^{-x} + 2$, (b) $y = (3 - 4x)e^{2x} + 2 \sin x$, (c) $y = 3e^{2x} + e^{-2x} + 2x e^{2x}$.

14. Solve the following differential equations with the method of variation of parameters:

(a) $y'' + y = \sec x$,

(b) $y'' + 4y' + 4y = \dfrac{e^{-2x}}{x^2}$.

Ans. (a) $y = c_1 \cos x + c_2 \sin x + \cos x \ln |\cos x| + x \sin x$,
(b) $y = c_1 e^{-2x} + c_2 x e^{-2x} - e^{-2x} \ln x$.

15. Solve the following set of simultaneous linear differential equation:

$$y'(x) - z'(x) - 2y(x) + 2z(x) = 1 - 2x,$$
$$y''(x) + 2z'(x) + y(x) = 0,$$
$$y(0) = z(0) = y'(0) = 0.$$

Ans. $y(x) = -2e^{-x} - 2xe^{-x} + 2; \quad z(x) = -2e^{-x} - 2xe^{-x} + 2 - x.$

16. Solve the following set of simultaneous linear differential equation:

$$y'(x) + z'(x) + y(x) + z(x) = 1,$$
$$y'(x) - y(x) - 2z(x) = 0,$$
$$y(0) = 1, \; z(0) = 0.$$

Ans. $y(x) = 2 - e^{-x}; \quad z(x) = e^{-x} - 1.$

17. In strength of materials, you will encounter the equation

$$\frac{d^4 y}{dx^4} = -4a^4 y$$

where a is a positive constant. Find the general solution of this equation. ($y(x)$ with four constants).

Ans. $y(x) = e^{ax}(c_1 \cos ax + c_2 \sin ax) + e^{-ax}(c_3 \cos ax + c_4 \sin ax)$

18. Solve

$$y''(t) + y(t) = \begin{cases} 1 - \frac{t^2}{\pi^2} & \text{if} \;\; 0 \le t \le \pi, \\ 0 & \text{if} \;\; t > \pi, \end{cases}$$
$$y(0) = y'(0) = 0.$$

This may be interpreted as an undamped system on which a force acts during some interval of time, for instance, the force acting on a gun barrel when a shell is fired, the barrel being braked by heavy springs (and then damped by a dashpot which we disregard for simplicity). Hint: at $t = \pi$ both y and y' must be continuous.

Ans. $y(t) = \begin{cases} -(1 + \frac{2}{\pi^2}) \cos t + (1 + \frac{2}{\pi^2}) - \frac{1}{\pi^2} t^2 & \text{for} \;\; 0 \le t \le \pi \\ [1 - 2(1 + \frac{2}{\pi^2})] \cos t + \frac{2}{\pi} \sin t & \text{for} \quad\; t \ge \pi. \end{cases}$

19. If $N_a(t)$, $N_b(t)$, $N_c(t)$ represent the number of nuclei of three radioactive substances which decay according to the scheme

$$a \to b \to c$$

with decay constants λ_a and λ_b, the substance c is considered stable. Then the functions are known to obey the system of differential equations

$$\frac{\mathrm{d}N_a}{\mathrm{d}t} = -\lambda_a N_a,$$

$$\frac{\mathrm{d}N_b}{\mathrm{d}t} = -\lambda_b N_b + \lambda_a N_a,$$

$$\frac{\mathrm{d}N_c}{\mathrm{d}t} = \lambda_b N_b.$$

Assuming that $N_a(0) = N_0$, and $N_b(0) = N_c(0) = 0$, find $N_a(t)$, $N_b(t)$ and $N_c(t)$ as functions of time t.

Ans. $N_a = N_0 e^{-\lambda_a t}$; $\quad N_b = N_0 \left[\frac{\lambda_a}{\lambda_b - \lambda_a} e^{-\lambda_a t} - \frac{\lambda_a}{\lambda_b - \lambda_a} e^{-\lambda_b t} \right]$; $\quad N_c = N_0 \left[1 - \frac{\lambda_b}{\lambda_b - \lambda_a} e^{-\lambda_a t} + \frac{\lambda_a}{\lambda_b - \lambda_a} e^{-\lambda_b t} \right]$.

20. Show that the particular solution of

$$m \frac{\mathrm{d}^2 x}{\mathrm{d}t^2} + c \frac{\mathrm{d}x}{\mathrm{d}t} + kx = F_0 \cos \omega t$$

can be written in the form of

$$x_\mathrm{p}(t) = C_1 \cos \omega t + C_2 \sin \omega t,$$

where

$$C_1 = \frac{(k - m\omega^2) F_0}{(k - m\omega^2)^2 + (c\omega)^2}, \quad C_2 = \frac{(c\omega) F_0}{(k - m\omega^2)^2 + (c\omega)^2}.$$

21. Show that the result of previous problem can be put in the form of

$$x_\mathrm{p}(t) = A \cos(\omega t - \phi)$$

where

$$A = \frac{F_0}{[(k - m\omega^2)^2 + (c\omega)^2]^{1/2}}, \quad \phi = \tan^{-1} \frac{c\omega}{m(\omega_0^2 - \omega^2)}.$$

22. For two identical undamped oscillators, A and B, each of mass m, and natural frequency ω_0, show that each of them is governed by the differential equation

$$m \frac{\mathrm{d}^2 x}{\mathrm{d}t^2} + m\omega_0^2 x = 0.$$

They are coupled in such a way that the coupling force exerted on A is $\alpha m(\mathrm{d}^2 x_B/\mathrm{d}t^2)$, and the coupling force on B is $\alpha m(\mathrm{d}^2 x_A/\mathrm{d}t^2)$, where α is the coupling constant with a magnitude less than one. Find the normal frequencies of the system.

Ans. $\omega = \omega_0 (1 \pm \alpha)^{-1/2}$.

6

Laplace Transforms

Among the tools that are very useful in solving linear differential equations is the Laplace transform method. The idea is to use an integral to transform the differential equation into an algebraic equation, from the solution of this algebraic equation we get the desired function through the inverse transform. The Laplace transform is named after the eminent French mathematician Pierre Simon Laplace (1749–1827), who is also remembered for the Laplace equation which is one of the most important equations in mathematical physics.

Laplace first studied this method in 1782. However, the power and usefulness of this method was not recognized until 100 years later. The techniques described in this chapter are mainly due to Oliver Heaviside (1850–1925), an innovative British electrical engineer, who also made significant contributions to electromagnetic theory.

The Laplace transform is especially useful in solving problems with nonhomogeneous terms of a discontinuous or impulsive nature. Such problems are common in physical sciences but are relatively awkward to handle by the methods previously discussed.

In this chapter certain properties of Laplace transforms are investigated and relevant formulas are tabulated in such a way that the solution of initial value problems involving linear differential equations can be conveniently obtained.

6.1 Definition and Properties of Laplace Transforms

6.1.1 Laplace Transform – A Linear Operator

The Laplace transform $\mathfrak{L}[f]$ of the function $f(t)$ is defined as

$$\mathfrak{L}[f] = \int_0^\infty e^{-st} f(t) \, \mathrm{d}t = F(s), \tag{6.1}$$

we assume that this integral exists. One of the reasons that Laplace transform is useful is that s can be chosen large enough that (6.1) converges even if $f(t)$ does not go to zero as $t \to \infty$. Of course, there are functions that diverge faster than e^{st}. For such functions, the Laplace transform does not exit. Fortunately such functions are of little physical interests.

Note that the transform is a function of s. The transforms of the functions of our concern not only exist, but also go to zero $(F(s) \to 0)$ as $s \to \infty$.

It follows immediately from the definition that the Laplace transform is a linear operator, that is

$$\mathcal{L}[af(t) + bg(t)] = \int_0^\infty e^{-st}[af(t) + bg(t)]dt$$

$$= a \int_0^\infty e^{-st} f(t)\, dt + b \int_0^\infty e^{-st} g(t)\, dt$$

$$= a\mathcal{L}[f] + b\mathcal{L}[g]. \tag{6.2}$$

For simple functions, the integral of the Laplace transform can be readily carried out. For example:

$$\mathcal{L}[1] = \int_0^\infty e^{-st}\, dt = \left[-\frac{1}{s}e^{-st}\right]_0^\infty = \frac{1}{s}. \tag{6.3}$$

It is also very easy to evaluate the transform of an exponential function

$$\mathcal{L}[e^{at}] = \int_0^\infty e^{-st} e^{at}\, dt = \int_0^\infty e^{-(s-a)t}\, dt = \left[-\frac{1}{s-a}e^{-(s-a)t}\right]_0^\infty.$$

As long as $s > a$, the upper limit vanishes and the lower limit gives $1/(s-a)$. Thus

$$\mathcal{L}[e^{at}] = \frac{1}{s-a}. \tag{6.4}$$

Similarly

$$\mathcal{L}[e^{-at}] = \frac{1}{s+a}. \tag{6.5}$$

With these relations, the Laplace transforms of the following hyperbolic functions

$$\cosh at = \frac{1}{2}(e^{at} + e^{-at}), \qquad \sinh at = \frac{1}{2}(e^{at} - e^{-at})$$

are easily obtained. Since Laplace transform is linear,

$$\mathcal{L}[\cosh at] = \frac{1}{2}\{\mathcal{L}[e^{at}] + \mathcal{L}[e^{-at}]\}$$

$$= \frac{1}{2}\left(\frac{1}{s-a} + \frac{1}{s+a}\right) = \frac{s}{s^2 - a^2}. \tag{6.6}$$

Similarly

$$\mathcal{L}\left[\sinh at\right] = \frac{a}{s^2 - a^2}. \tag{6.7}$$

Now the parameter a does not have to be restricted to real numbers. If a is purely imaginary $a = i\omega$, we will have

$$\mathcal{L}[e^{i\omega t}] = \frac{1}{s - i\omega}.$$

Since

$$\frac{1}{s - i\omega} = \frac{1}{s - i\omega} \times \frac{s + i\omega}{s + i\omega} = \frac{s}{s^2 + \omega^2} + i\frac{\omega}{s^2 + \omega^2},$$

and

$$\mathcal{L}[e^{i\omega t}] = \mathcal{L}[\cos \omega t + i \sin \omega t] = \mathcal{L}[\cos \omega t] + i\mathcal{L}[\sin \omega t],$$

equating the real part to real part and imaginary part to imaginary part we have

$$\mathcal{L}[\cos \omega t] = \frac{s}{s^2 + \omega^2}, \tag{6.8}$$

$$\mathcal{L}[\sin \omega t] = \frac{\omega}{s^2 + \omega^2}. \tag{6.9}$$

The definition of $\mathcal{L}[\cos \omega t]$ is, of course, still

$$\mathcal{L}[\cos \omega t] = \int_0^\infty e^{-st} \cos \omega t \, dt. \tag{6.10}$$

With integration by parts, we can evaluate this integral directly,

$$\int_0^\infty e^{-st} \cos \omega t \, dt = \left[-\frac{1}{s} e^{-st} \cos \omega t \right]_0^\infty - \int_0^\infty \frac{1}{s} e^{-st} \omega \sin \omega t \, dt$$

$$= \frac{1}{s} - \frac{\omega}{s} \int_0^\infty e^{-st} \sin \omega t \, dt,$$

$$\int_0^\infty e^{-st} \sin \omega t \, dt = \left[-\frac{1}{s} e^{-st} \sin \omega t \right]_0^\infty + \int_0^\infty \frac{1}{s} e^{-st} \omega \cos \omega t \, dt$$

$$= \frac{\omega}{s} \int_0^\infty e^{-st} \cos \omega t \, dt.$$

Combine these two equations,

$$\int_0^\infty e^{-st} \cos \omega t \, dt = \frac{1}{s} - \frac{\omega^2}{s^2} \int_0^\infty e^{-st} \cos \omega t \, dt.$$

Move the last term to the left-hand side,

$$\left(1 + \frac{\omega^2}{s^2} \right) \int_0^\infty e^{-st} \cos \omega t \, dt = \frac{1}{s}.$$

or

$$\int_0^\infty e^{-st} \cos\omega t \; dt = \frac{s}{s^2 + \omega^2},$$

which is exactly the same as (6.8), as it should be.

In principle, the Laplace transform can be obtained directly by carrying out the integral. However, very often it is much simpler to use the properties of the Laplace transform, rather than direct integration, to obtain the transform, as shown in the last example.

The Laplace transform has many interesting properties, they are the reasons that the Laplace transform is a powerful tool of mathematical analysis. We will now discuss some of them, and use them to generate more transforms as illustrations.

6.1.2 Laplace Transforms of Derivatives

The Laplace transform of a derivative is by definition

$$\mathcal{L}[f'] = \int_0^\infty e^{-st} \frac{df(t)}{dt} dt = \int_0^\infty e^{-st} df(t).$$

If we let $u = e^{-st}$ and $dv = df(t)$, then $du = -s\,e^{-st}dt$ and $v = f$. With integration by parts, we have $u\,dv = d(uv) - v\,du$, so

$$\mathcal{L}[f'] = \int_0^\infty \{d[(e^{-st}f(t)] + f(t)s\,e^{-st}dt\}$$

$$= \left[e^{-st}f(t)\right]_0^\infty + s\int_0^\infty e^{-st}f(t)\;dt = -f(0) + s\mathcal{L}[f]. \qquad (6.11)$$

Clearly

$$\mathcal{L}[f''] = \mathcal{L}[(f')'] = -f'(0) + s\mathcal{L}[f']$$
$$= -f'(0) + s\,(-f(0) + s\mathcal{L}[f]) = -f'(0) - sf(0) + s^2\mathcal{L}[f]. \quad (6.12)$$

Naturally this result can be extended to higher derivatives

$$\mathcal{L}[f^{(n)}] = -f^{(n-1)}(0) - \cdots - s^{n-1}f(0) + s^n\mathcal{L}[f]. \qquad (6.13)$$

These properties are crucial in solving differential equations. Here we will use them to generate $\mathcal{L}[t^n]$.

First let $f(t) = t$, then $f' = 1$ and $f(0) = 0$. By (6.11)

$$\mathcal{L}[1] = -0 + s\mathcal{L}[t],$$

rearranging and using (6.3), we have

$$\mathcal{L}[t] = \frac{1}{s}\mathcal{L}[1] = \frac{1}{s^2}. \qquad (6.14)$$

If we let $f(t) = t^2$, then $f' = 2t$ and $f(0) = 0$. Again by (6.11)

$$\mathcal{L}[2t] = -0 + s\mathcal{L}[t^2].$$

Thus, with (6.14)

$$\mathcal{L}[t^2] = \frac{1}{s}\mathcal{L}[2t] = \frac{2}{s}\mathcal{L}[t] = \frac{2}{s^3}. \tag{6.15}$$

Clearly this process can be repeated

$$\mathcal{L}[t^n] = \frac{n!}{s^{n+1}}. \tag{6.16}$$

6.1.3 Substitution: s-Shifting

If we know the Laplace transform $F(s)$ of the function $f(t)$, we can get the transform of $e^{at}f(t)$ by replacing s with $s - a$ in $F(s)$. This can be easily shown. By definition

$$F(s) = \int_0^\infty e^{-st} f(t)\ dt = \mathcal{L}[f(t)],$$

clearly

$$F(s - a) = \int_0^\infty e^{-(s-a)t} f(t)\ dt = \int_0^\infty e^{-st}e^{at} f(t)\ dt = \mathcal{L}\left[e^{at} f(t)\right]. \tag{6.17}$$

This simple relation is sometimes known as s-shifting (or first shifting) theorem.

With the help of the s-shifting theorem, we can derive the transforms of many more functions without carrying out the integration. For example, it follows from (6.16) and (6.17) that

$$\mathcal{L}[e^{-at}t^n] = \frac{n!}{(s+a)^{n+1}}. \tag{6.18}$$

It can also be easily shown that

$$\mathcal{L}[e^{-at}\cos\omega t] = \int_0^\infty e^{-st}e^{-at}\cos\omega t\ dt = \int_0^\infty e^{-(s+a)t}\cos\omega t\ dt. \tag{6.19}$$

Compare the integrals in (6.10) and (6.19), the only difference is that s is replaced by $s + a$. Therefore the last integral must equal to the right-hand side of (6.8) with s changed to $s + a$, that is

$$\mathcal{L}[e^{-at}\cos\omega t] = \frac{s+a}{(s+a)^2 + \omega^2}. \tag{6.20}$$

Similarly

$$\mathcal{L}[e^{-at}\sin\omega t] = \frac{\omega}{(s+a)^2 + \omega^2}. \tag{6.21}$$

6.1.4 Derivative of a Transform

If we differentiate the Laplace transform $F(s)$ with respect to s, we get

$$\frac{d}{ds}F(s) = \frac{d}{ds}\mathcal{L}\left[f(t)\right] = \frac{d}{ds}\int_0^\infty e^{-st}f(t)\,dt$$

$$= \int_0^\infty \frac{de^{-st}}{ds}f(t)\,dt = \int_0^\infty e^{-st}(-t)f(t)\,dt = \mathcal{L}\left[-tf(t)\right]. \quad (6.22)$$

Continuing this process, we have

$$\frac{d^n}{ds^n}\mathcal{L}\left[f(t)\right] = \mathcal{L}\left[(-t)^n f(t)\right]. \quad (6.23)$$

Many more formulas can be derived by taking advantage of this relation. For example, differentiating both sides of (6.9) with respect to s, we have

$$\frac{d}{ds}\mathcal{L}[\sin \omega t] = \frac{d}{ds}\frac{\omega}{s^2+\omega^2}.$$

Since

$$\frac{d}{ds}\mathcal{L}[\sin \omega t] = \frac{d}{ds}\int_0^\infty e^{-st}\sin \omega t\,dt = -\int_0^\infty t\,e^{-st}\sin \omega t\,dt = -\mathcal{L}[t\sin \omega t],$$

$$\frac{d}{ds}\frac{\omega}{s^2+\omega^2} = -\frac{2s\omega}{(s^2+\omega^2)^2},$$

therefore

$$\mathcal{L}[t\sin \omega t] = \frac{2s\omega}{(s^2+\omega^2)^2}. \quad (6.24)$$

Similarly we can show

$$\mathcal{L}[t\cos \omega t] = \frac{s^2-\omega^2}{(s^2+\omega^2)^2}. \quad (6.25)$$

6.1.5 A Short Table of Laplace Transforms

Since Laplace transform is a linear operator, two transforms can be combined to form a new one. For example:

$$\mathcal{L}[1-\cos \omega t] = \frac{1}{s} - \frac{s}{s^2+\omega^2} = \frac{\omega^2}{s(s^2+\omega^2)}, \quad (6.26)$$

$$\mathcal{L}[\omega t - \sin \omega t] = \frac{\omega}{s^2} - \frac{\omega}{s^2+\omega^2} = \frac{\omega^3}{s^2(s^2+\omega^2)}, \quad (6.27)$$

$$\mathcal{L}[\sin \omega t - \omega t\cos \omega t] = \frac{\omega}{s^2+\omega^2} - \frac{\omega(s^2-\omega^2)}{(s^2+\omega^2)^2} = \frac{2\omega^3}{(s^2+\omega^2)^2}, \quad (6.28)$$

Table 6.1. A short table of Laplace transforms, in each case s is assumed to be sufficiently large that the transform exists

$f(t)$	$F(s) = \mathfrak{L}[f(t)]$	$f(t)$	$F(s) = \mathfrak{L}[f(t)]$
1	$\dfrac{1}{s}$	$\delta(t)$	1
t	$\dfrac{1}{s^2}$	$\delta(t-c)$	e^{-sc}
t^n	$\dfrac{n!}{s^{n+1}}$	$\delta'(t-c)$	$s\,\mathrm{e}^{-sc}$
e^{at}	$\dfrac{1}{s-a}$	$u(t-c)$	$\dfrac{1}{s}\mathrm{e}^{-sc}$
$t\,\mathrm{e}^{at}$	$\dfrac{1}{(s-a)^2}$	$(t-c)^n u(t-c)$	$\dfrac{n!}{s^{n+1}}\mathrm{e}^{-sc}$
$t^n \mathrm{e}^{at}$	$\dfrac{n!}{(s-a)^{n+1}}$	$(t-c)^n \mathrm{e}^{a(t-c)} u(t-c)$	$\dfrac{n!}{(s-a)^{n+1}}\mathrm{e}^{-sc}$
$\sin \omega t$	$\dfrac{\omega}{s^2 + \omega^2}$	$\sin \omega(t-c) u(t-c)$	$\dfrac{\omega}{s^2 + \omega^2}\mathrm{e}^{-sc}$
$\cos \omega t$	$\dfrac{s}{s^2 + \omega^2}$	$\cosh a(t-c) u(t-c)$	$\dfrac{s}{s^2 - a^2}\mathrm{e}^{-sc}$
$\sinh at$	$\dfrac{a}{s^2 - a^2}$	$\sin \omega t$ of period $\dfrac{\pi}{\omega}$	$\dfrac{\omega}{s^2 + \omega^2}\coth\dfrac{s\pi}{2\omega}$
$\cosh at$	$\dfrac{s}{s^2 - a^2}$	t of period p	$\dfrac{1-(1+ps)\mathrm{e}^{-ps}}{ps^2(1-\mathrm{e}^{-ps})}$
$\mathrm{e}^{at}\sin \omega t$	$\dfrac{\omega}{(s-a)^2 + \omega^2}$	$\dfrac{1}{t}(\mathrm{e}^{bt} - \mathrm{e}^{at})$	$\ln\dfrac{s-a}{s-b}$
$\mathrm{e}^{at}\cos \omega t$	$\dfrac{s-a}{(s-a)^2 + \omega^2}$	$\dfrac{2}{t}(1 - \cosh at)$	$\ln\dfrac{s^2 - a^2}{s^2}$
$t\sin \omega t$	$\dfrac{2\omega s}{(s^2 + \omega^2)^2}$	$\dfrac{2}{t}(1 - \cos \omega t)$	$\ln\dfrac{s^2 + \omega^2}{s^2}$
$1 - \cos \omega t$	$\dfrac{\omega^2}{s(s^2 + \omega^2)}$	$\dfrac{\sin \omega t}{t}$	$\tan^{-1}\dfrac{\omega}{s}$
$\omega t - \sin \omega t$	$\dfrac{\omega^3}{s^2(s^2 + \omega^2)}$	$t^a \ (a > -1)$	$\dfrac{\Gamma(a+1)}{s^{a+1}}$
$\sin \omega t - \omega t \cos \omega t$	$\dfrac{2\omega^3}{(s^2 + \omega^2)^2}$	$t^{-1/2}$	$\sqrt{\dfrac{\pi}{s}}$
$\sin \omega t + \omega t \cos \omega t$	$\dfrac{2\omega s^2}{(s^2 + \omega^2)^2}$	$t^{1/2}$	$\dfrac{1}{2}\dfrac{\sqrt{\pi}}{s^{3/2}}$
$\cos at - \cos bt$	$\dfrac{(b^2 - a^2)\,s}{(s^2 + a^2)(s^2 + b^2)}$	$J_0(at)$	$\dfrac{1}{(s^2 + a^2)^{1/2}}$

$$\mathcal{L}[\sin \omega t + \omega t \cos \omega t] = \frac{\omega}{s^2 + \omega^2} + \frac{\omega(s^2 - \omega^2)}{(s^2 + \omega^2)^2} = \frac{2\omega s^2}{(s^2 + \omega^2)^2}, \qquad (6.29)$$

$$\mathcal{L}[\cos at - \cos bt] = \frac{s}{s^2 + a^2} - \frac{s}{s^2 + b^2} = \frac{(b^2 - a^2)s}{(s^2 + a^2)(s^2 + b^2)}. \qquad (6.30)$$

There are extensive tables of Laplace transforms (For example, F. Oberherttinger and E. Badii, Tables of Laplace Transforms, Springer, New York, 1973). A short list of some simple Laplace transforms is given in Table 6.1. The items in the left-hand side of the table are the ones we have shown so far. Items in the right-hand side are relations we are going to derive in the following sections.

6.2 Solving Differential Equation with Laplace Transform

6.2.1 Inverse Laplace Transform

In solving differential equation with Laplace transform, we encounter the inverse problem of determining the unknown function $f(t)$ which has a given transform $F(s)$. The notation of $\mathcal{L}^{-1}[F(s)]$ is conventionally used for the inverse Laplace transform of $F(s)$. That is, if

$$F(s) = \mathcal{L}[f(t)] = \int_0^\infty e^{-st} f(t) \, dt, \qquad (6.31)$$

then

$$f(t) = \mathcal{L}^{-1}[F(s)]. \qquad (6.32)$$

Since

$$f(t) = \mathcal{L}^{-1}[F(s)] = \mathcal{L}^{-1}[\mathcal{L}[f(t)]] = I[f(t)],$$

it follows that $\mathcal{L}^{-1}\mathcal{L}$ is the identity operator I. The inverse transforms are of great practical importance and there are a variety of ways to get them. In this section, we will first study the transform in the form of a quotient of two polynomials

$$F(s) = \frac{p(s)}{q(s)},$$

where $p(s)$ and $q(s)$ have real coefficients and no common factors. Since

$$\lim_{s \to \infty} F(s) = \lim_{s \to \infty} \int_0^\infty e^{-st} f(t) \, dt \to 0,$$

it is clear that the degree of $p(s)$ is lower than that of $q(s)$. There are several closely related methods to get the inverse of such a transform. For the sake of clarity, we list them separately.

By Inspection. If the expression is simple enough, one can get the inverse directly from the table. This is illustrated in the following examples.

Example 6.2.1. Find (a) $\mathcal{L}^{-1}\left[\dfrac{1}{s^4}\right]$; (b) $\mathcal{L}^{-1}\left[\dfrac{4}{(s+4)^3}\right]$; (c) $\mathcal{L}^{-1}\left[\dfrac{1}{s^2+4}\right]$.

Solution 6.2.1.
(a) Since

$$\mathcal{L}[t^3] = \frac{3!}{s^4} = \frac{6}{s^4}, \qquad t^3 = \mathcal{L}^{-1}\left[\frac{6}{s^4}\right],$$

we have

$$\mathcal{L}^{-1}\left[\frac{1}{s^4}\right] = \frac{1}{6}\mathcal{L}^{-1}\left[\frac{6}{s^4}\right] = \frac{1}{6}t^3.$$

(b) Since

$$\mathcal{L}[e^{-4t}t^2] = \frac{2}{(s+4)^3}, \qquad e^{-4t}t^2 = \mathcal{L}^{-1}\left[\frac{2}{(s+4)^3}\right],$$

so

$$\mathcal{L}^{-1}\left[\frac{4}{(s+4)^3}\right] = 2\mathcal{L}^{-1}\left[\frac{2}{(s+4)^3}\right] = 2e^{-4t}t^2.$$

(c) Since

$$\mathcal{L}[\sin 2t] = \frac{2}{s^2+4}, \qquad \sin 2t = \mathcal{L}^{-1}\left[\frac{2}{s^2+4}\right],$$

so

$$\mathcal{L}^{-1}\left[\frac{1}{s^2+4}\right] = \frac{1}{2}\mathcal{L}^{-1}\left[\frac{2}{s^2+4}\right] = \frac{1}{2}\sin 2t.$$

Example 6.2.2. Find (a) $\mathcal{L}^{-1}\left[\dfrac{1}{s^2+2s+5}\right]$;   (b) $\mathcal{L}^{-1}\left[\dfrac{2s+1}{s^2+2s+5}\right]$.

Solution 6.2.2. (a) First we note that

$$\frac{1}{s^2+2s+5} = \frac{1}{(s+1)^2+4} = \frac{1}{2} \times \frac{2}{(s+1)^2+4}.$$

Since

$$\mathcal{L}[e^{-t}\sin 2t] = \frac{2}{(s+1)^2+4}, \qquad e^{-t}\sin 2t = \mathcal{L}^{-1}\left[\frac{2}{(s+1)^2+4}\right],$$

so

$$\mathcal{L}^{-1}\left[\frac{1}{s^2+2s+5}\right] = \frac{1}{2}\mathcal{L}^{-1}\left[\frac{2}{(s+1)^2+4}\right] = \frac{1}{2}e^{-t}\sin 2t.$$

(b) Recall

$$\mathcal{L}[e^{-t}\sin 2t] = \frac{2}{(s+1)^2 + 4},$$

$$\mathcal{L}[e^{-t}\cos 2t] = \frac{s+1}{(s+1)^2 + 4},$$

so we write

$$\frac{2s+1}{s^2+2s+5} = \frac{2(s+1)-1}{(s+1)^2+4} = 2\frac{(s+1)}{(s+1)^2+4} - \frac{1}{2}\frac{2}{(s+1)^2+4}.$$

Thus

$$\mathcal{L}^{-1}\left[\frac{2s+1}{s^2+2s+5}\right] = 2\mathcal{L}^{-1}\left[\frac{(s+1)}{(s+1)^2+4}\right] - \frac{1}{2}\mathcal{L}^{-1}\left[\frac{2}{(s+1)^2+4}\right]$$

$$= 2\mathcal{L}^{-1}[\mathcal{L}[e^{-t}\cos 2t]] - \frac{1}{2}\mathcal{L}^{-1}[\mathcal{L}[e^{-t}\sin 2t]]$$

$$= 2e^{-t}\cos 2t - \frac{1}{2}e^{-t}\sin 2t.$$

Partial Fraction Decomposition. Take the partial fractions of $F(s)$ and then take the inverse of each term. Most probably you are familiar with partial fractions. We will use the following examples for review.

Example 6.2.3. Find $\mathcal{L}^{-1}\left[\dfrac{s-1}{s^2-s-2}\right]$.

Solution 6.2.3. First we note

$$\frac{s-1}{s^2-s-2} = \frac{s-1}{(s-2)(s+1)}$$

$$= \frac{a}{(s-2)} + \frac{b}{(s+1)} = \frac{a(s+1)+b(s-2)}{(s-2)(s+1)}.$$

The following are three different ways to determine a and b.

- First note that

$$s - 1 = a(s+1) + b(s-2)$$

  must hold for all s. One way is to set $s = 2$, then it follows that $a = \frac{1}{3}$. Similarly if we set $s = -1$, we see immediately that $b = \frac{2}{3}$.
- Another way is to collect the terms with the same powers in s, and require the coefficients of the corresponding terms on both sides of the equation be equal to each other. That is,

$$s - 1 = (a+b)s + (a-2b).$$

  This means $a + b = 1$ and $a - 2b = -1$. Thus $a = \frac{1}{3}$ and $b = \frac{2}{3}$.

- Still another way is to note that

$$\lim_{s\to 2}\left\{(s-2)\frac{s-1}{(s-2)(s+1)}\right\} = \lim_{s\to 2}\left\{(s-2)\left[\frac{a}{(s-2)}+\frac{b}{(s+1)}\right]\right\},$$

this means

$$\lim_{s\to 2}\left\{\frac{s-1}{(s+1)}\right\} = \lim_{s\to 2}\left\{a+(s-2)\frac{b}{(s+1)}\right\} = a.$$

We see immediately that $a = \frac{1}{3}$. Similarly

$$\lim_{s\to -1}\left\{(s+1)\frac{s-1}{(s-2)(s+1)}\right\} = \lim_{s\to -1}\left\{(s+1)\left[\frac{a}{(s-2)}+b\right]\right\} = b$$

gives $b = \frac{2}{3}$.

In some problems, one way is much simpler than others. Anyway, in this problem

$$\mathcal{L}^{-1}\left[\frac{s-1}{s^2-s-2}\right] = \mathcal{L}^{-1}\left[\frac{1}{3}\frac{1}{(s-2)}+\frac{2}{3}\frac{1}{(s+1)}\right]$$

$$= \frac{1}{3}\mathcal{L}^{-1}[\mathcal{L}[e^{2t}]] + \frac{2}{3}\mathcal{L}^{-1}[\mathcal{L}[e^{-t}]] = \frac{1}{3}e^{2t} + \frac{2}{3}e^{-t}.$$

Example 6.2.4. Find $\mathcal{L}^{-1}\left[\frac{1}{s(s^2+4)}\right]$.

Solution 6.2.4. There are two ways for us to take partial fractions.

- If we use complex roots,

$$\frac{1}{s(s^2+4)} = \frac{a}{s}+\frac{b}{s-2i}+\frac{c}{s+2i}.$$

Multiplying by s and taking the limit with $s \to 0$, we have

$$a = \lim_{s\to 0}\frac{1}{s^2+4} = \frac{1}{4}.$$

Multiplying by $s - 2i$ and taking the limit with $s \to 2i$, we have

$$b = \lim_{s\to 2i}\frac{1}{s(s+2i)} = -\frac{1}{8}.$$

Multiplying by $s + 2i$ and taking the limit with $s \to -2i$, we have

$$c = \lim_{s\to -2i}\frac{1}{s(s-2i)} = -\frac{1}{8}.$$

Therefore

$$\mathcal{L}^{-1}\left[\frac{1}{s\,(s^2+4)}\right] = \frac{1}{4}\mathcal{L}^{-1}\left[\frac{1}{s}\right] - \frac{1}{8}\mathcal{L}^{-1}\left[\frac{1}{s-2i}\right] - \frac{1}{8}\mathcal{L}^{-1}\left[\frac{1}{s+2i}\right]$$

$$= \frac{1}{4} - \frac{1}{8}e^{2it} - \frac{1}{8}e^{-2it} = \frac{1}{4} - \frac{1}{4}\cos 2t.$$

- Another way to take partial fractions is to note that

$$\frac{b}{s-2i} + \frac{c}{s+2i} = \frac{b(s+2i)+c(s-2i)}{(s-2i)(s+2i)} = \frac{(b+c)s+2i(b-c)}{s^2+4}.$$

If we let $b+c = b'$ and $2i(b-c) = c'$, then

$$\frac{1}{s\,(s^2+4)} = \frac{a}{s} + \frac{b's+c'}{s^2+4}.$$

An important point we should note is that if the denominator is second order in s, the numerator must be allowed the possibility of being first order in s. In other words, it will not be possible for us to get the correct answer if b' term is missing. With this understanding, the partial fractions can be taken directly as

$$\frac{1}{s\,(s^2+4)} = \frac{a}{s} + \frac{bs+c}{s^2+4} = \frac{a\,(s^2+4)+(bs+c)s}{s\,(s^2+4)}$$

$$= \frac{as^2+4a+bs^2+cs}{s\,(s^2+4)} = \frac{(a+b)s^2+cs+4a}{s\,(s^2+4)}.$$

Therefore

$$1 = (a+b)s^2 + cs + 4a.$$

The coefficients of s must be equal term by term. That is, $a+b=0$, $c=0$, $4a=1$. This gives $a=1/4$, $b=-1/4$, $c=0$. Thus

$$\mathcal{L}^{-1}\left[\frac{1}{s\,(s^2+4)}\right] = \mathcal{L}^{-1}\left[\frac{1}{4s} - \frac{1}{4}\frac{s}{s^2+4}\right] = \frac{1}{4} - \frac{1}{4}\cos 2t.$$

Example 6.2.5. Find $\mathcal{L}^{-1}\left[\dfrac{1}{s^3(s-1)}\right]$.

Solution 6.2.5.

$$\frac{1}{s^3(s-1)} = \frac{a}{s} + \frac{b}{s^2} + \frac{c}{s^3} + \frac{d}{(s-1)}$$

$$= \frac{as^2(s-1)+bs(s-1)+c(s-1)+ds^3}{s^3(s-1)}$$

$$= \frac{(a+d)s^3+(b-a)s^2+(c-b)s-c}{s^3(s-1)}.$$

This requires $a+d=0$, $b-a=0$, $c-b=0$, $-c=1$. Thus $c=-1$, $b=-1$, $a=-1$, $d=1$. Hence

$$\mathcal{L}^{-1}\left[\frac{1}{s^3(s-1)}\right] = \mathcal{L}^{-1}\left[\frac{-1}{s} - \frac{1}{s^2} - \frac{1}{s^3} + \frac{1}{(s-1)}\right]$$

$$= -1 - t - \frac{1}{2}t^2 + e^t.$$

Example 6.2.6. Find $\mathcal{L}^{-1}\left[\dfrac{2\omega}{(s^2+\omega^2)^2}\right]$.

Solution 6.2.6.

$$\frac{2\omega}{(s^2+\omega^2)^2} = \frac{2\omega}{[(s-i\omega)(s+i\omega)]^2}$$

$$= \frac{a}{(s-i\omega)} + \frac{b}{(s-i\omega)^2} + \frac{c}{(s+i\omega)} + \frac{d}{(s+i\omega)^2}.$$

Multiplying both sides by $(s-i\omega)^2$ and take the limit as $s \to i\omega$, we have

$$\lim_{s\to i\omega}\left\{\frac{2\omega}{(s+i\omega)^2}\right\} = \lim_{s\to i\omega}\left\{(s-i\omega)a + b + \frac{(s-i\omega)^2c}{(s+i\omega)} + \frac{(s-i\omega)^2d}{(s+i\omega)^2}\right\}.$$

Clearly

$$b = \frac{2\omega}{(2i\omega)^2} = -\frac{1}{2\omega}.$$

If after multiplying both sides by $(s-i\omega)^2$, we take the derivative first and then go to the limit $s \to i\omega$, we have

$$\lim_{s\to i\omega}\left\{\frac{d}{ds}\frac{2\omega}{(s+i\omega)^2}\right\} = \lim_{s\to i\omega}\left\{a + \frac{d}{ds}\left[\frac{(s-i\omega)^2c}{(s+i\omega)} + \frac{(s-i\omega)^2d}{(s+i\omega)^2}\right]\right\}.$$

This leads to

$$a = \lim_{s\to i\omega}\left\{\frac{-4\omega}{(s+i\omega)^3}\right\} = \frac{1}{2\omega^2 i}.$$

Similarly, we can show

$$d = -\frac{1}{2\omega}, \quad c = -\frac{1}{2\omega^2 i}.$$

Thus we have

$$\mathcal{L}^{-1}\left[\frac{2\omega}{(s^2+\omega^2)^2}\right] =$$

$$\mathcal{L}^{-1}\left[\frac{1}{2\omega^2 i}\frac{1}{(s-i\omega)} - \frac{1}{2\omega}\frac{1}{(s-i\omega)^2} - \frac{1}{2\omega^2 i}\cdot\frac{1}{(s+i\omega)} - \frac{1}{2\omega}\frac{1}{(s+i\omega)^2}\right] =$$

$$\frac{1}{2\omega^2 i}\mathcal{L}^{-1}\left[\frac{1}{(s-i\omega)} - \frac{1}{(s+i\omega)}\right] - \frac{1}{2\omega}\mathcal{L}^{-1}\left[\frac{1}{(s-i\omega)^2} + \frac{1}{(s+i\omega)^2}\right] =$$

$$\frac{1}{2\omega^2 i}\left(e^{i\omega t} - e^{-i\omega t}\right) - \frac{1}{2\omega}\left(t\, e^{i\omega t} + t\, e^{-i\omega t}\right) = \frac{1}{\omega^2}\sin\omega t - \frac{1}{\omega}t\cos\omega t.$$

The Heaviside Expansion. The Heaviside expansion is essentially a systematic way of taking partial fractions. In the partial fraction decomposition of $p(s)/q(s)$, an unrepeated factor $(s - a)$ of $q(s)$ gives rise to a single fraction of the form $A/(s - a)$. Thus $F(s)$ can be written as

$$F(s) = \frac{p(s)}{q(s)} = \frac{A}{s - a} + G(s), \tag{6.33}$$

where $G(s)$ is simply the rest of the expression. Multiplication by $(s - a)$ gives

$$\frac{(s - a)p(s)}{q(s)} = A + (s - a)G(s).$$

If we let s approach a, the second term in the right-hand side vanishes, since $G(s)$ has no factor that could cancel $(s - a)$. Therefore

$$A = \lim_{s \to a} \frac{(s - a)p(s)}{q(s)}. \tag{6.34}$$

Since $q(a) = 0$, because a is an unrepeated root of $q(s) = 0$, the limit in (6.34) is an indeterminant of the form of $0/0$. With the L'Hospital's rule, we have

$$A = \lim_{s \to a} \frac{p(s) + (s - a)p'(s)}{q'(s)} = \frac{p(a)}{q'(a)}. \tag{6.35}$$

Thus the constants in the partial fraction decomposition can be quickly determined.

Example 6.2.7. Use the Heaviside expansion to find $\mathcal{L}^{-1}\left[\dfrac{s - 1}{s^2 - s - 2}\right]$.

Solution 6.2.7. The roots of $s^2 - s - 2 = 0$ are $s = 2$ and $s = -1$, and $\frac{d}{ds}(s^2 - s - 2) = 2s - 1$. Therefore

$$\frac{s - 1}{s^2 - s - 2} = \frac{a}{(s - 2)} + \frac{b}{(s + 1)},$$

$$a = \lim_{s \to 2} \frac{s - 1}{2s - 1} = \frac{1}{3}, \quad b = \lim_{s \to -1} \frac{s - 1}{2s - 1} = \frac{2}{3}.$$

Thus

$$\mathcal{L}^{-1}\left[\frac{s - 1}{s^2 - s - 2}\right] = \frac{1}{3}\mathcal{L}^{-1}\left[\frac{1}{s - 2}\right] + \frac{2}{3}\mathcal{L}^{-1}\left[\frac{1}{s + 1}\right] = \frac{1}{3}e^{2t} + \frac{2}{3}e^{-t}.$$

Example 6.2.8. Use the Heaviside expansion to find $\mathcal{L}^{-1}\left[\dfrac{2s + 1}{s^2 + 2s + 5}\right]$.

Solution 6.2.8. The roots of $s^2 + 2s + 5 = 0$ are $s = -1 \pm 2i$, and $\frac{d}{ds}(s^2 + 2s + 2) = 2s + 2$. Thus

$$\frac{2s+1}{s^2+2s+5} = \frac{a}{s-(-1+2i)} + \frac{b}{s-(-1-2i)},$$

$$a = \lim_{s\to -1+2i} \frac{2s+1}{2s+2} = 1 + \frac{i}{4}, \qquad b = \lim_{s\to -1-2i} \frac{2s+1}{2s+2} = 1 - \frac{i}{4}.$$

Therefore

$$\mathcal{L}^{-1}\left[\frac{2s+1}{s^2+2s+5}\right]$$

$$= \left(1+\frac{1}{4}i\right)\mathcal{L}^{-1}\left[\frac{1}{s-(-1+2i)}\right] + \left(1-\frac{1}{4}i\right)\mathcal{L}^{-1}\left[\frac{1}{s-(-1-2i)}\right].$$

Recall

$$\mathcal{L}^{-1}\left[\frac{1}{s-c}\right] = e^{ct},$$

we have

$$\mathcal{L}^{-1}\left[\frac{1}{s-(-1+2i)}\right] = e^{(-1+2i)t} = e^{-t}e^{i2t} = e^{-t}\left(\cos 2t + i\sin 2t\right),$$

and

$$\mathcal{L}^{-1}\left[\frac{1}{s-(-1-2i)}\right] = e^{-t}e^{-i2t} = e^{-t}\left(\cos 2t - i\sin 2t\right).$$

Hence

$$\mathcal{L}^{-1}\left[\frac{2s+1}{s^2+2s+5}\right] = \left(1+\frac{1}{4}i\right)e^{-t}\left(\cos 2t + i\sin 2t\right)$$

$$+ \left(1-\frac{1}{4}i\right)e^{-t}\left(\cos 2t - i\sin 2t\right)$$

$$= 2e^{-t}\cos 2t - \frac{1}{2}e^{-t}\sin 2t.$$

In general, if $q(s)$ is a polynomial with unrepeated roots, the Heaviside expansion is the most efficient way in partial fraction decomposition. If $q(s)$ is already in the form of a product of factors $(s-a_1)(s-a_2)\cdots(s-a_n)$, then other methods of partial fraction may be equally or more efficient. In any case, if the complex roots are used, it is useful to keep in mind that if the original function is real, the final result must also be real. If there is an imaginary term in the final result, then there must be a mistake somewhere.

If $q(s)$ has repeated roots, we can write it as

$$\frac{p(s)}{q(s)} = \frac{A_m}{(s-a)^m} + \frac{A_{m-1}}{(s-a)^{m-1}} + \cdots + \frac{A_1}{(s-a)}.$$

With a similar argument, one can show that

$$A_k = \frac{1}{(m-k)!}\lim_{s\to a}\frac{d^{m-k}}{ds^{m-k}}\left[\frac{(s-a)^m p(s)}{q(s)}\right], \qquad k=1,\ldots,m. \qquad (6.36)$$

Unfortunately, in practice this formula is not necessarily simpler than other partial fraction methods, such as the one shown in Example 6.2.6. In fact, problems of that nature are best solved by using the derivatives of a transform. *Using Derivatives of the Transform.* In Example 6.2.6, we used the partial fraction to find $\mathcal{L}^{-1}[1/(s^2 + a^2)^2]$. A simpler way to handle such problems is to make use of the properties of derivatives. The procedures are illustrated in the following examples.

Example 6.2.9. Find   (a) $\mathcal{L}^{-1}\left[\dfrac{1}{(s^2 + a^2)^2}\right]$,   (b) $\mathcal{L}^{-1}\left[\dfrac{s}{(s^2 + a^2)^2}\right]$,

(c) $\mathcal{L}^{-1}\left[\dfrac{s^2}{(s^2 + a^2)^2}\right]$, (d) $\mathcal{L}^{-1}\left[\dfrac{s^3}{(s^2 + a^2)^2}\right]$.

Solution 6.2.9. (a) Taking the derivative

$$\frac{d}{da}\frac{a}{s^2 + a^2} = \frac{1}{s^2 + a^2} - \frac{2a^2}{(s^2 + a^2)^2},$$

we can write

$$\frac{1}{(s^2 + a^2)^2} = \frac{1}{2a^2}\left(\frac{1}{s^2 + a^2} - \frac{d}{da}\frac{a}{s^2 + a^2}\right).$$

Since

$$\mathcal{L}[\sin at] = \frac{a}{s^2 + a^2},$$

we have

$$\frac{1}{(s^2 + a^2)^2} = \frac{1}{2a^2}\left(\frac{1}{a}\mathcal{L}[\sin at] - \frac{d}{da}\mathcal{L}[\sin at]\right)$$

$$= \frac{1}{2a^3}\mathcal{L}[\sin at] - \frac{1}{2a^2}\frac{d}{da}\int_0^\infty e^{-st}\sin at\, dt$$

$$= \frac{1}{2a^3}\mathcal{L}[\sin at] - \frac{1}{2a^2}\int_0^\infty e^{-st}t\cos at\, dt$$

$$= \frac{1}{2a^3}\mathcal{L}[\sin at] - \frac{1}{2a^2}\mathcal{L}[t\cos at].$$

Therefore

$$\mathcal{L}^{-1}\left[\frac{1}{(s^2 + a^2)^2}\right] = \mathcal{L}^{-1}\left[\frac{1}{2a^3}\mathcal{L}[\sin at] - \frac{1}{2a^2}\mathcal{L}[t\cos at]\right]$$

$$= \frac{1}{2a^3}\sin at - \frac{1}{2a^2}t\cos at.$$

(b) Take derivative with respect to s

$$\frac{d}{ds}\frac{a}{s^2 + a^2} = \frac{-2as}{(s^2 + a^2)^2},$$

so

$$\frac{s}{(s^2 + a^2)^2} = -\frac{1}{2a}\frac{d}{ds}\frac{a}{s^2 + a^2} = -\frac{1}{2a}\frac{d}{ds}\mathcal{L}[\sin at]$$

$$= -\frac{1}{2a}\frac{d}{ds}\int_0^\infty e^{-st}\sin at\, dt = \frac{1}{2a}\int_0^\infty e^{-st}t\sin at\, dt$$

$$= \frac{1}{2a}\mathcal{L}[t\sin at].$$

Therefore

$$\mathcal{L}^{-1}\left[\frac{s}{(s^2 + a^2)^2}\right] = \mathcal{L}^{-1}\left[\frac{1}{2a}\mathcal{L}[t\sin at]\right] = \frac{1}{2a}t\sin at.$$

(c) It follows from the result of (b),

$$s\mathcal{L}\left[\frac{1}{2a}t\sin at\right] = s\frac{s}{(s^2 + a^2)^2} = \frac{s^2}{(s^2 + a^2)^2}.$$

Recall

$$\mathcal{L}\left[\frac{df}{dt}\right] = s\mathcal{L}[f] - f(0), \qquad s\mathcal{L}[f] = \mathcal{L}\left[\frac{df}{dt}\right] + f(0).$$

Let $f = \dfrac{1}{2a}t\sin at$, so $\dfrac{df}{dt} = \dfrac{1}{2a}\sin at + \dfrac{1}{2}t\cos at$; $f(0) = 0$,

we have

$$s\mathcal{L}\left[\frac{1}{2a}t\sin at\right] = \mathcal{L}\left[\frac{1}{2a}\sin at + \frac{1}{2}t\cos at\right].$$

Therefore

$$\mathcal{L}^{-1}\left[\frac{s^2}{(s^2 + a^2)^2}\right] = \mathcal{L}^{-1}\left[s\mathcal{L}\left[\frac{1}{2a}t\sin at\right]\right] = \frac{1}{2a}\sin at + \frac{1}{2}t\cos at.$$

(d) From the result of (c)

$$\frac{s^3}{(s^2 + a^2)^2} = s\frac{s^2}{(s^2 + a^2)^2} = s\mathcal{L}\left[\frac{1}{2a}\sin at + \frac{1}{2}t\cos at\right].$$

This time, let

$$f = \frac{1}{2a}\sin at + \frac{1}{2}t\cos at, \quad so \quad \frac{df}{dt} = \cos at - \frac{a}{2}t\sin at;\ f(0) = 0,$$

thus

$$s\mathcal{L}\left[\frac{1}{2a}\sin at + \frac{1}{2}t\cos at\right] = \mathcal{L}\left[\cos at - \frac{a}{2}t\sin at\right],$$

$$\mathcal{L}^{-1}\left[\frac{s^3}{(s^2 + a^2)^2}\right] = \mathcal{L}^{-1}\left[\mathcal{L}\left[\cos at - \frac{a}{2}t\sin at\right]\right] = \cos at - \frac{a}{2}t\sin at.$$

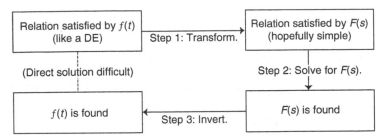

Fig. 6.1. Steps of using Laplace transform to solve differential equations

6.2.2 Solving Differential Equations

The idea of using Laplace transform to solve differential equation is expressed in Fig. 6.1. Suppose we have a differential equation in which the unknown function is $f(t)$. The first step is to apply the Laplace transform to this differential equation. The result is a relation satisfied by $F(s) = \mathfrak{L}[f]$. Generally this is an algebraic equation. The second step is to find $F(s)$ by solving this algebraic equation. The third and final step is to find the unknown function $f(t)$ by taking the inverse of the Laplace transform $F(s)$.

A few example will make this procedure clear.

Example 6.2.10. Find the solution of the differential equation

$$y'' + y = \sin 2t,$$

satisfying the initial conditions

$$y(0) = 0, \quad y'(0) = 1.$$

Solution 6.2.10. Applying Laplace transform to the equation ,

$$\mathfrak{L}[y'' + y] = \mathfrak{L}[\sin 2t],$$

we have

$$s^2 \mathfrak{L}[y] - sy(0) - y'(0) + \mathfrak{L}[y] = \frac{2}{s^2 + 4}.$$

With the initial values of $y(0)$ and $y'(0)$, this equation can be written as

$$(s^2 + 1)\mathfrak{L}[y] = 1 + \frac{2}{s^2 + 4}.$$

This algebraic equation can be easily solved to give

$$\mathfrak{L}[y] = \frac{s^2 + 6}{(s^2 + 1)(s^2 + 4)}.$$

Thus
$$y(t) = \mathcal{L}^{-1}\left[\frac{s^2 + 6}{(s^2 + 1)(s^2 + 4)}\right].$$

Using methods of the last section, we find
$$y(t) = \frac{5}{3}\sin t - \frac{1}{3}\sin 2t.$$

Example 6.2.11. Find the solution of the differential equation
$$y'' + 4y = \sin 2t, \qquad y(0) = 10, \ y'(0) = 0.$$

Solution 6.2.11. Applying the Laplace transform to both sides of the equation
$$\mathcal{L}[y'' + 4y] = \mathcal{L}[\sin 2t],$$

we have
$$s^2\mathcal{L}[y] - sy(0) - y'(0) + 4\mathcal{L}[y] = \frac{2}{s^2 + 4}.$$

With the initial values of $y(0)$ and $y'(0)$, this equation can be written as
$$(s^2 + 4)\mathcal{L}[y] = 10s + \frac{2}{s^2 + 4}.$$

Therefore
$$y = \mathcal{L}^{-1}\left[\frac{10s}{(s^2 + 4)} + \frac{2}{(s^2 + 4)^2}\right].$$

This leads to
$$y = 10\cos 2t + \frac{1}{8}\sin 2t - \frac{1}{4}t\cos 2t.$$

Example 6.2.12. Find the solution of the differential equation
$$y'' + 4y' + 4y = t^2 e^{-2t}, \qquad y(0) = 0, \ y'(0) = 0.$$

Solution 6.2.12. Applying the Laplace transform to the equation
$$\mathcal{L}[y'' + 4y' + 4y] = \mathcal{L}[t^2 e^{-2t}],$$

With the initial values of $y(0)$ and $y'(0)$, we have
$$s^2\mathcal{L}[y] + 4s\mathcal{L}[y] + 4\mathcal{L}[y] = \frac{2}{(s + 2)^3}.$$

Collecting terms
$$(s^2 + 4s + 4)\mathcal{L}[y] = (s + 2)^2\mathcal{L}[y] = \frac{2}{(s + 2)^3},$$

or

$$\mathcal{L}[y] = \frac{2}{(s+2)^5}.$$

The solution is the inverse transform

$$y = \frac{2}{4!}\mathcal{L}^{-1}\left[\frac{4!}{(s+2)^5}\right] = \frac{1}{12}t^4 e^{-2t}.$$

Example 6.2.13. Find the solution of the set of the differential equations

$$y' - 2y + z = 0,$$
$$z' - y - 2z = 0,$$

satisfying the initial conditions

$$y(0) = 1, \quad z(0) = 0.$$

Solution 6.2.13. Applying the Laplace transform to each of the equations

$$\mathcal{L}[y' - 2y + z] = \mathcal{L}[0],$$
$$\mathcal{L}[z' - y - 2z] = \mathcal{L}[0],$$

we obtain

$$s\mathcal{L}[y] - y(0) - 2\mathcal{L}[y] + \mathcal{L}[z] = 0,$$
$$s\mathcal{L}[z] - z(0) - \mathcal{L}[y] - 2\mathcal{L}[z] = 0.$$

After substituting the initial conditions and collecting terms, we have

$$(s - 2)\mathcal{L}[y] + \mathcal{L}[z] = 1,$$
$$\mathcal{L}[y] - (s - 2)\mathcal{L}[z] = 0.$$

This set of algebraic equations can be easily solved to give

$$\mathcal{L}[y] = \frac{s - 2}{(s - 2)^2 + 1},$$
$$\mathcal{L}[z] = \frac{1}{(s - 2)^2 + 1}.$$

Thus

$$y = \mathcal{L}^{-1}\left[\frac{s - 2}{(s - 2)^2 + 1}\right] = e^{2t}\cos t,$$
$$z = \mathcal{L}^{-1}\left[\frac{1}{(s - 2)^2 + 1}\right] = e^{2t}\sin t.$$

6.3 Laplace Transform of Impulse and Step Functions

Some of the most useful and interesting applications of the Laplace transform method occur in the solution of linear differential equations with discontinuous or impulsive nonhomogeneous functions. Equations of this type frequently arise in the analysis of the flow of current in electric circuits or the vibrations of mechanical systems, where voltages or forces of large magnitude act over very short time intervals.

To deal effectively with functions having jump discontinuities, we first introduce two functions known as delta function and step function.

6.3.1 The Dirac Delta Function

The delta function, $\delta(t)$, was first proposed in 1930 by Dirac in the development of the mathematical formalism of quantum mechanics. He required a function which is zero everywhere, except at a single point, where it is discontinuous and behaved like an infinitely high and infinitely narrow spike of unit area. Mathematicians were quick to point out that, strictly speaking, there is no function which has these properties. But Dirac supposed there was, and proceeded to use it so successfully that a new branch of mathematics was developed to justify its use. This area of mathematics is called the theory of distribution or of generalized functions. While it is nice to know that the mathematical foundation of the delta function has been established in complete details, for applications in physical sciences we need only to know its operational definition.

Definition of δ Function. The delta function is a sharply peaked function defined as

$$\delta(t - t_0) = \begin{cases} 0 & t \neq t_0 \\ \infty & t = t_0, \end{cases} \tag{6.37}$$

but such that the integral of $\delta(t - t_0)$ is normalized to unity:

$$\int_{-\infty}^{+\infty} \delta(t - t_0)\, dt = 1. \tag{6.38}$$

Clearly the limits $-\infty$ and ∞ may be replaced by $t_0 - \epsilon$ and $t_0 + \epsilon$ as long as $\epsilon > 0$, since $\delta(t - t_0)$ is equal to zero for $t \neq t_0$. We can think of it as an infinitely high and infinitely narrow function shown in Fig. 6.2, where $h \to \infty$ and $\tau \to 0$ in such a way that the area under it is equal to one.

Mathematically, the δ function is defined by how it behaves inside an integral. In fact the first operation where Dirac used the delta function is the integration

$$\int_{-\infty}^{+\infty} f(t)\delta(t - t_0)\, dt,$$

where $f(t)$ is a continuous function. This integral can be evaluated by the following argument. Since $\delta(t - t_0)$ is zero for $t \neq t_0$, the limit of integration

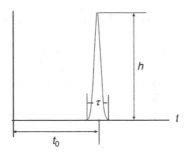

Fig. 6.2. A sharply peaked function. If $h \to \infty$ and $\tau \to 0$ in such a way that the area under it is equal to 1, then this function becomes a delta function $\delta(t - t_0)$

may be changed to $t_0 - \epsilon$ and $t_0 + \epsilon$, where ϵ is a small positive number. Moreover, since $f(x)$ is continuous at $t = t_0$, its values within the interval $(t_0 - \epsilon, t_0 + \epsilon)$ will not differ much from $f(t_0)$ and we can claim, approximately, that

$$\int_{-\infty}^{+\infty} f(t)\delta\left(t - t_0\right) dt = \int_{t_0-\epsilon}^{t_0+\epsilon} f(t)\delta\left(t - t_0\right) dt \approx f(t_0) \int_{t_0-\epsilon}^{t_0+\epsilon} \delta\left(t - t_0\right) dt$$

with the approximation improving as ϵ approaches zero. However,

$$\int_{t_0-\epsilon}^{t_0+\epsilon} \delta\left(t - t_0\right) dt = 1$$

for all values of ϵ. It appears then that letting $\epsilon \to 0$, we have exactly

$$\int_{-\infty}^{+\infty} f(t)\delta\left(t - t_0\right) dt = f(t_0). \tag{6.39}$$

This integral is sometimes referred to as the shifting property of the delta function: $\delta\left(t - t_0\right)$ acts as a sieve, selecting from all possible values of $f(t)$ its value at the point $t = t_0$.

Delta Function with Complicated Arguments. In general the argument of the delta function can be any function of the independent variable. It turns out that such a function can always be rewritten as a sum of delta functions of simple argument. Here are some examples.

- $\delta(-t)$

  Let $t' = -t$, then $dt = -dt'$. We can write

$$\int_{-\infty}^{+\infty} f(t)\delta\left(-t\right) dt = -\int_{+\infty}^{-\infty} f(-t')\delta\left(t'\right) dt' = \int_{-\infty}^{+\infty} f(-t')\delta\left(t'\right) dt' = f(0).$$

Since

$$\int_{-\infty}^{+\infty} f(t)\delta(t)\, dt = f(0),$$

therefore

$$\delta(-t) = \delta(t). \tag{6.40}$$

This result is almost self evident.

- $\delta(at)$

Let $t' = at$, then $dt = dt'/a$. Hence, if $a > 0$,

$$\int_{-\infty}^{+\infty} f(t)\delta(at)\, dt = \int_{-\infty}^{+\infty} f\left(\frac{t'}{a}\right)\delta(t')\frac{1}{a}dt' = \frac{1}{a}\int_{-\infty}^{+\infty} f\left(\frac{t'}{a}\right)\delta(t')\, dt'$$

$$= \frac{1}{a}f\left(\frac{0}{a}\right) = \frac{1}{a}f(0).$$

Since

$$\int_{-\infty}^{+\infty} f(t)\frac{1}{a}\delta(t)\, dt = \frac{1}{a}\int_{-\infty}^{+\infty} f(t)\delta(t)\, dt = \frac{1}{a}f(0),$$

therefore

$$\delta(at) = \frac{1}{a}\delta(t).$$

Since

$$\delta(-at) = \delta(at),$$

we can write

$$\delta(at) = \frac{1}{|a|}\delta(t). \tag{6.41}$$

- $\delta(t^2 - a^2)$

The argument of this function goes to zero when $t = a$ and $t = -a$, which seems to imply two δ functions. There can be contributions to the integral

$$\int_{-\infty}^{+\infty} f(t)\delta(t^2 - a^2)\, dt = \int_{-\infty}^{+\infty} f(t)\delta[(t-a)(t+a)]dt$$

only at the zeros of the argument of the delta function. That is

$$\int_{-\infty}^{+\infty} f(t)\delta(t^2 - a^2)\, dt = \int_{-a-\epsilon}^{-a+\epsilon} f(t)\delta(t^2 - a^2)\, dt + \int_{a-\epsilon}^{a+\epsilon} f(t)\delta(t^2 - a^2)\, dt.$$

Near the two zeros, $t^2 - a^2$ can be approximated as

$$t^2 - a^2 = (t-a)(t+a) = \begin{cases} (-2a)(t+a) & t \to -a \\ (+2a)(t-a) & t \to +a \end{cases}.$$

In the limit as $\epsilon \to 0$, the integral becomes

$$\int_{-\infty}^{+\infty} f(t)\delta\left(t^2 - a^2\right) dt = \int_{-a-\epsilon}^{-a+\epsilon} f(t)\delta\left((-2a)\left(t + a\right)\right) dt$$

$$+ \int_{a-\epsilon}^{a+\epsilon} f(t)\delta\left((2a)\left(t - a\right)\right) dt = \frac{1}{|2a|} \int_{-a-\epsilon}^{-a+\epsilon} f(t)\delta(t + a)dt$$

$$+ \frac{1}{|2a|} \int_{a-\epsilon}^{a+\epsilon} f(t)\delta(t - a)dt = \int_{-\infty}^{+\infty} f(t)\frac{1}{|2a|}[\delta(t + a) + \delta(t - a)]dt.$$

Therefore

$$\delta\left(t^2 - a^2\right) = \frac{1}{|2a|}[\delta(t + a) + \delta(t - a)]. \tag{6.42}$$

The Laplace Transform of the Delta Function and Its Derivative. It follows from the definitions of the Laplace transform and the delta function that the Laplace transform of the delta function is given by

$$\mathcal{L}[\delta(t - a)] = \int_0^\infty e^{-st}\delta(t - a)dt = e^{-sa}. \tag{6.43}$$

The Laplace transform of the derivative of a delta function can be evaluated using integration by parts:

$$\mathcal{L}[\delta'(t - a)] = \int_0^\infty e^{-st}\frac{d}{dt}\delta(t - a)dt = \int_0^\infty e^{-st}d(\delta(t - a))$$

$$= \left[e^{-st}\delta(t - a)\right]_0^\infty - \int_0^\infty \delta(t - a)\frac{d}{dt}e^{-st}dt.$$

Since $\delta(t - a)$ vanishes everywhere except at $t = a$, at both upper and lower limits the integrated part is equal to zero. Therefore

$$\mathcal{L}[\delta'(t - a)] = s\int_0^\infty \delta(t - a)e^{-st}dt = s\,e^{-sa}. \tag{6.44}$$

In dealing with phenomena of an impulsive nature, this is a very useful expression.

6.3.2 The Heaviside Unit Step Function

Definition of the Step Function. The Heaviside unit step function $u(t - c)$ can be defined from the integration of the delta function $\delta(t' - c)$

$$u(t - c) = \int_{-\infty}^t \delta(t' - c)dt'. \tag{6.45}$$

The delta function is identically equal to zero if $t' < c$. The upper limit of the integration variable t' is t. If t is less than c, then all t' will be less than c.

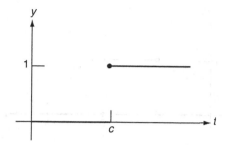

Fig. 6.3. The Heaviside unit step function $u(t - c)$

The integral is equal to zero. If t is greater than c, the integral is equal to one by the definition of the delta function. Thus

$$u(t - c) = \begin{cases} 0 & t < c \\ 1 & t > c \end{cases}.$$
(6.46)

The step function can be defined directly with (6.46) without referring to (6.45). However, with (6.45), it is immediately clear that

$$\frac{d}{dt} u(t - c) = \delta(t - c).$$
(6.47)

A plot of the Heaviside unit step function $y = u(t - c)$ is shown in Fig. 6.3. Interestingly, the function does not get its name because it is heavy on one side, but, rather from the British engineer Oliver Heaviside. Very often this function is simply called step function.

Very often we have to deal with a pulse of a finite duration. These step functions are very convenient in such situation. For example, the square pulse

$$y(t) = \begin{cases} 0 & 0 < t < \pi \\ 1 & \pi < t < 2\pi \\ 0 & 2\pi < t < \infty \end{cases}$$

can be expressed as

$$y(t) = u(t - \pi) - u(t - 2\pi).$$

The sketch of this function is shown in Fig. 6.4.

Shifting Operation. In some problems a system which becomes active at $t = 0$, because of some initial disturbance, is subsequently acted upon by another disturbance beginning at a later time $t = c$. In this situation, the analytical description is greatly facilitated by the function

$$y = f(t - c)u(t - c),$$

which represent a shifting operation. First, $f(t - c)$ represents a translation of $f(t)$ by a distance c in the positive t-direction. Multiplying $u(t - c)$ has the

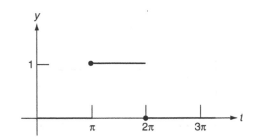

Fig. 6.4. The square impulse $u(t - \pi) - u(t - 2\pi)$

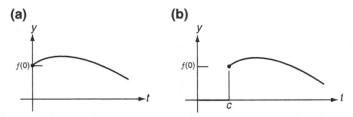

Fig. 6.5. A translation of a given function. (**a**) $y = f(t)$; (**b**) $y = f(t - c)u(t - c)$

effect of "cutting off" or making everything vanish to the left of c. This is shown in Fig. 6.5.

Laplace Transform Involving Step Function. The Laplace transform of the step function is easily determined:

$$\mathfrak{L}[u(t - c)] = \int_0^\infty e^{-st} u(t - c) dt = \int_c^\infty e^{-st} dt$$

$$= \frac{1}{s} e^{-sc}. \tag{6.48}$$

The step function is particularly important in transform theory because of the following relationship between the transform of $f(t)$ and that of its translation $f(t - c)u(t - c)$.

$$\mathfrak{L}[f(t - c)u(t - c)] = \int_0^\infty e^{-st} f(t - c)u(t - c) dt = \int_c^\infty e^{-st} f(t - c) dt.$$

Making a change of variable $t' = t - c$, we have

$$\int_c^\infty e^{-st} f(t - c) dt = \int_0^\infty e^{-s(t'+c)} f(t') dt' = e^{-sc} \int_0^\infty e^{-st'} f(t') dt'$$

$$= e^{-sc} \mathfrak{L}[f(t)].$$

Therefore

$$\mathfrak{L}[f(t - c)u(t - c)] = e^{-sc} \mathfrak{L}[f(t)]. \tag{6.49}$$

Its inverse is of considerable importance.

$$f(t - c)u(t - c) = \mathcal{L}^{-1}[e^{-sc}\mathcal{L}[f(t)]].\tag{6.50}$$

This relationship is sometimes referred as t-shifting (or second shifting) theorem.

6.4 Differential Equations with Discontinuous Forcing Functions

In this section we turn our attention to some examples in which the nonho-mogeneous term, or forcing function, is discontinuous.

We start with the simplest case. A particle of mass m initially at rest is set into motion by a sudden blow at $t = t_0$. Assuming no friction, we wish to find the position as a function of time. Such a common every-day occurrence is rather "awkward" for "ordinary" mathematics to deal with. However, with Laplace transform and delta function, it becomes very easy.

In the Newton's dynamic equation

$$m\frac{d^2x}{dt^2} = F,\tag{6.51}$$

let us express the force of the sudden blow by the delta function

$$F = P\delta(t - t_0).\tag{6.52}$$

The initial conditions are

$$x(0) = 0, \quad x'(0) = 0.\tag{6.53}$$

Applying the Laplace transform to the differential equation

$$\mathcal{L}[mx''] = \mathcal{L}[P\delta(t - t_0)],\tag{6.54}$$

we obtain

$$ms^2\mathcal{L}[x] = P\,e^{-st_0}.\tag{6.55}$$

So

$$x(t) = \frac{P}{m}\mathcal{L}^{-1}\left[\frac{e^{-st_0}}{s^2}\right] = \frac{P}{m}\mathcal{L}^{-1}[e^{-st_0}\mathcal{L}[t]]$$

$$= \frac{P}{m}(t - t_0)u(t - t_0).\tag{6.56}$$

This means

$$x(t) = \begin{cases} 0 & t < t_0 \\ \frac{P}{m}(t - t_0) & t > t_0 \end{cases}.\tag{6.57}$$

The result says that the particle will stay put until t_0, after that the distance will increase linearly with time. The velocity of the particle is given by

$$v = \frac{dx}{dt} = \frac{P}{m}, \tag{6.58}$$

which is a constant. In fact we see that the amplitude P of the delta function is equal to mv which is the momentum. This shows that what the sudden blow did is to impart a momentum P to the particle. This momentum stays the same with the particle thereafter.

Example 6.4.1. Let us consider the damped, driven, harmonic oscillator. The mass m is driven by an applied force $F(t)$. It also experiences the spring force $-kx(t)$ and a friction force $-bx'(t)$, proportional to its velocity. The differential equation describing the motion is

$$mx'' + bx' + kx = F(t).$$

If it is at rest initially

$$x(0) = 0, \quad x'(0) = 0,$$

and the force function is an ideal impulse peaked at t_0, that is

$$F(t) = P_0 \delta(t - t_0),$$

find the displacement x as a function of time t.

Solution 6.4.1. Applying the Laplace transform to both sides of the equation

$$\mathcal{L}\left[x'' + \frac{b}{m}x' + \frac{k}{m}x\right] = \frac{P_0}{m}\mathcal{L}[\delta(t - t_0)]$$

leads to

$$s^2 \mathcal{L}[x] + \frac{b}{m}s\mathcal{L}[x] + \frac{k}{m}\mathcal{L}[x] = \frac{P_0}{m}e^{-st_0}.$$

Therefore

$$\mathcal{L}[x] = \frac{P_0}{m}\frac{1}{s^2 + \frac{b}{m}s + \frac{k}{m}}e^{-st_0}.$$

Let us write

$$s^2 + \frac{b}{m}s + \frac{k}{m} = s^2 + \frac{b}{m}s + \left(\frac{b}{2m}\right)^2 - \left(\frac{b}{2m}\right)^2 + \frac{k}{m}$$

$$= \left(s + \frac{b}{2m}\right)^2 + \frac{k}{m} - \left(\frac{b}{2m}\right)^2,$$

and simplify the notation with

$$\alpha = \frac{b}{2m}, \quad \omega^2 = \frac{k}{m} - \left(\frac{b}{2m}\right)^2,$$

so we have

$$\mathcal{L}[x] = \frac{P_0}{m\omega} \frac{\omega}{(s+\alpha)^2 + \omega^2} e^{-st_0}.$$

Three different cases arise:
(a) The oscillatory case, $\omega^2 > 0$.

$$x(t) = \mathcal{L}^{-1}\left[\frac{P_0}{m\omega} \frac{\omega}{(s+\alpha)^2 + \omega^2} e^{-st_0}\right] = \frac{P_0}{m\omega} \mathcal{L}^{-1}\left[e^{-st_0} \frac{\omega}{(s+\alpha)^2 + \omega^2}\right]$$

$$= \frac{P_0}{m\omega} \mathcal{L}^{-1}\left[e^{-st_0} \mathcal{L}[e^{-\alpha t} \sin \omega t]\right] = \frac{P_0}{m\omega} e^{-\alpha(t-t_0)} \sin \omega(t - t_0) u(t - t_0).$$

(b) The over-damped case $\omega^2 < 0$. Let $\beta^2 = -\omega^2$,

$$x(t) = \mathcal{L}^{-1}\left[\frac{P_0}{m\beta} \frac{\beta}{(s+\alpha)^2 - \beta^2} e^{-st_0}\right] = \frac{P_0}{m\beta} \mathcal{L}^{-1}\left[e^{-st_0} \frac{\beta}{(s+\alpha)^2 - \beta^2}\right]$$

$$= \frac{P_0}{m\beta} \mathcal{L}^{-1}[e^{-st_0} \mathcal{L}[e^{-\alpha t} \sinh \beta t]] = \frac{P_0}{m\beta} e^{-\alpha(t-t_0)} \sinh \beta(t - t_0) u(t - t_0).$$

(c) The critically damped case $\omega^2 = 0$.

$$x(t) = \mathcal{L}^{-1}\left[\frac{P_0}{m} \frac{1}{(s+\alpha)^2} e^{-st_0}\right] = \frac{P_0}{m} \mathcal{L}^{-1}\left[e^{-st_0} \frac{1}{(s+\alpha)^2}\right]$$

$$= \frac{P_0}{m} \mathcal{L}^{-1}[e^{-st_0} \mathcal{L}[e^{-\alpha t} t]] = \frac{P_0}{m} e^{-\alpha(t-t_0)} (t - t_0) u(t - t_0).$$

Notice in all three cases, $x(t)$ is equal to zero before $t = t_0$, as you would expect, because the system cannot respond until after the impulse has occurred. This kind of behavior is often referred to as being causal. Causality, a characteristic of solutions involving time, requires that there can be no response before the application of a drive.

It is interesting to note that Newton's equation is invariant under the transformation of $t \to -t$. Thus clearly causality is not implied by Newton's equation. The causality shown here is the result of the definition of the Laplace transform. The fact that this important physical requirement is built in the Laplace transformation is another reason that this method is so useful.

Example 6.4.2. A mass $m = 1$ is attached to spring with constant $k = 4$, and there is no friction, $b = 0$. The mass is released from rest with $x(0) = 3$. At the instant $t = 2\pi$ the mass is struck with a hammer, providing an impulse $P_0 = 8$. Determine the motion of the mass.

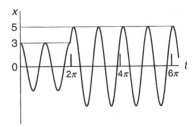

Fig. 6.6. The plot of $x(t) = 3\cos 2t + 4\sin 2(t - 2\pi)u(t - 2\pi)$

Solution 6.4.2. We need to solve the initial value problem

$$x'' + 4x = 8\delta(t - 2\pi); \quad x(0) = 3, \quad x'(0) = 0.$$

Apply the Laplace transform to get

$$s^2 \mathcal{L}[x] - 3s + 4\mathcal{L}[x] = 8e^{-2\pi s},$$

so

$$(s^2 + 4)\mathcal{L}[x] = 3s + 8e^{-2\pi s}.$$

Therefore

$$\mathcal{L}[x] = \frac{3s}{(s^2 + 4)} + \frac{8e^{-2\pi s}}{(s^2 + 4)},$$

hence

$$x(t) = \mathcal{L}^{-1}\left[\frac{3s}{s^2 + 4}\right] + \mathcal{L}^{-1}\left[\frac{8e^{-2\pi s}}{s^2 + 4}\right] = 3\mathcal{L}^{-1}\left[\frac{s}{s^2 + 4}\right] + 4\mathcal{L}^{-1}\left[\frac{2e^{-2\pi s}}{s^2 + 4}\right]$$

$$= 3\mathcal{L}^{-1}[\mathcal{L}[\cos 2t]] + 4\mathcal{L}^{-1}[e^{-2\pi s}\mathcal{L}[\sin 2t]]$$

$$= 3\cos 2t + 4\sin 2(t - 2\pi)u(t - 2\pi)$$

or

$$x(t) = \begin{cases} 3\cos 2t & t < 2\pi \\ 3\cos 2t + 4\sin 2(t - 2\pi) & t > 2\pi \end{cases}.$$

As $3\cos 2t + 4\sin 2t = 5\cos(2t - \theta)$ and $\theta = \tan^{-1}(4/3)$, we see the effect of the impulse at $t = 2\pi$. It instantaneously increases the amplitude of the oscillations from 3 to 5. Although the frequency is still the same, there is a discontinuity in velocity. The plot of $x(t)$ is shown in Fig. 6.6.

Example 6.4.3. Consider the RLC series circuit shown in Fig. 6.7 with $R = 110\ \Omega$, $L = 1$ H, $C = 0.001$ F, and a battery supplying an emf of 90 V. Initially there is no current in the circuit and no charge on the capacitor. At $t = 0$ the switch is closed and at $t = T$ $(T = 1$ s$)$ the battery is removed from the circuit in such a way that the RLC circuit is still closed but without emf. Find the current $i(t)$ as a function of time.

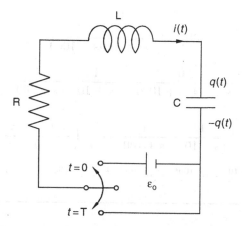

Fig. 6.7. A RLC circuit. The open circuit without charge on the capacitor is closed at $t = 0$. At $t = T$, the battery is removed from the circuit in such a way that the circuit is closed but without emf

Solution 6.4.3. The circuit equation is given by

$$Li' + Ri + \frac{1}{C}q = e(t)$$

$$i = \frac{dq}{dt}$$

and the initial conditions are

$$i(0) = 0, \quad q(0) = 0.$$

In this problem

$$e(t) = 90[u(t) - u(t-1)].$$

Apply the Laplace transform to the two differential equations to get

$$Ls\mathcal{L}[i] + R\mathcal{L}[i] + \frac{1}{C}\mathcal{L}[q] = \mathcal{L}[e(t)]$$

$$\mathcal{L}[i] = s\mathcal{L}[q].$$

Combine the two equations we obtain

$$Ls\mathcal{L}[i] + R\mathcal{L}[i] + \frac{1}{Cs}\mathcal{L}[i] = \mathcal{L}[e(t)].$$

Putting in the R, L, C, and $e(t)$ values in, we have

$$s\mathcal{L}[i] + 110\mathcal{L}[i] + \frac{1}{0.001s}\mathcal{L}[i] = \mathcal{L}[90[u(t) - u(t-1)]]$$

$$= 90\frac{1 - e^{-s}}{s}.$$

Therefore

$$\mathcal{L}[i] = 90\frac{1 - \mathrm{e}^{-s}}{s^2 + 110s + 1000}.$$

Since

$$\frac{90}{s^2 + 110s + 1000} = \frac{1}{s + 10} - \frac{1}{s + 100},$$

so we have

$$i(t) = \mathcal{L}^{-1}\left[\frac{1}{s + 10} - \frac{1}{s + 100} - \mathrm{e}^{-s}\left(\frac{1}{s + 10} - \frac{1}{s + 100}\right)\right]$$

$$= \mathrm{e}^{-10t} - \mathrm{e}^{-100t} - (\mathrm{e}^{-10(t-1)} - \mathrm{e}^{-100(t-1)})u(t - 1).$$

6.5 Convolution

Another important general property of the Laplace transform has to do with the products of transforms. It often happens that we are given two transforms $F(s)$ and $G(s)$ whose inverses $f(t)$ and $g(t)$ we know, and we would like to calculate the inverse of the product $F(s)G(s)$ from those known inverses $f(t)$ and $g(t)$. The inverse is called the convolution of $f(t)$ and $g(t)$. In order to understand the meaning of the mathematical formulation, we will first consider a specific example.

6.5.1 The Duhamel Integral

Let us once again consider the damped driven oscillator

$$mx'' + bx' + kx = f(t) \tag{6.59}$$

with $x(0) = 0$, $x'(0) = 0$. Applying the Laplace transform we get

$$\mathcal{L}[x] = \frac{\mathcal{L}[f(t)]}{m[(s + \alpha)^2 + \omega^2]}, \tag{6.60}$$

where $\alpha = \frac{b}{2m}$, $\omega^2 = \frac{k}{m} - (\frac{b}{2m})^2$. If $f(t)$ is a unit impulse at time τ,

$$f(t) = \delta(t - \tau), \tag{6.61}$$

then we have

$$\mathcal{L}[x] = \frac{\mathrm{e}^{-s\tau}}{m[(s + \alpha)^2 + \omega^2]}. \tag{6.62}$$

Therefore

$$x(t) = \mathcal{L}^{-1}[\mathrm{e}^{-s\tau}\mathcal{L}[\frac{1}{m\omega}\mathrm{e}^{-\alpha t}\sin\omega t]]$$

$$= \frac{1}{m\omega}\mathrm{e}^{-\alpha(t-\tau)}\sin\omega(t - \tau)u(t - \tau). \tag{6.63}$$

For $t > \tau$,

$$x(t) = \frac{1}{m\omega}e^{-\alpha(t-\tau)}\sin\omega(t-\tau)u(t-\tau). \qquad (6.64)$$

If we designate the solution as $g(t)$ in the particular case where τ is equal to zero,

$$g(t) = \frac{1}{m\omega}e^{-\alpha t}\sin\omega t, \qquad (6.65)$$

so in general if τ is not equal to zero,

$$x(t) = g(t-\tau). \qquad (6.66)$$

If the force function is

$$f(t) = P\delta(t-\tau), \qquad (6.67)$$

the solution (or the response function) is clearly

$$x(t) = Pg(t-\tau). \qquad (6.68)$$

Now we consider the response of the system under a general external force function shown in Fig. 6.8.

This force may be assumed to be made up of a series of impulses of varying magnitude. As we have discussed, the impulse is actually momentum P imparted. Since the change of momentum is equal to force $(\Delta P/\Delta t = f)$, the impulse imparted during a short time interval is equal to force multiplied by the time duration.

Assuming that at time τ, the force $f(\tau)$ acts on the system for a short period of time $\Delta\tau$, the impulse acting at $t = \tau$ is given by $f(\tau)\Delta\tau$. At any time t, the elapsed time since the impulse is $t - \tau$, so the response of the system at time t due to this impulse is

$$\Delta x(t) = f(\tau)\Delta\tau g(t-\tau). \qquad (6.69)$$

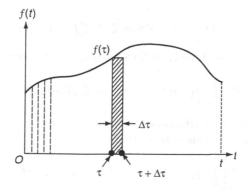

Fig. 6.8. An arbitrary forcing function

The total response at time t can be found by summing all the responses due to the elementary impulses acting all times

$$x(t) = \sum f(\tau)g(t - \tau)\Delta\tau. \tag{6.70}$$

Letting $\Delta\tau \to 0$ and replacing the summation by the integration, we obtain

$$x(t) = \int_0^t f(\tau)g(t - \tau)\mathrm{d}\tau \tag{6.71}$$

or

$$x(t) = \frac{1}{m\omega}\int_0^t f(\tau)\mathrm{e}^{-\alpha(t-\tau)}\sin\omega(t - \tau)\mathrm{d}\tau. \tag{6.72}$$

This result is known as the Duhamel integral. In many cases the function $f(\tau)$ has a form that permits an explicit integration. In the case such integration is not possible, it can be evaluated numerically without much difficulty.

6.5.2 The Convolution Theorem

The Duhamel integral can also be viewed in the following way. Since

$$g(t) = \frac{1}{m\omega}\mathrm{e}^{-\alpha t}\sin\omega t = \mathcal{L}^{-1}\left[\frac{1}{m[(s + \alpha)^2 + \omega^2]}\right], \tag{6.73}$$

and by (6.60)

$$\begin{aligned}
\mathcal{L}[x(t)] &= \mathcal{L}[f(t)]\frac{1}{m[(s + \alpha)^2 + \omega^2]} \\
&= \mathcal{L}[f(t)]\mathcal{L}[g(t)],
\end{aligned} \tag{6.74}$$

it follows that

$$x(t) = \mathcal{L}^{-1}[\mathcal{L}[f(t)]\mathcal{L}[g(t)]]. \tag{6.75}$$

On the other hand

$$x(t) = \int_0^t f(\tau)g(t - \tau)\mathrm{d}\tau, \tag{6.76}$$

therefore

$$\int_0^t f(\tau)g(t - \tau)\mathrm{d}\tau = \mathcal{L}^{-1}[\mathcal{L}[f(t)]\mathcal{L}[g(t)]]. \tag{6.77}$$

It turns out, as long as these transforms exist, this relationship

$$\mathcal{L}\left[\int_0^t f(\tau)g(t - \tau)\mathrm{d}\tau\right] = \mathcal{L}[f(t)]\mathcal{L}[g(t)]] \tag{6.78}$$

is generally true for any arbitrary functions of f and g. It is known as the convolution theorem. If this is true, then

$$\mathcal{L}\left[\int_0^t f(t - \lambda)g(\lambda)\mathrm{d}\lambda\right] = \mathcal{L}[f(t)]\mathcal{L}[g(t)]] \tag{6.79}$$

must also be true, since the roles played by f and g in the equation are symmetric. This can be easily demonstrated directly by a change of variable. Let $\lambda = t - \tau$, then

$$\int_0^t f(\tau)g(t-\tau)\mathrm{d}\tau = \int_t^0 f(t-\lambda)g(\lambda)\mathrm{d}(-\lambda)$$

$$= \int_0^t f(t-\lambda)g(\lambda)\mathrm{d}\lambda. \tag{6.80}$$

The proof of the convolution theorem goes as follows.
By definition

$$\mathfrak{L}\left[\int_0^t f(t-\lambda)g(\lambda)\mathrm{d}\lambda\right] = \int_0^\infty \mathrm{e}^{-st}\left[\int_0^t f(t-\lambda)g(\lambda)\mathrm{d}\lambda\right]\mathrm{d}t. \tag{6.81}$$

Now with

$$u(t-\lambda) = \begin{cases} 1 & \lambda < t \\ 0 & \lambda > t \end{cases} \tag{6.82}$$

and

$$f(t-\lambda)g(\lambda)u(t-\lambda) = \begin{cases} f(t-\lambda)g(\lambda) & \lambda < t \\ 0 & \lambda > t. \end{cases} \tag{6.83}$$

We can write

$$\int_0^\infty f(t-\lambda)g(\lambda)u(t-\lambda)\mathrm{d}\lambda = \int_0^t f(t-\lambda)g(\lambda)u(t-\lambda)\mathrm{d}\lambda$$

$$+ \int_t^\infty f(t-\lambda)g(\lambda)u(t-\lambda)\mathrm{d}\lambda, \tag{6.84}$$

the second term on the right-hand side is equal to zero because the lower limit of λ is t, so $\lambda > t$. In the first term on the right-hand side, the range of λ is between 0 and t, so $\lambda < t$, using (6.83) we have

$$\int_0^\infty f(t-\lambda)g(\lambda)u(t-\lambda)\mathrm{d}\lambda = \int_0^t f(t-\lambda)g(\lambda)\mathrm{d}\lambda. \tag{6.85}$$

Putting (6.85) into (6.81)

$$\mathfrak{L}\left[\int_0^t f(t-\lambda)g(\lambda)\mathrm{d}\lambda\right] = \int_0^\infty \mathrm{e}^{-st}\left[\int_0^\infty f(t-\lambda)g(\lambda)u(t-\lambda)\mathrm{d}\lambda\right]\mathrm{d}t, \tag{6.86}$$

and changing the order of integration

$$\int_0^\infty \mathrm{e}^{-st}\left[\int_0^\infty f(t-\lambda)g(\lambda)u(t-\lambda)\mathrm{d}\lambda\right]\mathrm{d}t$$

$$= \int_0^\infty g(\lambda)\left[\int_0^\infty \mathrm{e}^{-st}f(t-\lambda)u(t-\lambda)\mathrm{d}t\right]\mathrm{d}\lambda, \tag{6.87}$$

we obtain

$$\mathcal{L}[\int_0^t f(t-\lambda)g(\lambda)d\lambda] = \int_0^\infty g(\lambda)\left[\int_0^\infty e^{-st}f(t-\lambda)u(t-\lambda)dt\right]d\lambda. \quad (6.88)$$

Because of the presence of $u(t-\lambda)$, the integrand of the inner integral is identically zero for all $t < \lambda$. Hence the inner integration effectively starts not at $t = 0$ but at $t = \lambda$. Therefore

$$\mathcal{L}\left[\int_0^t f(t-\lambda)g(\lambda)d\lambda\right] = \int_0^\infty g(\lambda)\left[\int_\lambda^\infty e^{-st}f(t-\lambda)dt\right]d\lambda. \quad (6.89)$$

Now in the inner integral on the right, let $t - \lambda = \tau$ and $dt = d\tau$. Then

$$\begin{aligned}
\mathcal{L}\left[\int_0^t f(t-\lambda)g(\lambda)d\lambda\right] &= \int_0^\infty g(\lambda)\left[\int_0^\infty e^{-s(\tau+\lambda)}f(\tau)d\tau\right]d\lambda \\
&= \int_0^\infty e^{-s\lambda}g(\lambda)\left[\int_0^\infty e^{-s\tau}f(\tau)d\tau\right]d\lambda \\
&= \left[\int_0^\infty e^{-s\tau}f(\tau)d\tau\right]\left[\int_0^\infty e^{-s\lambda}g(\lambda)d\lambda\right] \\
&= \mathcal{L}[f(t)]\mathcal{L}[g(t)] \quad (6.90)
\end{aligned}$$

as asserted.

A common notation is to designate the convolution integral as

$$\int_0^t f(t-\lambda)g(\lambda)d\lambda = f(t) * g(t). \quad (6.91)$$

So the convolution theorem is often written as

$$\mathcal{L}[f]\mathcal{L}[g] = \mathcal{L}[f * g]. \quad (6.92)$$

Example 6.5.1. Use convolution to find

$$\mathcal{L}^{-1}\left[\frac{1}{s^2(s-a)}\right].$$

Solution 6.5.1. Since

$$\mathcal{L}[t] = \frac{1}{s^2}, \quad \mathcal{L}[e^{at}] = \frac{1}{s-a},$$

we can write

$$\mathcal{L}^{-1}\left[\frac{1}{s^2(s-a)}\right] = \mathcal{L}^{-1}\left[\frac{1}{s^2}\cdot\frac{1}{s-a}\right] = \mathcal{L}^{-1}[\mathcal{L}[t]\mathcal{L}[e^{at}]].$$

Therefore

$$\mathcal{L}^{-1}\left[\frac{1}{s^2(s-a)}\right] = t\,e^{at} = \int_0^t \tau e^{a(t-\tau)}d\tau = \frac{1}{a^2}\left(e^{at} - at - 1\right).$$

6.6 Further Properties of Laplace Transforms

6.6.1 Transforms of Integrals

From the property of the transform of a derivative, one can derive a formula
for the transform of an integral.

$$\mathcal{L}\left[\int_0^t f(x)\mathrm{d}x\right] = \int_0^\infty \mathrm{e}^{-st}\left[\int_0^t f(x)\mathrm{d}x\right]\mathrm{d}t.$$

Let

$$g(t) = \int_0^t f(x)\mathrm{d}x,$$

then

$$g'(t) = f(t) \quad\text{and}\quad g(0) = 0.$$

Since

$$\mathcal{L}\left[g'(t)\right] = s\mathcal{L}\left[g(t)\right] - g(0),$$

we have

$$\mathcal{L}\left[f(t)\right] = s\mathcal{L}\left[\int_0^t f(x)\mathrm{d}x\right]. \tag{6.93}$$

Thus if $F(s) = \mathcal{L}\left[f(t)\right]$,

$$\mathcal{L}\left[\int_0^t f(x)\mathrm{d}x\right] = \frac{\mathcal{L}\left[f(t)\right]}{s} = \frac{1}{s}F(s). \tag{6.94}$$

This formula is very useful in finding the inverse transform of a fraction
that has the form of $p(s)/\left[s^n q(s)\right]$. For example:

$$\mathcal{L}^{-1}\left[\frac{1}{s(s-a)}\right] = \mathcal{L}^{-1}\left[\frac{1}{s}\mathcal{L}\left[\mathrm{e}^{at}\right]\right] = \mathcal{L}^{-1}\mathcal{L}\left[\int_0^t \mathrm{e}^{ax}\mathrm{d}x\right]$$

$$= \int_0^t \mathrm{e}^{ax}\mathrm{d}x = \frac{1}{a}(\mathrm{e}^{at} - 1).$$

Similarly

$$\mathcal{L}^{-1}\left[\frac{1}{s^2(s-a)}\right] = \mathcal{L}^{-1}\left[\frac{1}{s}\mathcal{L}\left[\frac{1}{a}(\mathrm{e}^{at} - 1)\right]\right]$$

$$= \int_0^t \frac{1}{a}(\mathrm{e}^{ax} - 1)\mathrm{d}x = \frac{1}{a^2}(\mathrm{e}^{at} - at - 1).$$

This method is often more convenient than the method of partial fractions.

6.6.2 Integration of Transforms

Differentiation of $F(s)$ corresponds to multiplication of $f(t)$ by $-t$. It is natural
to expect that integration of $F(s)$ will correspond to division of $f(t)$ by t. This
is indeed the case, provided the limits of integration are appropriately chosen.

If $F(s')$ is the Laplace transform of $f(t)$, then

$$\int_s^\infty F(s')ds' = \int_s^\infty \left[\int_0^\infty e^{-s't}f(t)dt\right]ds' = \int_0^\infty \left[\int_s^\infty e^{-s't}f(t)ds'\right]dt$$

$$= \int_0^\infty f(t)\left[\int_s^\infty e^{-s't}ds'\right]dt = \int_0^\infty f(t)\left[-\frac{1}{t}e^{-s't}\right]_{s'=s}^\infty dt$$

$$= \int_0^\infty f(t)\frac{1}{t}e^{-st}dt = \int_0^\infty e^{-st}\frac{1}{t}f(t)dt = \mathcal{L}\left[\frac{f(t)}{t}\right]. \qquad (6.95)$$

This relationship, namely

$$\mathcal{L}\left[\frac{f(t)}{t}\right] = \int_s^\infty \mathcal{L}[f(t)]\,ds'$$

is useful if $\mathcal{L}[f(t)]$ is known.

Example 6.6.1. Find (a) $\mathcal{L}\left[\frac{1}{t}(e^{-at} - e^{-bt})\right]$,   (b) $\mathcal{L}\left[\frac{\sin t}{t}\right]$.

Solution 6.6.1. (a)

$$\mathcal{L}\left[\frac{1}{t}(e^{-at} - e^{-bt})\right] = \int_s^\infty \mathcal{L}\left[e^{-at} - e^{-bt}\right]ds' = \int_s^\infty \left(\frac{1}{s'+a} - \frac{1}{s'+b}\right)ds'$$

$$= [\ln(s'+a) - \ln(s'+b)]_{s'=s}^\infty = \left[\ln\frac{s'+a}{s'+b}\right]_{s'=s}^\infty$$

$$= \ln 1 - \ln\frac{s+a}{s+b} = \ln\frac{s+b}{s+a}.$$

(b)

$$\mathcal{L}\left[\frac{\sin t}{t}\right] = \int_s^\infty \mathcal{L}[\sin t]\,ds' = \int_s^\infty \frac{1}{s'^2+1}ds' = [\tan^{-1} s']_{s'=s}^\infty$$

$$= \frac{\pi}{2} - \tan^{-1} s = \cot^{-1} s = \tan^{-1}\frac{1}{s}.$$

6.6.3 Scaling

If $F(s) = \mathcal{L}[f(t)] = \int_0^\infty e^{-st}f(t)dt$ is already known, then $\mathcal{L}[f(at)]$ can be easily obtained by a change of scale. By definition

$$\mathcal{L}[f(at)] = \int_0^\infty e^{-st}f(at)dt = \frac{1}{a}\int_0^\infty e^{-(s/a)at}f(at)d(at). \qquad (6.96)$$

Let $t' = at$, the integral becomes

$$\int_0^\infty e^{-(s/a)at}f(at)d(at) = \int_0^\infty e^{-(s/a)t'}f(t')d(t'),$$

which is just the Laplace transform of f with the parameter s replaced by s/a. Therefore

$$\mathcal{L}[f(at)] = \frac{1}{a} F\left(\frac{s}{a}\right). \tag{6.97}$$

Example 6.6.2. If $\mathcal{L}[f(t)]$ is known to be $\dfrac{1}{s(1+2s)}$, Find $\mathcal{L}[f(2t)]$.

Solution 6.6.2.

$$\mathcal{L}[f(2t)] = \frac{1}{2} \frac{1}{(s/2)[1 + 2(s/2)]} = \frac{1}{s(1+s)}.$$

Example 6.6.3. Find $\mathcal{L}\left[\dfrac{\sin \omega t}{t}\right]$.

Solution 6.6.3. Since $\mathcal{L}\left[\dfrac{\sin t}{t}\right] = \tan^{-1}\dfrac{1}{s}$, then

$$\mathcal{L}\left[\frac{\sin \omega t}{\omega t}\right] = \frac{1}{\omega} \tan^{-1}\frac{\omega}{s}.$$

Therefore

$$\mathcal{L}\left[\frac{\sin \omega t}{t}\right] = \tan^{-1}\frac{\omega}{s}.$$

6.6.4 Laplace Transforms of Periodic Functions

Very often the input functions in physical systems are periodic functions. A function is said to be periodic if there is a number p such that

$$f(t+p) = f(t).$$

The least value of p is called the period of f. A periodic function is one that has the characteristic

$$f(t) = f(t+p) = f(t+2p) = \cdots f(t+np) \cdots. \tag{6.98}$$

The Laplace transform of $f(t)$ is a series of integrals

$$\mathcal{L}[f] = \int_0^\infty e^{-st} f(t)dt$$

$$= \int_0^p e^{-st} f(t)dt + \int_p^{2p} e^{-st} f(t)dt \cdots \int_{np}^{(n+1)p} e^{-st} f(t)dt \cdots. \tag{6.99}$$

It follows from a change of variable $t = \tau + np$ that

$$\int_{np}^{(n+1)p} e^{-st} f(t)dt = \int_0^p e^{-s(\tau+np)} f(\tau + np)d\tau = e^{-snp} \int_0^p e^{-s\tau} f(\tau)d\tau.$$

The dummy integration variable τ can be set equal to t, thus

$$\mathcal{L}[f] = \int_0^p e^{-st} f(t)dt + e^{-sp} \int_0^p e^{-st} f(t)dt + \cdots + e^{-snp} \int_0^p e^{-st} f(t)dt + \cdots$$

$$= (1 + e^{-sp} + e^{-2sp} + \cdots + e^{-nsp} + \cdots) \int_0^p e^{-st} f(t)dt. \qquad (6.100)$$

With the series expansion, $1/(1-x) = 1 + x + x^2 + \cdots$, this equation becomes

$$\mathcal{L}[f] = \frac{1}{1 - e^{-sp}} \int_0^p e^{-st} f(t)dt. \qquad (6.101)$$

Example 6.6.4. Half-wave rectifier: Find the Laplace transform of the periodic function (shown in Fig. 6.9) whose definition over one period is:

$$f(t) = \begin{cases} \sin \omega t & \text{if } 0 < t < \dfrac{\pi}{\omega} \\[2mm] 0 & \text{if } \dfrac{\pi}{\omega} < t < \dfrac{2\pi}{\omega} \end{cases}.$$

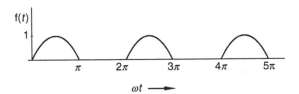

Fig. 6.9. Half-wave rectifier. The definition over one period is $f(t) = \sin \omega t$ for $0 < t < \pi/\omega$ and $f(t) = 0$ for $\pi/\omega < t < 2\pi/\omega$

Solution 6.6.4.

$$\mathcal{L}[f] = \frac{1}{1 - e^{-s2\pi/\omega}} \int_0^{2\pi/\omega} e^{-st} f(t)dt = \frac{1}{1 - e^{-s2\pi/\omega}} \int_0^{\pi/\omega} e^{-st} \sin \omega t \, dt.$$

The integral can be evaluated with integration by parts. However, it is easier to note that the integral is the imaginary part of

$$\int_0^{\pi/\omega} e^{-st} e^{i\omega t} dt = \left[\frac{1}{-s + i\omega} e^{-st+i\omega t} \right]_0^{\pi/\omega} = \frac{1}{-s + i\omega} \left(e^{-s\pi/\omega + i\pi} - 1 \right)$$

$$= \frac{-s - i\omega}{s^2 + \omega^2} \left(-e^{-s\pi/\omega} - 1 \right).$$

Thus

$$\mathcal{L}[f] = \frac{1}{1 - e^{-s2\pi/\omega}} \frac{\omega(1 + e^{-s\pi/\omega})}{s^2 + \omega^2} = \frac{\omega}{(s^2 + \omega^2)(1 - e^{-s\pi/\omega})}.$$

Example 6.6.5. Full-wave rectifier: Find the Laplace transform of the periodic function (shown in Fig. 6.10) whose definition over one period is:

$$f(t) = |\sin \omega t| \qquad 0 < t < \frac{\pi}{\omega}.$$

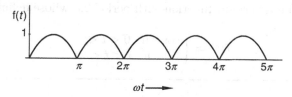

Fig. 6.10. Full-wave rectifier. The definition over one period is $f(t) = |\sin \omega t|$, for $0 < t < \pi/\omega$

Solution 6.6.5. In this case, the period is just π/ω. Therefore

$$\mathcal{L}[f] = \frac{1}{1 - e^{-s\pi/\omega}} \int_0^{\pi/\omega} e^{-st} \sin \omega t \, dt = \frac{\omega(1 + e^{-s\pi/\omega})}{(s^2 + \omega^2)(1 - e^{-s\pi/\omega})}.$$

This is a perfectly good result. It can be simplified somewhat by multiplying both numerator and denominator $\exp(\frac{s\pi}{2\omega})$

$$\mathcal{L}[f] = \frac{\omega(e^{s\pi/(2\omega)} + e^{-s\pi/(2\omega)})}{(s^2 + \omega^2)(e^{s\pi/(2\omega)} - e^{-s\pi/(2\omega)})} = \frac{\omega}{(s^2 + \omega^2)} \coth \frac{s\pi}{2\omega}.$$

6.6.5 Inverse Laplace Transforms Involving Periodic Functions

Any Laplace transform $F(s)$ that either has a factor $(1 - e^{-sp})^{-1}$, or can be written in a form with such a factor as in the last example, indicates that its inverse transform is a periodic function. However, its period may be a multiple of p. This is illustrated in the following example.

Example 6.6.6. Find the inverse and its period of the Laplace transform

$$F(s) = \frac{s}{(s^2 + 1)(1 - e^{-s\pi})}.$$

Solution 6.6.6.

$$f(t) = \mathcal{L}^{-1}\left[\frac{s}{(s^2+1)(1-e^{-s\pi})}\right]$$

$$= \mathcal{L}^{-1}\left[\frac{s}{(s^2+1)}(1+e^{-s\pi}+e^{-2s\pi}+e^{-3s\pi}+\cdots)\right]$$

$$= \cos t + u(t-\pi)\cos(t-\pi) + u(t-2\pi)\cos(t-2\pi)$$
$$+u(t-3\pi)\cos(t-3\pi) + u(t-4\pi)\cos(t-4\pi)\cdots$$

$$= \cos t - u(t-\pi)\cos t + u(t-2\pi)\cos t - u(t-3\pi)\cos t + \cdots$$

$$= [1-u(t-\pi)]\cos t + [u(t-2\pi)-u(t-3\pi)]\cos t + \cdots.$$

Therefore $f(t)$ is a periodic function with period 2π, whose definition over one period is

$$f(t) = \begin{cases} \cos t & \text{if } 0 < t < \pi \\ 0 & \text{if } \pi < t < 2\pi \end{cases}.$$

6.6.6 Laplace Transforms and Gamma Functions

The Laplace transform of t^n is defined as

$$\mathcal{L}\left[t^n\right] = \int_0^\infty e^{-st}t^n \, dt. \tag{6.102}$$

If we make a change of variable and let $st = x$, the integral becomes

$$\mathcal{L}\left[t^n\right] = \int_0^\infty e^{-x}\left(\frac{x}{s}\right)^n d\left(\frac{x}{s}\right) = \frac{1}{s^{n+1}}\int_0^\infty e^{-x}x^n dx. \tag{6.103}$$

The last integral is known as the *gamma function* of $n+1$, written as $\Gamma(n+1)$. Gamma function occurs frequently in practice. It is given by

$$\Gamma(n) = \int_0^\infty e^{-x}x^{n-1}dx, \tag{6.104}$$

this is well defined as long as n is not zero or a negative integer. For $n = 1$,

$$\Gamma(1) = \int_0^\infty e^{-x}dx = \left[e^{-x}\right]_0^\infty = 1. \tag{6.105}$$

With integration by parts, one can easily show

$$\Gamma(n+1) = \int_0^\infty e^{-x}x^n dx = \left[-x^n e^{-x}\right]_0^\infty + n\int_0^\infty e^{-x}x^{n-1}dx = n\Gamma(n). \tag{6.106}$$

Thus if n is a positive integer,

$$\Gamma(n+1) = n\Gamma(n) = n(n-1)\cdots 1\Gamma(1) = n!, \qquad (6.107)$$

and according to (6.103)

$$\mathcal{L}[t^n] = \frac{\Gamma(n+1)}{s^{n+1}} = \frac{n!}{s^{n+1}} \qquad (6.108)$$

in agreement with the result we obtained before. Since $\Gamma(n)$ is a tabulated function, as long as $n > -1$, $\mathcal{L}[t^n]$ can still be evaluated even if n is not an integer. For example

$$\mathcal{L}\left[\frac{1}{\sqrt{t}}\right] = \frac{\Gamma(\frac{1}{2})}{s^{\frac{1}{2}}} = \frac{\sqrt{\pi}}{s^{1/2}}, \qquad (6.109)$$

$$\mathcal{L}\left[\sqrt{t}\right] = \frac{\Gamma(1+\frac{1}{2})}{s^{1+\frac{1}{2}}} = \frac{\frac{1}{2}\Gamma(\frac{1}{2})}{s^{\frac{3}{2}}} = \frac{1}{2}\frac{\sqrt{\pi}}{s^{3/2}}, \qquad (6.110)$$

where we have used the well-known result $\Gamma(\frac{1}{2}) = \sqrt{\pi}$.

Example 6.6.7. Find the Laplace transform of $e^{at}(1+2at)/\sqrt{\pi t}$.

Solution 6.6.7. Use s-shifting property

$$\mathcal{L}\left[\frac{e^{at}}{\sqrt{\pi t}}\right] = \frac{1}{(s-a)^{1/2}},$$

$$\mathcal{L}\left[\frac{e^{at}2at}{\sqrt{\pi t}}\right] = \frac{2a}{\sqrt{\pi}}\mathcal{L}\left[e^{at}\sqrt{t}\right] = \frac{a}{(s-a)^{3/2}}.$$

Thus

$$\mathcal{L}\left[\frac{e^{at}(1+2at)}{\sqrt{\pi t}}\right] = \frac{1}{(s-a)^{1/2}} + \frac{a}{(s-a)^{3/2}} = \frac{s}{(s-a)^{3/2}}.$$

6.7 Summary of Operations of Laplace Transforms

The properties of the Laplace transform are not difficult to understand. However, because there are so many of them, it is not easy to decide which one to use for a specific problem. In Table 6.2 we summarize these operations. In the last column we give a simple example and in the first column we give a name to characterize the operation. This classification is helpful in remembering the details of each operation.

In Sect. 6.2.1, we discussed the inverse of the transform $F(s)$ in the form of a quotient of two polynomials. If $F(s)$ is not in that form, sometimes we can use the properties of the Laplace transforms to obtain the inverse.

Table 6.2. Summary of Laplace transform operations

Name	$h(t)$	$\mathfrak{L}\,[h(t)]$	Example: Let $f(t) = t$
Definition	$f(t)$	$F(s)$	$\mathfrak{L}\,[t] = \int_0^\infty e^{-st}t \; dt = \dfrac{1}{s^2} = F(s)$
Multiply t	$tf(t)$	$-\dfrac{d}{ds}F(s)$	$\mathfrak{L}\,[t \cdot t] = -\dfrac{d}{ds}\dfrac{1}{s^2} = \dfrac{2}{s^3}$
Divide t	$\dfrac{f(t)}{t}$	$\int_s^\infty F(\varsigma)d\varsigma$	$\mathfrak{L}\left[\dfrac{t}{t}\right] = \int_s^\infty \dfrac{1}{\varsigma^2}d\varsigma = \dfrac{1}{s}$
Derivative	$f'(t)$	$sF(s) - f(0)$	$\mathfrak{L}\left[\dfrac{dt}{dt}\right] = s\dfrac{1}{s^2} - 0 = \dfrac{1}{s}$
Integral	$\int_0^t f(\tau)d\tau$	$\dfrac{F(s)}{s}$	$\mathfrak{L}\left[\int_0^t \tau \; d\tau\right] = \dfrac{1/s^2}{s} = \dfrac{1}{s^3}$
Shifting $-s$	$e^{at}f(t)$	$F(s-a)$	$\mathfrak{L}\left[e^{at}t\right] = \dfrac{1}{(s-a)^2}$
Shifting $-t$	$u(t-a)f(t-a)$	$e^{-sa}F(s)$	$\mathfrak{L}\,[u(t-a)(t-a)] = e^{-sa}\dfrac{1}{s^2}$
Scaling	$f(at)$	$\dfrac{1}{a}F\left(\dfrac{s}{a}\right)$	$\mathfrak{L}\,[at] = \dfrac{1}{a}\dfrac{1}{(s/a)^2} = a\dfrac{1}{s^2}$
Period $-p$	periodic $f(t)$	$\dfrac{\int_0^P e^{-st}f(t)\,dt}{1 - e^{-ps}}$	$\mathfrak{L}\,[f] = \dfrac{1 - (1+ps)e^{-ps}}{s^2(1 - e^{-ps})}$
Convolution	$\int_0^t f(\tau)g(t-\tau)d\tau$	$F(s) \cdot G(s)$	Let $g(t) = f(t),\;\; G(s) = F(s)$ $\mathfrak{L}\left[\int_0^t \tau(t-\tau)d\tau\right] = \dfrac{1}{s^2}\dfrac{1}{s^2} = \dfrac{1}{s^4}$

Example 6.7.1. Find $\mathfrak{L}^{-1}\left[\ln\dfrac{s+a}{s-b}\right]$.

Solution 6.7.1. The transform is not in the form of a quotient of two polynomials, but its derivative is. Let

$$\mathfrak{L}\,[f(t)] = \ln\dfrac{s+a}{s-b}, \qquad f(t) = \mathfrak{L}^{-1}\left[\ln\dfrac{s+a}{s-b}\right].$$

Since

$$\mathfrak{L}\,[tf(t)] = -\dfrac{d}{ds}\mathfrak{L}\,[f(t)] = -\dfrac{d}{ds}\ln\dfrac{s+a}{s-b}$$

$$= -\dfrac{d}{ds}\ln(s+a) + \dfrac{d}{ds}\ln(s-b) = \dfrac{1}{s-b} - \dfrac{1}{s+a},$$

and

$$tf(t) = \mathcal{L}^{-1}\left[-\frac{d}{ds}\mathcal{L}[f(t)]\right] = \mathcal{L}^{-1}\left[\frac{1}{s-b} - \frac{1}{s+a}\right],$$

therefore

$$\mathcal{L}^{-1}\left[\ln\frac{s+a}{s-b}\right] = f(t) = \frac{1}{t}\mathcal{L}^{-1}\left[\frac{1}{s-b} - \frac{1}{s+a}\right]$$

$$= \frac{1}{t}\left(e^{bt} - e^{-at}\right).$$

Example 6.7.2. Find $\mathcal{L}^{-1}\left[\ln\frac{s^2-a^2}{s^2}\right]$.

Solution 6.7.2.

$$\mathcal{L}^{-1}\left[\ln\frac{s^2-a^2}{s^2}\right] = \frac{1}{t}\mathcal{L}^{-1}\left[-\frac{d}{ds}\ln\frac{s^2-a^2}{s^2}\right] = -\frac{1}{t}\mathcal{L}^{-1}\left[\frac{2s}{s^2-a^2} - \frac{2s}{s^2}\right]$$

$$= \frac{2}{t}\left(1 - \cosh at\right).$$

Example 6.7.3. Find $\mathcal{L}^{-1}\left[\ln\frac{s^2+\omega^2}{s^2}\right]$.

Solution 6.7.3.

$$\mathcal{L}^{-1}\left[\ln\frac{s^2+\omega^2}{s^2}\right] = \frac{1}{t}\mathcal{L}^{-1}\left[-\frac{d}{ds}\ln\frac{s^2+\omega^2}{s^2}\right] = -\frac{1}{t}\mathcal{L}^{-1}\left[\frac{2s}{s^2+\omega^2} - \frac{2s}{s^2}\right]$$

$$= \frac{2}{t}\left(1 - \cos\omega t\right).$$

Example 6.7.4. Find $\mathcal{L}^{-1}\left[\tan^{-1}\frac{1}{s}\right]$.

Solution 6.7.4.

$$\mathcal{L}^{-1}\left[\tan^{-1}\frac{1}{s}\right] = \frac{1}{t}\mathcal{L}^{-1}\left[-\frac{d}{ds}\tan^{-1}\frac{1}{s}\right] = -\frac{1}{t}\mathcal{L}^{-1}\left[\frac{1}{(1/s)^2+1}\frac{d}{ds}\frac{1}{s}\right]$$

$$= \frac{1}{t}\mathcal{L}^{-1}\left[\frac{1}{1+s^2}\right] = \frac{1}{t}\sin t.$$

6.8 Additional Applications of Laplace Transforms

6.8.1 Evaluating Integrals

Many integrals from 0 *to* ∞ can be evaluated by the Laplace transform method.

By direct substitution. Integrals involving e^{-at} can be obtained from the Laplace transformation with a simple substitution.

$$\int_0^\infty e^{-at} f(t)dt = \left\{ \int_0^\infty e^{-st} f(t)dt \right\}_{s=a} = \{ \mathfrak{L}\,[f(t)] \}_{s=a}\,. \qquad (6.111)$$

Example 6.8.1. Find $\int_0^\infty e^{-3t} \sin t\; dt$.

Solution 6.8.1.

$$\int_0^\infty e^{-3t} \sin t\; dt = \{ \mathfrak{L}\,[\sin t] \}_{s=3} = \left\{ \frac{1}{s^2+1} \right\}_{s=3} = \frac{1}{10}$$

Example 6.8.2. Find $\int_0^\infty e^{-2t} t \cos t\; dt$.

Solution 6.8.2. Since

$$\int_0^\infty e^{-2t} t \cos t\; dt = \{ \mathfrak{L}[t \cos t] \}_{s=2}\,,$$

$$\{ \mathfrak{L}[t \cos t] \} = -\frac{d}{ds}\mathfrak{L}[\cos t] = -\frac{d}{ds}\frac{s}{s^2+1} = \frac{s^2-1}{(s^2+1)^2}\,,$$

therefore

$$\int_0^\infty e^{-2t} t \cos t\; dt = \left\{ \frac{s^2-1}{(s^2+1)^2} \right\}_{s=2} = \frac{3}{25}\,.$$

Use integral of the transform. In Sect. 6.6.2 we have shown

$$\mathfrak{L}\left[\frac{f(t)}{t}\right] = \int_s^\infty F(s')ds'\,,$$

where

$$\mathfrak{L}\left[\frac{f(t)}{t}\right] = \int_0^\infty e^{-st}\frac{f(t)}{t}dt\,, \qquad F(s) = \mathfrak{L}\,[f(t)] = \int_0^\infty e^{-st} f(t)dt\,.$$

Setting $s = 0$, we obtain an equally important formula

$$\int_0^\infty \frac{f(t)}{t}dt = \int_0^\infty \mathfrak{L}\,[f(t)]\,ds\,. \qquad (6.112)$$

This formula can be used if the integral on left side is difficult to do directly.

Example 6.8.3. Find $\displaystyle\int_0^\infty \left[\frac{e^{-t} - e^{-3t}}{t}\right] dt.$

Solution 6.8.3.

$$\int_0^\infty \left[\frac{e^{-t} - e^{-3t}}{t}\right] dt = \int_0^\infty \mathcal{L}\left[e^{-t} - e^{-3t}\right] ds = \int_0^\infty \left(\frac{1}{s+1} - \frac{1}{s+3}\right) ds$$

$$= \left[\ln(s+1) - \ln(s+3)\right]_0^\infty = \left[\ln\frac{s+1}{s+3}\right]_0^\infty$$

$$= \ln 1 - \ln\frac{1}{3} = \ln 3.$$

Example 6.8.4. Find $\displaystyle\int_0^\infty \frac{\sin t}{t} dt.$

Solution 6.8.4.

$$\int_0^\infty \frac{\sin t}{t} dt = \int_0^\infty \mathcal{L}\left[\sin t\right] ds = \int_0^\infty \frac{1}{s^2 + 1} ds$$

$$= \left[\tan^{-1} s\right]_0^\infty = \frac{\pi}{2}. \tag{6.113}$$

Use double integrals. We can solve the problem of the last example by a double integral. Starting with

$$\mathcal{L}[1] = \int_0^\infty e^{-st} dt = \frac{1}{s}, \tag{6.114}$$

if we rename t as x, and s as t, we have

$$\int_0^\infty e^{-tx} dx = \frac{1}{t}. \tag{6.115}$$

So

$$\int_0^\infty \frac{\sin t}{t} dt = \int_0^\infty \sin t \left[\frac{1}{t}\right] dt = \int_0^\infty \sin t \left[\int_0^\infty e^{-tx} dx\right] dt. \tag{6.116}$$

Interchanging the order of integration, we have

$$\int_0^\infty \frac{\sin t}{t} dt = \int_0^\infty \left[\int_0^\infty e^{-tx} \sin t \, dt\right] dx. \tag{6.117}$$

The integral in the bracket is recognized as the Laplace transform of $\sin t$ with the parameter s replaced by x, thus

$$\int_0^\infty \frac{\sin t}{t} dt = \int_0^\infty \frac{1}{1 + x^2} dx = \left[\tan^{-1} x\right]_0^\infty = \frac{\pi}{2}. \tag{6.118}$$

This method can be applied to more complicated cases.

Example 6.8.5. Find $\displaystyle\int_0^\infty \frac{\sin^2 t}{t^2}\,\mathrm{d}t$.

Solution 6.8.5. First we note

$$\sin^2 t = \frac{1}{2}(1 - \cos 2t),$$

then write

$$\int_0^\infty \frac{\sin^2 t}{t^2}\,\mathrm{d}t = \frac{1}{2}\int_0^\infty (1 - \cos 2t)\left[\frac{1}{t^2}\right]\mathrm{d}t.$$

With

$$\frac{1}{t^2} = \int_0^\infty \mathrm{e}^{-tx} x\,\mathrm{d}x,$$

we have

$$\int_0^\infty \frac{\sin^2 t}{t^2}\,\mathrm{d}t = \frac{1}{2}\int_0^\infty (1 - \cos 2t)\left[\int_0^\infty \mathrm{e}^{-tx} x\,\mathrm{d}x\right]\mathrm{d}t$$

$$= \frac{1}{2}\int_0^\infty \left[\int_0^\infty \mathrm{e}^{-tx}(1 - \cos 2t)\mathrm{d}t\right] x\,\mathrm{d}x.$$

Since

$$\int_0^\infty \mathrm{e}^{-tx}(1 - \cos 2t)\mathrm{d}t = [\mathcal{L}(1 - \cos 2t)]_{s=x}$$

$$= \frac{1}{x} - \frac{x}{x^2 + 4} = \frac{4}{x(x^2 + 4)},$$

$$\int_0^\infty \frac{\sin^2 t}{t^2}\,\mathrm{d}t = \frac{1}{2}\int_0^\infty \left[\frac{4}{x(x^2+4)}\right] x\,\mathrm{d}x$$

$$= \int_0^\infty \frac{2}{(x^2+4)}\mathrm{d}x = \left[\tan^{-1}\frac{x}{2}\right]_0^\infty = \frac{\pi}{2}.$$

Use inverse Laplace transform. If the integral is difficult to do, we can first find its Laplace transform and then take the inverse.

Example 6.8.6. Find $\displaystyle\int_0^\infty \frac{\cos x}{x^2 + b^2}\,\mathrm{d}x$

Solution 6.8.6. In order to use Laplace transform to evaluate this integral, we change $\cos x$ to $\cos tx$, and at the end we set $t = 1$. Let

$$I(t) = \int_0^\infty \frac{\cos tx}{x^2 + b^2}\,\mathrm{d}x.$$

$$\mathcal{L}\left[I(t)\right] = \int_0^\infty \frac{1}{x^2 + b^2} \mathcal{L}\left[\cos tx\right] dx = \int_0^\infty \frac{1}{x^2 + b^2} \frac{s}{s^2 + x^2} dx$$

$$= \frac{s}{s^2 - b^2} \int_0^\infty \left[\frac{1}{x^2 + b^2} - \frac{1}{s^2 + x^2}\right] dx$$

$$= \frac{s}{s^2 - b^2} \left\{ \left[\frac{1}{b} \tan^{-1} \frac{x}{b}\right]_0^\infty - \left[\frac{1}{s} \tan^{-1} \frac{x}{s}\right]_0^\infty \right\}$$

$$= \frac{s}{s^2 - b^2} \left\{ \frac{\pi}{2b} - \frac{\pi}{2s} \right\} = \frac{\pi}{2b} \frac{1}{s + b}.$$

$$I(t) = \mathcal{L}^{-1}\left[\frac{\pi}{2b} \frac{1}{s + b}\right] = \frac{\pi}{2b} e^{-bt}.$$

Thus

$$\int_0^\infty \frac{\cos x}{x^2 + b^2} dx = I(1) = \frac{\pi}{2b} e^{-b}.$$

6.8.2 Differential Equation with Variable Coefficients

If $f(t)$ in the formula

$$\mathcal{L}\left[tf(t)\right] = -\frac{d}{ds} \mathcal{L}\left[f(t)\right]$$

is taken to be the nth derivative of $y(t)$, then

$$\mathcal{L}\left[ty^{(n)}(t)\right] = -\frac{d}{ds} \mathcal{L}\left[y^{(n)}(t)\right]$$

$$= -\frac{d}{ds} \left\{ s^n \mathcal{L}\left[y(t)\right] - s^{n-1} y(0) \cdots -y^{n-1}(0) \right\}. \quad (6.119)$$

This equation can be used to transform a linear differential equation with variable coefficients into a differential equation involving the transform. This procedure is useful if the new equation can be readily solved.

Example 6.8.7. Find the solution of

$$ty''(t) - ty'(t) - y(t) = 0, \quad y(0) = 0, \ y'(0) = 2.$$

Solution 6.8.7.

$$\mathcal{L}\left[ty''(t)\right] = -\frac{d}{ds} \left\{ s^2 \mathcal{L}\left[y(t)\right] - sy(0) - y'(0) \right\}.$$

Let $\mathcal{L}\left[y(t)\right] = F(s)$, with $y(0) = 0$ we have

$$\mathcal{L}\left[ty''(t)\right] = -2sF(s) - s^2 F'(s),$$
$$\mathcal{L}\left[ty'(t)\right] = -F(s) - sF'(s).$$

Therefore

$$\mathcal{L}[ty''(t) - ty'(t) - y(t)] = -s(s-1)F'(s) - 2sF(s) = 0.$$

It follows

$$\frac{dF(s)}{F(s)} = -2\frac{ds}{s-1},$$

$$\ln F(s) = \ln(s-1)^{-2} + \ln C$$

$$F(s) = \frac{C}{(s-1)^2}$$

$$y(t) = \mathcal{L}^{-1}[F(s)] = \mathcal{L}^{-1}\left[\frac{C}{(s-1)^2}\right] = C\,e^t t.$$

Since

$$y'(t) = C\,e^t t + C\,e^t,$$

$$y'(0) = C = 2,$$

therefore

$$y(t) = 2e^t t.$$

It can be easily verified that this is indeed the solution, since it satisfies both the equation and the initial conditions.

Example 6.8.8. Zeroth Order Bessel Function: Find the solution of

$$ty''(t) + y'(t) + ty(t) = 0, \qquad y(0) = 1,\ y'(0) = 0.$$

Solution 6.8.8. With $\mathcal{L}[y(t)] = F(s)$ and $y(0) = 1, y'(0) = 0$,

$$\mathcal{L}[ty''(t) + y'(t) + ty(t)] = -\frac{d}{ds}\{s^2 F(s) - s\} + sF(s) - 1 - \frac{d}{ds}F(s) = 0.$$

Collecting terms

$$(s^2 + 1)\frac{d}{ds}F(s) + sF(s) = 0,$$

or

$$\frac{dF(s)}{F(s)} = -\frac{s\,ds}{s^2 + 1} = -\frac{1}{2}\frac{ds^2}{s^2 + 1}.$$

It follows

$$\ln F(s) = -\frac{1}{2}\ln(s^2 + 1) + \ln C,$$

$$F(s) = \frac{C}{(s^2 + 1)^{1/2}}.$$

To find the inverse of this Laplace transform, we expand it in a series in the case $s > 1$,

$$F(s) = \frac{C}{s}\left[1 + \frac{1}{s^2}\right]^{-1/2}$$

$$= \frac{C}{s}\left[1 - \frac{1}{2s^2} + \frac{1\cdot 3}{2^2\cdot 2!}\frac{1}{s^4} \cdots + \frac{(-1)^n(2n)!}{(2^n n!)^2}\frac{1}{s^{2n}} + \cdots\right].$$

Inverting term by term, we have

$$y(t) = \mathcal{L}^{-1}\left[F(s)\right] = C\sum_{n=0}^{\infty}\frac{(-1)^n t^{2n}}{(2^n n!)^2}.$$

Since $y(0) = 1$, therefore $C = 1$. It turned out this series with $C = 1$ is known as the Bessel function of zeroth order $J_0(t)$, that is,

$$J_0(t) = \sum_{n=0}^{\infty}\frac{(-1)^n t^{2n}}{(2^n n!)^2},$$

which we will discuss in more detail in the chapter on Bessel functions. Thus the solution of the equation is

$$y(t) = J_0(t).$$

Furthermore, this development shows that

$$\mathcal{L}\left[J_0(t)\right] = \frac{1}{(s^2+1)^{1/2}}.$$

With the scaling property of the Laplace transform, we also have (for $a > 0$)

$$\mathcal{L}\left[J_0(at)\right] = \frac{1}{a}\frac{1}{[(s/a)^2 + 1]^{1/2}} = \frac{1}{(s^2 + a^2)^{1/2}}.$$

6.8.3 Integral and Integrodifferential Equations

Equations in which the unknown function appears under the integral are called integral equations. If the derivatives are also in the equation, then they are called integrodifferential equations. They are often difficult to solve. But if the integrals are in the form of a convolution, then Laplace transform can be used to solve them. The following example will make the procedure clear.

Example 6.8.9. Solve the integral equation

$$y(t) = t + \int_0^t y(\tau)\sin(t-\tau)d\tau.$$

Solution 6.8.9.

$$\mathcal{L}[y(t)] = \mathcal{L}[t] + \mathcal{L}\left[\int_0^t y(\tau)\sin(t-\tau)d\tau\right]$$

$$= \mathcal{L}[t] + \mathcal{L}[y(t)]\mathcal{L}[\sin t] = \frac{1}{s^2} + \mathcal{L}[y(t)]\frac{1}{s^2+1}.$$

Solving for $\mathcal{L}[y(t)]$

$$\left(1 - \frac{1}{s^2+1}\right)\mathcal{L}[y(t)] = \frac{1}{s^2}.$$

So

$$\mathcal{L}[y(t)] = \frac{s^2+1}{s^4} = \frac{1}{s^2} + \frac{1}{s^4}$$

and

$$y(t) = \mathcal{L}^{-1}\left[\frac{1}{s^2} + \frac{1}{s^4}\right] = t + \frac{1}{6}t^3.$$

Example 6.8.10. Find the solution of

$$y'(t) - 3\int_0^t e^{-2(t-\tau)}y(\tau)d\tau = z(t), \qquad y(0) = 4,$$

$$z(t) = \begin{cases} 4e^{-2t} & 1 < t \\ 0 & 0 < t < 1 \end{cases}.$$

Solution 6.8.10. First note

$$\mathcal{L}[z(t)] = \mathcal{L}\left[4e^{-2t}u(t-1)\right] = \mathcal{L}\left[4e^{-2}e^{-2(t-1)}u(t-1)\right] = 4e^{-2}\frac{1}{s+2}e^{-s},$$

$$\mathcal{L}\left[3\int_0^t e^{-2(t-\tau)}y(\tau)d\tau\right] = 3\frac{1}{s+2}\mathcal{L}[y].$$

Applying the Laplace transform to both sides of the equation leads to

$$s\mathcal{L}[y] - 4 - 3\frac{1}{s+2}\mathcal{L}[y] = 4e^{-2}\frac{1}{s+2}e^{-s},$$

collecting terms

$$\frac{s^2+2s-3}{s+2}\mathcal{L}[y] = 4 + 4e^{-2}\frac{1}{s+2}e^{-s},$$

or

$$\mathcal{L}[y] = \frac{4(s+2) + 4e^{-2}e^{-s}}{s^2+2s-3}$$

$$= \frac{3}{s-1} + \frac{1}{s+3} + e^{-2}\left(\frac{1}{s-1} - \frac{1}{s+3}\right)e^{-s}.$$

Thus

$$y(t) = 3e^t + e^{-3t} + e^{-2}\left(e^{(t-1)} - e^{-3(t-1)}\right)u(t-1).$$

6.9 Inversion by Contour Integration

As we have seen that to be able to invert a given transform $F(s)$ to find the function $f(t)$ is the key of solving differential equations with the Laplace Transform. For those familiar with the complex contour integration, we present in this section a universal technique of finding the inverse of the Laplace Transform. The derivation is highly imaginative, but the result is elegantly simple.

First let us extend the Laplace transform to the complex domain. The function $F(z)$ is the same function as $F(s)$ except with s replaced by z. In the complex plane, $F(z)$ will have some singular points. Let us choose a line $x = b$ in the complex plane such that all singular points of $F(z)$ are in the left-hand side of this line. Then $F(z)$ is analytic on the line $x = b$ and in the entire half plane to the right of this line. If s is any point in this half plane, we can choose a semicircular contour $C = C_1 + C_2$, as shown in Fig. 6.11, and apply the Cauchy's integral formula,

$$F(s) = \frac{1}{2\pi i} \oint_C \frac{F(z)}{z-s} dz$$
$$= \frac{1}{2\pi i} \int_{b+iR}^{b-iR} \frac{F(z)}{z-s} dz + \frac{1}{2\pi i} \int_{C_1} \frac{F(z)}{z-s} dz. \qquad (6.120)$$

Now if we let R go to infinity, (6.120) is still valid, but all values of z on the semicircle C_1 are infinitely large. Since $F(z) \to 0$ as $z \to \infty$,

$$\lim_{R \to \infty} \int_{C_1} \frac{F(z)}{z-s} dz = 0.$$

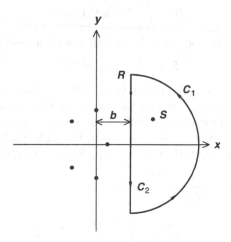

Fig. 6.11. The first contour used to obtain the complex inversion of the Laplace transform

Therefore in this limit (6.120) becomes

$$F(s) = \frac{1}{2\pi i} \int_{b+i\infty}^{b-i\infty} \frac{F(z)}{z-s} dz = \frac{1}{2\pi i} \int_{b-i\infty}^{b+i\infty} \frac{F(z)}{s-z} dz.$$

In the last step we have changed the sign of the integrand and interchanged the upper and lower limits of the integral.

Taking the inverse Laplace transform, we have

$$\mathcal{L}^{-1}[F(s)] = \mathcal{L}^{-1} \left[\frac{1}{2\pi i} \int_{b-i\infty}^{b+i\infty} \frac{F(z)}{s-z} dz \right].$$

Since the inverse Laplace operator $\mathcal{L}^{-1}$ refers only to the variable s, we can write

$$\mathcal{L}^{-1}[F(s)] = \frac{1}{2\pi i} \int_{b-i\infty}^{b+i\infty} F(z) \mathcal{L}^{-1} \left[\frac{1}{s-z} \right] dz.$$

Since

$$\mathcal{L}^{-1} \left[\frac{1}{s-z} \right] = e^{zt}$$

we have

$$\mathcal{L}^{-1}[F(s)] = \frac{1}{2\pi i} \int_{b-i\infty}^{b+i\infty} F(z) e^{zt} dz. \qquad (6.121)$$

This procedure is called the Mellin inversion. This integral is from $b - i\infty$ to $b + i\infty$ along C_2. Usually the evaluation of this integral is accomplished by the residue theorem. To use the residue theorem, we must have a closed contour. To close the contour, we have to add a returning line integral from $b + i\infty$ to $b - i\infty$ in such a way that the value of the integral is not changed. This can be done with the semicircular contour C_3 in the left half-plane as shown in Fig. 6.12, since

$$\lim_{R\to\infty} \int_{C_3} F(z) e^{zt} dz = 0. \qquad (6.122)$$

This can be understood from the fact that with a positive t, the integrand

$$F(z)e^{zt} = F(z)e^{xt+iyt}$$

goes to zero as z goes to infinity. The factor e^{iyt} is oscillatory with a maximum value of 1. For x to change from b to $-\infty$ as on C_3, the factor e^{xt} is always less than e^{bt}. Therefore $F(z)e^{zt}$ will go to zero as long as $F(z)$ is going to zero. Note that this will not be the case in the right half plane where x will go to positive infinite and e^{zt} will blow up. Thus with C_3, we have

$$\mathcal{L}^{-1}[F(s)] = \frac{1}{2\pi i} \int_{b-i\infty}^{b+i\infty} F(z) e^{zt} dz.$$

$$= \frac{1}{2\pi i} \lim_{R\to\infty} \left[\int_{C_2} F(z) e^{zt} dz + \int_{C_3} F(z) e^{zt} dz \right]$$

$$= \frac{1}{2\pi i} \oint_C F(z) e^{zt} dz, \qquad (6.123)$$

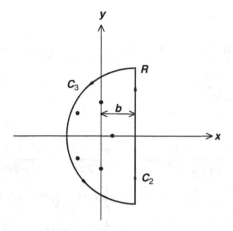

Fig. 6.12. The contour used in evaluating the complex inversion integral

where $C = C_2 + C_3$ as shown in Fig. 6.12 with $R \to \infty$. This contour is also called the Bromwich contour. Since b is on the right of all singular points of $F(z)$, the contour C encloses all singular points of $e^{zt}F(z)$. Therefore by the residue theorem

$$\mathcal{L}^{-1}[F(s)] = \frac{1}{2\pi i} \oint_C F(z)e^{zt}dz$$
$$= \sum \text{all residues of } F(z)e^{zt}. \qquad (6.124)$$

Example 6.9.1. Using the complex inversion integral to find

$$\mathcal{L}^{-1}\left[\frac{1}{(s+a)^2 + b^2}\right].$$

Solution 6.9.1. Since

$$\frac{1}{(s+a)^2 + b^2} = \frac{1}{[s-(-a+ib)][s-(-a-ib)]},$$

the residues of

$$\frac{e^{zt}}{(z+a)^2 + b^2}$$

at the two singular points are

$$r_1 = \lim_{z \to -a+ib} [z-(-a+ib)] \frac{e^{zt}}{[z-(-a+ib)][z-(-a-ib)]}$$
$$= \frac{e^{(-a+ib)t}}{2ib}$$

and

$$r_2 = \lim_{z \to -a-ib} [z - (-a - ib)] \frac{e^{zt}}{[z - (-a + ib)][z - (-a - ib)]}$$

$$= \frac{e^{(-a-ib)t}}{-2ib}.$$

Therefore

$$\mathcal{L}^{-1}\left[\frac{1}{(s + a)^2 + b^2}\right] = \frac{e^{(-a+ib)t}}{2ib} + \frac{e^{(-a-ib)t}}{-2ib}$$

$$= \frac{1}{b}e^{-at}\frac{1}{2i}\left(e^{ibt} - e^{-ibt}\right)$$

$$= \frac{1}{b}e^{-at}\sin bt.$$

This is a familiar result. This example shows that the complex inversion integral is indeed another way of finding the inverse Laplace transform. In more difficult applications, the use of the complex inversion integral and the contour integration is either the only way or the simplest way of finding the inverse Laplace transform.

6.10 Computer Algebraic Systems for Laplace Transforms

Before closing this chapter, we should mention that a number of commercial computer packages are available to perform algebraic manipulations, including Laplace transforms. They are called computer algebraic systems, some prominent ones are Matlab, Maple, Mathematica, MathCad, and MuPAD.

This book is written with the software "Scientific WorkPlace," which also provides an interface to MuPAD. (Before version 5, it also came with Maple). Instead of requiring the user to adhere to a rigid syntax, the user can use natural mathematical notations. For example, to compute $\mathcal{L}(\cos t - 2\sin t)$, all you have to do is (1) type $\cos t - 2\sin t$ in the math-mode, (2) click on the "Compute" button, (3) click on the "Transforms" button in the pull-down menu, and (4) click on the "Laplace" button in the submenu. The program will return with

$$\cos t - 2\sin t, \text{ Laplace transform is} : \frac{s}{s^2 + 1} - \frac{2}{s^2 + 1}.$$

The program also recognizes the Laplace transform symbol. An alternative way to do the same problem is to choose, from the "Miscellaneous Symbols" panel, the Laplace transform symbol and type

$$\mathcal{L}(cos\,t - 2sin\,t)$$

in the math-mode, and click on "Compute" and then choose "Evaluate." The program will return with

$$\mathcal{L}(cos\,t - 2sin\,t) = \frac{s}{s^2 + 1} - \frac{2}{s^2 + 1}$$

Similarly, one can compute the inverse of the Laplace transform. For example, type

$$\mathcal{L}^{-1}\left(\frac{s - 2}{s^2 + 1}\right)$$

and click on "Compute" and then on "Evaluate," the program will return with

$$\mathcal{L}^{-1}\left(\frac{s - 2}{s^2 + 1}\right) = cos\,t - 2sin\,t$$

To compute the Laplace transform of a derivative, one has to define the function first. For example, to find $\mathfrak{L}(y''')$, first type

$$y\,(t)$$

and click on "Compute", then on "Definition" in the menu, then on "New Definition" in the submenu. Then type

$$\mathcal{L}(y''')$$

and click on "Compute" and then on "Evaluate". The program will return with

$$\mathcal{L}(y''') = s^3 \mathcal{L}(y) - sy'(0) - s^2 y(0) - y''(0).$$

The program can also use Laplace transform to solve ordinary differential equations. For example, to solve the equation

$$y' + y = x + \sin x,$$

first type this equation in math-mode, then click on "Compute." In the pull-down menu, click on "Solve ODE," then click on "Laplace" in the submenu. The program returns with a question asking which is the independent variable. Type x and click on "OK." The program will return with

$$\text{Laplace solution is}: x - \frac{1}{2}\cos x + \frac{1}{2}\sin x + e^{-x}\left(y\,(0) + \frac{3}{2}\right) - 1.$$

Unfortunately, not every problem can be solved by a computer algebraic system. Sometimes it fails to find the solution. Even worse, for a variety of reasons, the intention of the user is sometimes misinterpreted, and the computer returns with an answer to a wrong problem without the user knowing it. One should be aware of these pitfalls.

Computer algebraic systems are no substitute for the knowledge of the subject matter, but they are useful supplements.

Exercises

1. Find the Laplace transformation of each of the following functions by direct integration.

   (a) $\frac{1}{2}t^2$,   (b) e^{3t},   (c) $3\sin(3t)$.

   Ans. (a) $\frac{1}{s^3}$,   (b) $\frac{1}{s-3}$,   (c) $\frac{9}{s^2+9}$.

2. Find the Laplace transformation of each of the following functions by using the "Multiply t" operation (see Table 6.2).

   (a) te^t,   (b) $t\cos t$,   (c) $t^2\cos t$.

   Ans. (a) $\frac{1}{(s-1)^2}$,   (b) $\frac{s^2-1}{(s^2+1)^2}$,   (c) $\frac{2s(s^2-3)}{(s^2+1)^3}$.

3. Find the Laplace transformation of each of the following functions by using the "Divide t" operation (see Table 6.2).

   (a) $\frac{1}{t}\left(e^{2t}-e^{-2t}\right)$,   (b) $\frac{2}{t}(1-\cos(2t))$,   (c) $\frac{1}{t}\sin(4t)$.

   Ans. (a) $\ln\left(\frac{s+2}{s-2}\right)$,   (b) $\ln\left(\frac{s^2+4}{s^2}\right)$,   (c) $\frac{\pi}{2}-\tan^{-1}\left(\frac{s}{4}\right)$.

4. Find the Laplace transformation of each of the following functions by using the "Shifting – s" operation (see Table 6.2).

   (a) $e^{at}\sin 3t$,   (b) $e^{-2t}t\sin at$,   (c) $\sinh t\cos t$.

   Ans. (a) $\frac{3}{(s-a)^2+9}$,   (b) $\frac{2a(s+2)}{[(s+2)^2+a^2]^2}$,   (c) $\frac{s^2-2}{s^4+4}$.

5. Use the definition of Laplace transformation to show

   (a) $\mathcal{L}[f'] = s\mathcal{L}[f] - f(0)$;

   use (a) to show

   (b) $\mathcal{L}[f''] = s^2\mathcal{L}[f] - sf(0) - f'(0)$.

6. Use the results of previous problem and the fact that $\frac{d^2}{dt^2}\cos at = -a^2\cos at$ and $\frac{d^2}{dt^2}\sin at = -a^2\sin at$ to show

   (a) $\mathcal{L}[\cos at] = \frac{s}{s^2+a^2}$,   (b) $\mathcal{L}[\sin at] = \frac{a}{s^2+a^2}$.

7. Differentiate both sides of part (b) of the previous problem with respect to a, and show that

   (a) $\mathcal{L}[t\cos at] = \frac{1}{s^2+a^2} - \frac{2a^2}{(s^2+a^2)^2}$,

   Differentiate both sides of part (a) of the previous problem with respect to s, and show that

   (b) $\mathcal{L}[-t\cos at] = \frac{1}{s^2+a^2} - \frac{2s^2}{(s^2+a^2)^2}$.

8. Use the results of problems 6 and 7 to show

(a) $\mathcal{L}^{-1}\left[\dfrac{1}{(s^2+a^2)^2}\right] = \dfrac{1}{2a^3}(\sin at - at\cos at),$

(b) $\mathcal{L}^{-1}\left[\dfrac{s^2}{(s^2+a^2)^2}\right] = \dfrac{1}{2}\left(t\cos at + \dfrac{1}{a}\sin at\right).$

9. Do problem 8 with convolution theorem.
   Hint: you may need the following integral

$$\int_0^t \sin a\tau \cos a\tau\, d\tau = \frac{1}{4a}(1-\cos 2at);$$

$$\int_0^t \sin^2 a\tau\, d\tau = \frac{1}{4a}(2at - \sin 2at).$$

10. If $f(t)=t^n$, $g(t)=t^m$, $n>-1,\ m>-1,$
    (a) show that

$$\int_0^t \tau^n(t-\tau)^m d\tau = t^{n+m+1}\int_0^1 y^n(1-y)^m dy.$$

    (b) By using the convolution theorem, show that

$$\int_0^1 y^n(1-y)^m dy = \frac{n!m!}{(n+m+1)!}.$$

    Hint: (a) let $\tau = yt$, (b) use convolution theorem to evaluate $\int_0^t \tau^n (t-\tau)^m d\tau$.

11. Find the Laplace transformation of each of the following functions by direct integration.
    (a) $\sin(t-a)\,u(t-a)$,

    (b) $f(t) = \begin{cases} \cos(t-\pi) & t>\pi \\ 0 & t<\pi \end{cases}$,

    (c) $f(t) = \begin{cases} 0 & 0\le t<5 \\ 1 & 5\le t<10 \\ 0 & 10\le t \end{cases}$.

    Ans. (a) $e^{-as}\frac{1}{s^2+1}$, (b) $e^{-\pi s}\frac{s}{s^2+1}$, (c) $\frac{1}{s}\left(e^{-5s}-e^{-10s}\right)$.

12. Do the previous problem by using the "Shifting $-t$" operation (see Table 6.2).

13. Use the partial fraction to find the inverse Laplace transform of the following expressions.
    (a) $\mathcal{L}^{-1}\left[\dfrac{4}{s^2-4s}\right]$, (b) $\mathcal{L}^{-1}\left[\dfrac{1}{s(s^2+1)}\right]$, (c) $\mathcal{L}^{-1}\left[\dfrac{1}{s^2(s^2+1)}\right]$.

    Ans. (a) $e^{4t}-1$, (b) $1-\cos t$, (c) $t-\sin t$.

14. Do the previous problem by using the formula

$$\mathcal{L}\left[\int_0^t f(\tau)\,d\tau\right] = \frac{1}{s}\mathcal{L}\left[f(t)\right].$$

15. Use the Heaviside expansion to solve the previous problem.

16. Use the Laplace transform to solve the following differential equations
    (a) $y'' + 2y' + y = 1$, $y(0) = 2$, $y'(0) = -2$,

    (b) (b) $y'' + y = \sin(3t)$, $y(0) = y'(0) = 0$.

    Ans. (a) $y(t) = 1 + (1-t)e^{-t}$, (b) $y(t) = \frac{3}{8}\sin t - \frac{1}{8}\sin 3t$.

17. Use the Laplace transform to solve the following set of equations

$$\frac{dy}{dt} = 2y - 3z,$$

$$\frac{dz}{dt} = -2y + z,$$

$$y(0) = 8, \quad z(0) = 3.$$

    Ans. $y(t) = 3e^{4t} + 5e^{-t}$, $\quad z(t) = 5e^{-t} - 2e^{4t}$.

18. Find the solution of the integrodifferential equation

$$y'(t) - \int_0^t y(\tau)\cos(t-\tau)\,d\tau = 0, \quad y(0) = 1.$$

    Ans. $y(t) = 1 + \frac{1}{2}t^2$.

19. Solve the following equations with the initial conditions at $t = 0$, both y and all its derivatives are equal to zero.
    (a) $y'' + 2y' + y = A\delta(t - t_0)$,
    (b) $y'''' - y = A\delta(t - t_0)$.
    Ans. (a) $y(t) = A(t - t_0)e^{-(t-t_0)}u(t - t_0)$,
    (b) $y(t) = \frac{1}{2}A\left[\sinh(t - t_0) - \sin(t - t_0)\right]u(t - t_0)$.

20. Consider a resistance R and an inductance L connected in series with a voltage $V(t)$. The equation governs the current is

$$L\frac{di}{dt} + Ri = V(t).$$

    Suppose $i(0) = 0$ and $V(t)$ is a voltage impulse at $t = t_0$ given by

$$V(t) = A\delta(t - t_0).$$

Find the current by the Laplace transform method.

Ans. $i(t) = \dfrac{A}{L} e^{-R(t-t_0)/L} u(t - t_0)$.

21. The damped harmonic oscillator is governed by

$$mx'' + bx' + kx = f(t); \quad \text{with } x(0) = x'(0) = 0.$$

(a) Find the solution by convolution. (Express $x(t)$ as an integral).
(b) If $f(t) = P\delta(t - t_0)$, find the solution by evaluating the convolution integral.
(c) If $b = 0$, and $f(t) = F_0 \sin \omega_0 t$ where $\omega_0 = \sqrt{k/m}$, solve the problem by Laplace transformation.
(d) If $b = 0$, and $f(t) = F_0 u(t - t_0)$ where $u(t - t_0)$ is the step function, solve the problem.

Ans. (a) $x(t) = \dfrac{1}{m\omega} \int_0^t f(\tau) e^{-\alpha(t-\tau)} \sin \omega (t - \tau) \, d\tau$
where $\alpha = \dfrac{b}{2m}$, $\omega^2 = \dfrac{k}{m} - \left(\dfrac{b}{2m}\right)^2$,
(b) $x(t) = \dfrac{P}{m\omega} e^{-\alpha(t-t_0)} \sin \omega (t - t_0) u(t - t_0)$,
(c) $x(t) = \dfrac{F_0}{2m\omega_0^2} (\sin \omega_0 t - \omega_0 t \cos \omega_0 t)$,
(d) $x(t) = \dfrac{F_0}{m\omega_0^2} [1 - \cos \omega_0 (t - t_0)] u(t - t_0)$.

22. Using the complex inversion integral, find the inverses of the following Laplace transforms

(a) $\dfrac{1}{(s+1)(s+3)}$, (b) $\dfrac{1}{(s+2)^2}$, (c) $\dfrac{1}{(s^2+9)(s^2+4)}$.

Ans. (a) $\frac{1}{2}(e^{-t} - e^{-3t})$, (b) $t e^{-2t}$, (c) $\frac{1}{30}(3 \sin 2t - 2 \sin 3t)$.

References

This bibliograph includes the references cited in the text and a few other books and tables that might be useful.

1. M. Abramowitz, I.A. Stegun: *Handbook of Mathematical Functions* (Dover, New York 1970)
2. G.B. Arfken, H.J. Weber: *Mathematical Methods for Physicists,* 5th edn. (Academic Press, San Diego, 2001)
3. M. L. Boas: *Mathematical Methods in the Physical Sciences,* 3rd edn. (Wiley, New York 2006)
4. W.E. Boyce, R.C. DiPrima: *Elementary Differential Equations and Boundary Value Problems,* 4th edn. (Wiley, New York 1986)
5. T.C. Bradbury: *Mathematical Methods with Applications to Problems in the Physical Sciences* (Wiley, New York 1984)
6. E. Butkov: *Mathematical Physics* (Addison-Wesley, Reading 1968)
7. F.W. Byron, Jr., R.W. Fuller: *Mathematics of Classical and Quantum Physics* (Dover, New York 1992)
8. T.L. Chow: *Mathematical Methods for Physicists: A Concise Introduction* (Cambridge University Press, Cambridge 2000)
9. R.V. Churchill: *Operational Mathematics,* 3rd edn. (McGraw-Hill, New York 1972)
10. H. Cohen: *Mathematics for Scientists and Engineeers* (Prentice-Hall, Englewood Cliffs 1992)
11. R.E. Collins: *Mathematical Methods for Physicists and Engineers* (Reinhold, New York 1968)
12. C.H. Edwards Jr., D.E. Penney: *Differential Equations and Boundary Value Problems* (Prentice-Hall, Englewood Cliffs 1996)
13. A. Erdélyi, W. Magnus, F. Oberhettinger, F. Tricomi: *Tables of Integral Transforms, Vol. 1* (McGraw-Hill, New York 1954)
14. L.R. Ford: *Differential Equations* (McGraw-Hill, New York 1955)
15. M.D. Greenberg: *Advanced Engineering Mathematics,* 2nd edn. (Prentice Hall, Upper Saddle River 1998)
16. I.S. Gradshteyn, I.M. Ryzhik: *Table of Integrals, Series and Products* (Academic Press, Orlando 1980)
17. D.W. Hardy, C.L. Walker: *Doing Mathematics with Scientific WorkPlace and Scientific Notebook,* Version 5 (MacKichan, Poulsbo 2003)

18. S. Hasssani: *Mathematical Methods: For Students of Physics and Related Fields* (Springer, New York 2000)
19. F.B. Hilderbrand: *Advanced Calculus for Applications,* 2nd edn. (Prentice-Hall, Englewood Cliffs 1976)
20. E.L. Ince: *Ordinary Differential Equations* (Dover, New York 1956)
21. H. Jeffreys, B.S. Jeffreys: *Mathematical Physics* (Cambridge University Press, Cambridge 1962)
22. D.E. Johnson and J.R. Johnson: *Mathematical Methods in Engineering Physics* (Prentice-Hall, Upper Sadddle River 1982)
23. D.W. Jordan, P. Smith: *Mathematical Techniques: An Introduction for the Engineering, Physical, and Mathematical Sciences,* 3rd edn. (Oxford University Press, Oxford 2002)
24. E. Kreyszig: *Advanced Engineering Mathematics,* 8th edn. (Wiley, New York 1999)
25. B.R. Kusse, E.A. Westwig: *Mathematical Physics: Applied Mathematics for Scientists and Engineers,* 2nd edn. (Wiley, New York 2006)
26. S.M. Lea: *Mathematics for Physicists* (Brooks/Cole, Belmont 2004)
27. T.D. Lee: "*Reminiscences*". In *Thirty Years Since Parity Nonconservation* ed. by R. Novick (Birkhäuser, Boston 1988) pp. 153–165
28. H. Margenau, G.M. Murphy: *Methods of Mathematical Physics* (Van Nostrand, Princeton 1956)
29. J. Mathew, R.L. Walker: *Mathematical Methods of Physics,* 2nd edn. (Benjamin, New York 1970)
30. P.C. Matthews: *Vector Calculus* (Springer, London 1998)
31. D.A. McQuarrie: *Mathematical Methods for Scientists and Engineers* (University Science Books, Sausalito 2003)
32. P.M. Morse, H. Feshbach: *Methods of Theoretical Physics* (McGraw-Hill, New York 1953)
33. G.M. Murphy: *Ordinary Differential Equations and Their Solutions* (Van Nostrand, Princeton 1960)
34. H.E. Newell, Jr.: *Vector Analysis* (McGraw-Hill, New York 1955)
35. F. Oberhettinger, E. Badii: *Tables of Laplace Transforms* (Springer, New York 1973)
36. A.D. Polyanin, V.F. Zaitsev: *Handbook of Exact Solutions for Ordinary Differential Equations* (CRC Press, Boca Raton 1995)
37. M.C. Potter, J.L. Goldber, E.F. Aboufadel: *Advanced Engineering Mathematics,* 3rd edn. (Oxford University Press, New York 2005)
38. W.H. Press, S.A. Teukolsky, W.T. Vettering, B.P. Flannery: *Numerical Recipes,* 2nd edn. (Cambridge University Press, Cambridge 1992)
39. R.J. Rice: *Numerical Methods, Software and Analysis* (McGraw-Hill, New York 1983)
40. K.F. Riley, M.P. Hobson, S.J. Bence: *Mathematical Methods for Physics and Engineering,* 2nd edn. (Cambridge University Press, Cambridge 2002)
41. H.M. Schey: *Div, Grad, Curl, and All That: An Informal Text on Vector Calculus,* 4th edn. (Norton, New York 2004)
42. K.A. Stroud, D.J. Booth: *Advanced Engineering Mathematics,* 4th edn. (Industrial Press, New York 2003)
43. C.R. Wylie, L.C. Barrett: *Advanced Engineering Mathematics,* 5th edn. (McGraw-Hill, New York 1982)
44. D. Zwillinger: *Handbook of Differential Equations* (Academic Press, San Diego 1998)

Index